Computational Methods for Molecular Modelling and Docking

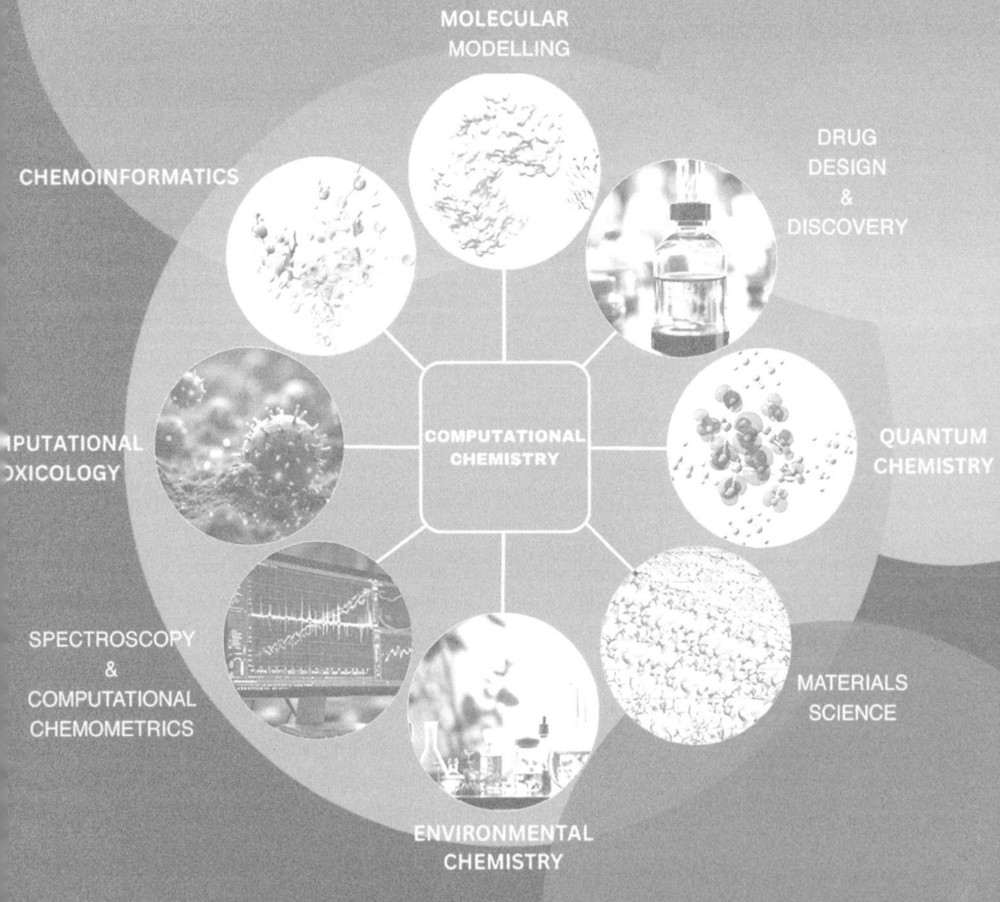

Computational Methods for Molecular Modelling and Docking

edited by

Arup Kumar Ghosh
Suban Kumar Sahoo

JENNY STANFORD
PUBLISHING

Published by

Jenny Stanford Publishing Pte. Ltd.
101 Thomson Road
#06-01, United Square
Singapore 307591

Email: editorial@jennystanford.com
Web: www.jennystanford.com

British Library Cataloguing-in-Publication Data
A catalogue record for this book is available from the British Library.

Computational Methods for Molecular Modelling and Docking

ISBN 978-981-5352-25-2 (Hardcover)
ISBN 978-1-042-00204-7 (eBook)

DOI: 10.1201/9781042002047

Contents

Preface

The tools of computational chemistry, such as molecular mechanics, density functional theory, ab initio, and semi-empirical methods, and molecular docking and dynamics, have been widely used for molecular modelling to calculate molecular geometries, electronic and thermodynamic properties, kinetics and transition states, spectra, and many other properties. Computational chemistry is a fast-growing field, and computational tools are used in diverse areas, such as drug design and discovery, catalysis, materials science, and bioinformatics.

This book, entitled *Computational Methods for Molecular Modelling and Docking*, includes a total of seven chapters contributed by reputed researchers. The first chapter presents a brief overview of computational chemistry and its development in recent years. The theories and methods that are used in various computational tools have been discussed along with their advantages and limitations. Chapter 2 presents molecular modelling approaches for studying chemical dynamics simulations in the condensed phase. Two models, viz., molecular mechanics/molecular mechanics (MM/MM) and quantum mechanics/molecular mechanics (QM/MM), are discussed. Furthermore, simulation results are used to perform a systematic investigation, with a focus on understanding the role of environmental molecules on chemical reactions and intermolecular energy transfer processes. Chapter 3 highlights the combined effectiveness of docking and molecular dynamics, demonstrating how their integration improves the accuracy of binding affinity predictions, fine-tunes ligand positioning, and aids in the optimization of lead compounds. Chapter 4 offers a comprehensive review of molecular docking techniques, detailing their development and applications in drug discovery. It also discusses the foundational theory, including sampling algorithms, scoring functions, and the software tools used in molecular docking. Some of the basic underlying protocols in computer-aided drug discovery through concise discussions on drug development against the SARS-CoV-2 spike (S) protein are presented in Chapter 5. This chapter also demonstrates a structure-based drug design (SBDD) approach that has been followed to screen a large library of peptidomimetic molecules and identify novel hits that may be taken for further clinical considerations. Chapter 6 exemplifies the molecular docking as a tool to test and predict the efficacies of the flavonoids as antivirals targeting SARS-CoV-2. It also covers the various databases available for searching the structures of proteins and ligands or drug-like molecules. Finally, Chapter 7 provides significant insights into the activity of different polyphenolic compounds for the treatment of several fatal diseases using a multi-targeted drug-repurposing approach by using different computational methodologies, e.g., conceptual density functional theory (DFT), molecular docking, molecular dynamics (MD) simulation, and quantitative structure-activity relationship (QSAR).

This book provides a basic understanding as well as relevant examples in the field of molecular modelling and docking, which will benefit a wide range of readers, including students (undergraduate, graduate, and postgraduate), new researchers, established scientists, and academicians.

Arup K. Ghosh, PhD
Suban K. Sahoo, PhD

Chapter 1

Introduction to Computational Chemistry

Srusty Jarupula, Suban K. Sahoo, and Arup Kumar Ghosh
Department of Chemistry, Sardar Vallabhbhai National Institute of Technology, Surat, Gujarat, India
arupchem87@gmail.com, akg@chem.svnit.ac.in

The tools of computational chemistry, such as molecular mechanics, density functional theory, ab initio and semi-empirical methods, molecular docking, and dynamics, have been widely used for molecular modelling to calculate molecular geometries, electronic and thermodynamic properties, kinetics and transition states, spectra, and many other properties. Computational chemistry is a fast-growing field, and computational tools are used in diversified areas, such as drug design and discovery, catalysis, materials science, and bioinformatics. This chapter presents a brief overview of computational chemistry and its development in recent years. The theories and methods that are used by computational chemists

Computational Methods for Molecular Modelling and Docking
Edited by Arup Kumar Ghosh and Suban Kumar Sahoo

ISBN 978-981-5352-25-2 (Hardcover), 978-1-042-00204-7 (eBook)
www.jennystanford.com

DOI: 10.1201/9781042002047-1

are discussed briefly, along with their advantages and limitations. Also, the applications of computational chemistry in different fields have been presented.

1.1 Introduction

The branch of chemistry that studies chemical systems and processes using mathematical models and computational simulations is known as *"computational chemistry."* It works by using the principles of theoretical chemistry and physics along with computer science to predict chemical properties and behaviors of systems and solve chemical problems. In chemistry, many aspects of molecules can be computed, including bond angles, bond lengths, dihedral angles, electron density distributions, activation energies, excited states, vibrational frequencies, chemical shifts, coupling constants, etc. [1, 2]. Also, the thermodynamic properties and kinetic parameters are calculated with the aid of computation [3]. Computational chemistry has also proven its worth in the field of materials science by predicting the properties of semiconductors, superconductors, ceramics, and other advanced materials [4]. One of the pivotal domains where computational chemistry finds wide use is the study of protein folding, protein dynamics, and drug delivery. Through molecular modelling and molecular dynamics (MD), theoretical studies related to protein folding and protein interaction with other molecules can be achieved [5]. The application of computational chemistry across various fields is illustrated in Figure 1.1. Most of these calculations are quite precise, and the calculated values are quite close to the experimental data, while the remaining aspects can be obtained approximately. Even in the approximate methods, sufficiently accurate data that are in accordance with practical experiments can be obtained [6]. Thus, computational chemistry plays an important role in chemical reaction simulations, studying the MD of atoms and molecules, molecular modelling by simulating the structure and behavior of molecules, analyzing the properties of materials at the molecular or atomic level, drug designing, and calculating the molecular properties by studying the interactions using quantum principles.

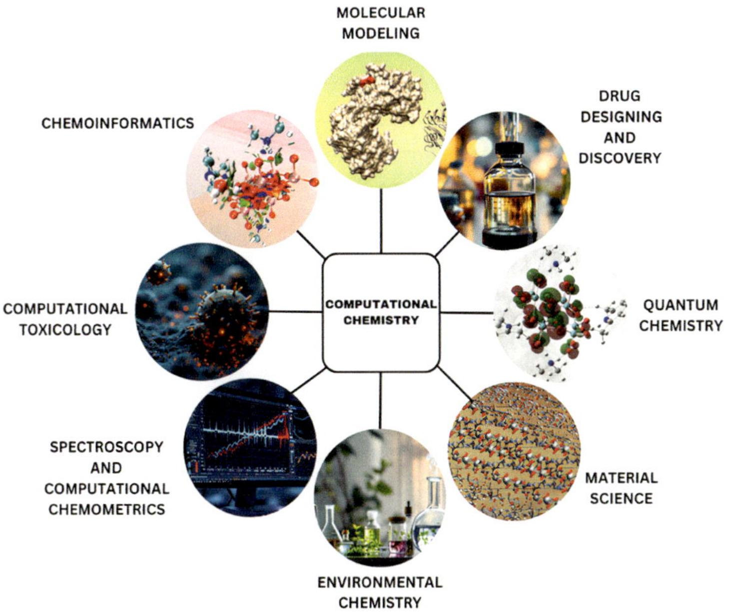

Figure 1.1 Illustration of applications of computational chemistry in various fields.

The use of computational methods dates back to the 1940s and 1950s. With the development of computers, the first semi-empirical atomic orbital calculations were performed [7]. Initially, the focus was on developing simple calculations and theoretical frameworks. During this period, the computational cost was very high, and only simple calculations could be performed. But in the 1950s, computational methods underwent significant theoretical advancements and developments, such as the introduction of Hartree-Fock (HF) calculation, configuration interaction (CI), and poly-atomic calculation using Gaussian [8]. During the 1960s, the empirical methods were replaced by semi-empirical methods such as CNDO [9]. Even after these developments, the study was still limited to small molecular systems with high computational costs. Also, these calculations often take weeks or days to complete, even on mainframe computers. During the 1970s and 1980s, the development of efficient ab initio computer programs, such as ATMOL, Gaussian, IBMOL, and POLYAYTOM, and the introduction of force fields such as

MM2 resulted in a rapid expansion of computational methods [10]. Also, as personal computers and workstations came into existence, computational costs began to decrease, and calculations that once took weeks could be done in a few hours or days. Some of the advancements of the 21st century include the development of time-dependent DFT methods, hybrid quantum mechanics and molecular mechanics, GPU-accelerated calculations, and multi-scale modelling techniques. Additionally, with the introduction of machine learning and artificial intelligence (AI), an exponential growth in the use of computers and computational methods has been witnessed, as evidenced by the 2024 Nobel Prize in computational protein design using neural networks. With all these developments and advancements, computational costs have decreased drastically, and calculations that once took months can now be achieved in minutes or maybe even seconds. Also, the introduction of supercomputers and cloud technology allows complex simulations with relative ease.

In recent times, other than theoretical chemists, researchers from different fields of interest have also started taking advantage of computational chemistry. For example, synthetic chemists often take the help of computational chemistry to model and study molecular systems prior to synthesizing them in the laboratory, which helps to reduce wastage of raw materials and labor and to stop the generation of unnecessary toxic waste. It also helps us study and analyze chemical properties and phenomena that are difficult to study experimentally. Again, with the advancement of computational technologies and the availability of user-friendly software, the ease with which computational calculations can be done even without any knowledge of the underlying principles has attracted many people, including theoreticians working in other fields, to start using computational chemistry. Nevertheless, it becomes very important for a researcher to understand what computational methods can and cannot predict. Hence, some basic theoretical knowledge is very much essential for every researcher utilizing various tools of computational chemistry to reduce computational time and search space and to refine their results.

In this chapter, an overview of the basic principles of the various types of computational methods has been presented. Initially, a summary of basic knowledge has been presented that a person must possess to smoothly assimilate the knowledge in the field of computational chemistry. Subsequently, the underlying principles, advantages, disadvantages, and applications of different theories used in the various dimensions of computational chemistry have been discussed. Finally, a summary of the various software developed for the different types of computations and visualizations has been included.

1.2 Elementary Terminologies

1.2.1 Theories, Models, and Approximations

In any scientific method, the first step is to form a "hypothesis." A hypothesis is an educated or rational guess from the already known facts, and then compared with all available data to develop its details. When such hypotheses are known to be consistent with known facts over a period, then they are usually called "theories" and published. It must be noted that "theories" are scientifically tested and proven. Normally, once established as theories, they are useful for explaining one or more observed phenomena and also help in predicting the results of experiments to be performed in the future. Nowadays, theories that have proven their worth over a significant period of time are also used to substantiate new experimental data. Generally, the theories are presented in mathematical form, which represents a particular phenomenon. However, the mathematical calculation difficulties might be so great that it may not be possible to solve them directly. Thus, to achieve results, often a small, insignificant part of the mathematical representation is left out of the overall calculation, or an average-based calculation is performed instead of an exact mathematical description.

Models are representations of the theories and can be constructed in a variety of forms. Hence, models may or may not be mathematical descriptions. As per Bailer-Jones (2009), models could be anything ranging from toys to abstract objects [11]. For example, Bohr's model of the atom produces an image of electrons revolving around the nucleus in the minds of people. In fact, models usually aid in the visualization of the whole theory, and thus, they can be very useful for understanding any phenomena without the need for doing complex mathematics related to the theories. At the same time, it is important to realize that these models will always have flaws and exceptions.

Approximations that are often used in physics and chemistry are generally a form of mathematical constructs derived from other complex mathematical expressions. In general, these complex mathematical expressions or functions are truncated or modified to generate simpler functions that yield results that are approximately equal to the solutions of the original functions. The simpler functions are known as approximate functions, and the process of generating such simpler functions with approximate solutions is known as approximation methods or simply approximations. Some of the approximate solutions may be very clumsy, while others are more accurate than any data from experiments. The entire field of computational chemistry is based on approximations. For example, the Born–Oppenheimer (BO) approximation is a well-known approximation method used for quantum mechanical studies. Nevertheless, computational chemists must acquire sufficient knowledge about the approximations they are dealing with and how accurate these approximate results will be for their investigation.

1.2.2 Atomic Units

The atomic unit, also referred to as the Hartree system, is a concept where some of the fundamental constants are considered as unity. The value of these fundamental constants in atomic units, as well as the most commonly used units, is listed in Table 1.1. The concept of atomic units eases the work of theoreticians and simplifies overall theoretical calculation, which is nowadays mostly performed with the aid of various software. It should be noted that most software yields results where values are predicted in atomic units, which need to be suitably converted prior to any comparison with macroscopic properties.

Table 1.1 Values of fundamental properties in atomic and relevant units

Property	Atomic unit	Value in relevant unit
Length	Bohr	1 Bohr = 0.529177249 Å
Weight/Mass	Atomic mass unit (amu)	1 amu = 1.6605402 $\times 10^{-27}$ kg
Energy	Hartree	1 Hartree = 27.2116 eV
Charge	Electron charge	1 electron = 1.602188 $\times 10^{-19}$ C
Charge separation	Bohr electron	1 Bohr electron = 2.541765 Debye
Time	Specialized time	1 specialized time = 2.418884326509 $\times 10^{-17}$ s
Action	Reduced-Plank Constant (h)	1 h = 1.054 $\times 10^{-34}$ Js

1.2.3 Computable Quantities

As mentioned in earlier sections, computational chemistry can be utilized as a tool to calculate and predict a variety of physical parameters. Thus, there are certain physical quantities that are commonly calculated and studied by computational chemists. These are known as computable quantities, and the most significant computable properties are discussed next.

1.2.3.1 Electronic structure or optimized geometry

Isolated molecules are the fundamental unit from which all materials are assembled. Some basic criteria are required to specify a molecule, such as the atoms of which it is composed and the manner in which those atoms are connected. In fact, the connections between the atoms of a molecule reveal the relative positions of all the atoms in space. This connectivity, or "bonding," is itself a property that can be determined. The determination of a connectivity or bonding pattern between the atoms is one of the most fundamental computable properties of molecules and is known as electronic structure or optimized geometry calculation. In this case, the "correct" or "best" structure is the one having the least possible energy for a particular given overall connectivity pattern decided by the theoretician, and is known as the optimized structure. This optimized geometry not only yields basic structural information, such as bond lengths, bond, and dihedral angles within a molecule, but can be further utilized for computation of other properties such as weak/strong interactions, transition, and excited-state calculations.

1.2.3.2 Energy

Systems or molecules having the least energy are the most stable. The potential energy of the system or molecule comprises bending energy, stretching energy, hydrogen bonds, and conformational energy, whereas the kinetic energy of the system comprises rotational, translational, and vibrational energy. Also, it is important to define a reference system (having zero energy) for formulating any mathematical representation of molecules. This reference or zero-energy varies for each approximation [12]. Certain molecular methods use stainless molecules as zero energy, whereas other methods assume chemical standard states for the same purpose. However, in any particular approximation, the total energy value relative to the methods of zero energy is very inaccurate. This inaccuracy is the result of systematic error [12]. Thus, it must be noted that the most accurate values or relative energies are obtained by subtracting total energies from separate calculations. In addition to various types of potential energies and kinetic energies, it is possible to compute thermodynamic parameters such as internal energy, enthalpy, free energy, and heat capacity [3]. In fact, electronic structure and energy calculations are often combined for the prediction of potential energy surfaces (PESs).

1.2.3.3 Electrostatics

Most interactions within a molecule are a consequence of electron-electron interactions that are commonly described by electron density or wavefunction. The study of the interaction between electrons or charged particles is termed electrostatics. According to Coulomb's law, the force of interaction and energy between two particles with charges q_1 and q_2, which are at a distance r_{12} is:

$$F = \frac{q_1 q_2}{r_{12}^2}$$

Another very useful function of electrostatics is the electrostatic potential (Φ). It is a function defined at every point in 3D space. The energy of placing an infinitesimal point charge at any point in space without any movement of existing charges (polarization of electron density) in the system is the electrostatic potential times the charge on the particle [12]. Also, in the above Coulomb's law, we assume that the charges are separated by a vacuum. If the charges are separated by any medium, the dielectric constant of that medium would also be included. The electrostatic potential (Φ) is related to charge density (ρ) by the "Poisson equation." It can be solved both numerically and analytically. The Poisson equation is expressed as:

$$\nabla^2 \Phi = -\rho$$

Poisson's equation is used to obtain the electrostatic properties of molecules.

1.2.3.4 Spectral properties

Once the optimized geometry of the molecule or system has been achieved, that structure can be further utilized for the calculation of a wide range of spectral properties that include nuclear magnetic resonance (NMR) spectra, chemical shifts, coupling constants, electron paramagnetic resonance (EPR) spectra, vibrational, and electronic spectra and electron affinities and ionization potentials [1, 2].

1.3 Methods

1.3.1 Ab Initio Methods

Ab initio is a term derived from Latin that means "from the beginning." Thus, ab initio methods are directly derived from the most fundamental principles or theories. Such methods do not use or incorporate any form of experimental data. It is an approximate quantum mechanical calculation. These approximations are usually mechanical, using a simpler form for a function or finding an approximate solution to a differential equation [12]. Ab initio methods include HF approximations, correlation, and methods that are discussed below.

1.3.1.1 HF approximations

The HF approximation is one of the fundamental principles of quantum chemistry that helps in approximating the behavior of a multi-electron system. This method helps us simplify complex many-body problems into sets of manageable one-electron equations, allowing calculations related to molecular energies and properties. It provides the foundational framework for understanding and calculating the electronic structures of molecules, while balancing computational feasibility with theoretical precision. This approximation consists of three important concepts, namely, the central-field approximation, the variational principle, and the wavefunctions and molecular orbitals of the system [12]. According to the central-field approximation, the HF method assumes that each electron moves independently in the average field created by all the other electrons, thus simplifying the process of solving the electron-electron interactions. This is obtained by integrating the coulombic repulsion term, thus giving an average effect rather than using the specific interactions. Also, the total wavefunction of the system of electrons is represented as Slater's determinant, ensuring that the wavefunction is asymmetric with reference to the exchange of electrons, thus meeting the requirements of fermions. This representation helps us with the formation of molecular orbitals from linear combinations of atomic orbitals (LCAO) using Gaussian-type orbitals (GTOs) as the basic functions.

The HF approximation has three methods. They are Restricted Hartree-Fock (RHF), Unrestricted Hartree-Fock (UHF), and Restricted Open Shell Hartree-Fock (ROHF). The RHF is used in systems where all electrons are paired, and both spin-up and spin-down electrons are in the same spatial orbital. Although this method is computationally systematic, it does not accurately describe systems with unpaired electrons. However, the UHF provides separate spatial orbitals for both spin-up and spin-down electrons, thus involving systems with unpaired electrons. However, this method can bring in contamination and affect accuracy depending on the type of chemical system. Hence, a new method of ROHF was formed, which involves both RHF and UHF. It can be used for both paired and unpaired electrons and minimizes the spin contamination [12].

HF calculations always give energies that are higher than the HF limits because of the approximations in the correlation of electrons [12]. This limitation often leads to errors in activation energies, bond lengths, and other properties. However, HF is a valuable tool, as it acts as a starting point for more sophisticated methods such as Density Functional Theory (DFT) and Post-Hartree-Fock methods.

1.3.1.2 Correlation

Many calculations in quantum chemistry initially begin with HF calculations and then adjust themselves for electron correlation. It includes various types of correlated calculations, such as Møller-Plesset Perturbation (MP) Theory, Configuration Interaction (CI), Multiconfigurational Self-Consistent Field (MCSCF), Multireference Configuration Interaction (MRCI), and Coupled Cluster (CC) Theory. These techniques collectively are known as Correlated calculations.

Incorporating correlation improves the accuracy of molecular geometries and computed energies. In organic compounds, correlation acts as an added refinement to achieve high accuracy (it is not required to achieve quantitative results) [12]. The exceptions to this rule include compounds that show Jahn-Teller distortion, which normally require correlation to get quantitatively accurate results. Also, in some extreme cases, transition-metal systems require correlation for achieving correct results [12].

1.3.1.2.1 *MP theory*

The MP theory is a widely used method in computational chemistry for adding electron correlation to HF wavefunctions. MP theory considers the wavefunction as a first-order perturbation, with correlation effects added through higher-order corrections. The second-order method, known as MP2, adds a minimal amount of correlation. Higher-order calculations, such as MP3 and MP4, are also common, with MP4 providing accuracy comparable to Configuration Interaction with Singles and Doubles (CISD) [13]. However, even when higher-order calculations like MP5 can be performed, they are rarely used due to the high computational cost, which often scales with the complexity of N10 or even worse.

The MP2 calculations are not variational in nature, and thus, the total energies obtained are generally lower than the exact total energy [12]. Depending on the chemical system, the behavior of energy across different orders varies. For some systems, energies decrease successively, whereas for others, MP2 may yield an energy below the exact value, with fluctuations in higher orders and a slight alteration around the exact energy. The advantage of Møller-Plesset is its extensive size. There is also a local version, known as local MP2 (LMP2), which requires less computational time compared to standard MP2 calculations. It is less vulnerable to basic set superimposition errors. However, LMP2 recovers about 98% of the energy correction provided by traditional MP2 [12, 13].

1.3.1.2.2 *CI*

CI improves the HF wavefunction by including electron correlation by a multiple-determinant approach. The CI wavefunction is a linear combination of multiple Slater determinants generated by promoting electrons from occupied to unoccupied orbitals [14]. This method gives a much more accurate representation of the electron correlation of a multi-electron system. There are four types of CI calculations. They are

- Configuration Interaction Singles (CIS): Involves single excitations with only one electron being promoted. It helps approximate excited states but does not alter the ground state energy.
- Configuration Interaction Singles and Doubles (CIDS): It involves both single and double excitations and provides a corrected ground state energy that is responsible for the correlation effects.
- Higher-order excitations such as CISDT and CISDTQ: CISDT involves triple excitation, while CISDTQ involves quadrupole excitations, which are generally used in cases where high accuracy is expected.
- Full configuration interaction (FCI): This method involves all possible excitations and gives an exact solution within the given set, but is rarely performed due to resource limitations and high cost.

The CI calculations have a high computational cost and often scale in the complexity of N8 or worse. Thus, it is more suitable for smaller molecular systems [14]. Also, the results from CI calculations vary for different packages due to differences in the HF reference state used. However, these differences are very small, and they diminish with higher-order calculations. CI methods are very useful for understanding electronic structures and excitation energy of various chemical systems.

1.3.1.2.3 *MCSCF*

MCSCF is an advanced computational method that extends far beyond the traditional single-determinant approaches. It uses multiple determinants to achieve the most accurate representation of electronic wavefunctions [15]. It optimizes orbitals, particularly for multi-determinant wavefunctions, and provides a few configurations with efficient correlation energy calculations. It is thus more efficient than CI calculations. However, CI calculations achieve a higher total correlation energy as they include more configurations.

MCSCF plays an important role in cases where the HF determinant's weight in single-reference CI calculations provides poor qualitative system descriptions [16]. But there are a few challenges faced while using it; thus, it requires experts due to the important process of active space selection [16]. Also, the user is required to choose the appropriate molecular orbitals, while the proper correlation needs to include both bonding and antibonding orbitals. The user also needs a proper orbital selection strategy, as poor selection can decrease accuracy when more orbitals are added, leading to inaccurate results. It is important to keep these points in mind, as the success of MCSCF calculations highly depends on the expertise and judgement of the user while selecting appropriate configurations and orbitals.

MCSCF generally has two variations, namely, the Complete Active Space Self-Consistent Field (CASSCF), which is famous for maximum valence-region correlation and includes all possible combinations of the active space [17]. The next one is the Restricted Active Space Self-Consistent Field (RASSCF), where smaller sets of Configuration State Functions are used by restricting the number of electrons in certain subspaces [18].

1.3.1.2.4 *MRCI*

MRCI is an advanced computational technique in which calculations start with an MCSCF wavefunction instead of an HF wavefunction, and the MCSCF wavefunction serves as the reference state. In this method, more CI determinants are included when compared to conventional CI calculations [19]. This method is computationally intensive and requires a large amount of computing resources. Also, this method is very costly. But, these methods have certain benefits, including the ability to provide optimal correlation for specific problems and to offer a more thorough electronic structure description.

This method uses specific abbreviations to indicate the type of calculation [12]. MCSCF +1+2 are the MRCI calculations with MCSCF references of single (+1) and double (+2) excitations. CASSCF+1+2 and GVB+1+2 are similar variations that use complete active space self-consistent fields (SCFs) and generalized valence bonds, respectively, as references. It is important to note that the notations +1 and +2 indicate the inclusion of single and double excitations from respective reference spaces.

1.3.1.2.5 *CC and related approaches*

The core concept of CC is quite similar to that of CI in cases related to multiple determinants, with wavefunctions having a linear combination of determinants [20]. However, the CC method has a much more sophisticated determinant selection process compared to that of CI. Hierarchically, there are various orders of CC expansion, such as CCSD, which includes single and double excitations; CCSDT, which includes single, double, and triple excitations; and CCSD(T), which includes triple excitations perturbatively.

The Quadratic Configuration Interaction (QCI) is an approach that includes both CI and CC algorithms [12]. Thus, the QCISD helps us approximate CCSD, while QCISD(T) includes the perturbative triple excitations. It also provides optimal correlations for organic molecules. Also, another variation of the CC method is the Symmetry Adapted Cluster (SAC), which is a size-consistent approach [21]. For excited states, a CI out of this space, called a SAC-CI, is done, which gives us the CI interactions from the SAC space. It is quite popular in the case of excited states as it improves the accuracy of electronic excited-state energies [22]. It is also computationally less intensive compared to FCI [21]. The use of the Brueckner Double technique nullifies the single excitation contribution by optimizing the orbitals. It also includes double excitations. It gives us accuracy and computational cost that is similar to CCSD [23]. The choice of method depends on the size of the system, the accuracy expected, the type of chemical problem, the availability of computational resources, and so on.

1.3.1.3 Quantum Monte Carlo (QMC) methods

The QMC method is an alternative approach that ignores the limitations of HF calculations by considering initial wavefunctions with explicit electron correlations. There are various types of QMC methods, such as the Variational Monte Carlo, Diffusion Monte Carlo, and Green's Function Monte Carlo [12]. The wavefunction for a variational QMC calculation might take the functional form:

$$\Psi = D_\alpha D_\beta \prod_{i<j} f(r_{ij})$$

where, D_α and D_β are the determinants for the and spin electrons, and $f(r_{ij})$ accounts for electron correlations. The presence of $f(r_{ij})$ in the wavefunction means that the correlations are explicitly taken into consideration, unlike the case of HF.

The variational QMC method searches for the wavefunction of the above form that has the minimum expectation value of energy, therefore approximating the ground state of the system [24].

Some important points regarding this method are that it does not introduce errors due to the dual determinant approach and is valid in cases where the Hamiltonian lacks spin terms (no spin coupling). It should be noted that QMC evaluates integrals over wavefunctions for calculating the expectation values of energy by using Monte Carlo (MC) integration. Hence, it is extremely time-consuming. However, it is one of the most accurate methods available to date.

1.3.1.4 Natural orbitals

Natural orbitals are a crucial tool used for calculating molecular properties efficiently, especially in cases involving multiple-determinant methods such as MCSCF, CC, CI, and MRCI [12]. These orbitals are defined as the eigenfunctions of the first-order reduced density matrix. The main advantage of natural orbitals is their ability to compress the complex wavefunction information. The information is condensed into a set of orbitals and their corresponding occupation numbers, representing the electron population in each natural orbital, instead of working with many determinants.

These numbers are close to 1 or 2 electrons for occupied orbitals, whereas higher-energy orbitals tend to have smaller occupation values [12]. The weight of these small occupations is slightly like the weights of excited configurations of the wavefunctions in higher orbitals, but they are not mathematically identical. One can compute the expected value of the property operator with the help of natural orbitals by weighing their respective occupation numbers, which are useful for calculating their molecular properties.

1.3.2 Density Functional Theory (DFT)

In recent years, DFT has become very popular because it is computationally less intensive compared to other methods with similar accuracy [55]. This theory has been developed recently compared to the ab initio methods. It was developed by Walter Kohn and Pierre Hohenberg in the 1960s. Walter Kohn then won the Nobel Prize in 1998 for his work in DFT. This method helps us to investigate the electronic structure of many-body systems, particularly the atoms, molecules, and the condensed phase, and is used to predict various properties of molecular systems and materials in physics, chemistry, and materials science.

The basic assumption behind DFT is that the energy of a molecule can be determined from the electron density instead of a wavefunction. This theory originated from the two theorems by Hohenberg-Kohn, in which the first theorem stated that the ground-state properties of a many-electron system are uniquely determined by the electron density, and the second theorem stated that the universal function for the energy (E_n) can be defined in terms of density, where the exact ground state is the global minimum of this functional. The original theorem was only applicable to finding the ground-state electronic energy of a molecule. Thus, the practical development of this method was done by Kohn and Sham, who formulated a method with a similar structure to the HF method. Thus, the electron density in this formulation is expressed as a linear combination of basis functions with a similar mathematical form to HF orbitals. So, the Kohn-Sham approach introduces a system of non-interacting electrons that generates the same density as the real systems. It also allows the kinetic energy to be computed exactly by improving accuracy. The SCF is used to solve the Kohn-Sham equations. The most challenging part of DFT is to approximate accurately, as various levels of approximation exist, such as LDA, GGA, meta-GGA, and hybrid functionals. Also, in DFT, pseudopotentials are often used to represent the effect of core electrons and reduce the computational cost. DFT can also treat systems with unpaired electrons by using separate functionals for up- and down-spin densities. Time-dependent DFT (TDDFT) is an extension of DFT that handles time-dependent phenomena and excited states. This method is generally accurate for ground-state properties and structures but faces challenges with certain excited states, strongly correlated systems, and long-range interactions.

Linear scaling techniques are an interesting development in DFT. The Coulomb terms are replaced by these algorithms in distant regions of the molecule with multipole expansions, which results in a method with time complexity of *N* for large molecules. Fast Multipole Method (FMM) and Continuous Fast Multipole Method (CFMM) are two common linear-scaling techniques [56, 57]. It uses the numeric integral accuracy and basis set. Linear scaling techniques are useful only in cases where the number of heavy atoms exceeds 30 or more, and depend on the shape of the molecule. Moreover, the linear systems are most effective, while the folded, globular, or planar systems are less efficient. CPU time may vary significantly for different conformers of the same molecule. Manual adjustment of multipole order can improve performance but might affect the accuracy of the result. Thus, it is necessary to take precautions when manually adjusting the multipole, considering its potential impacts on accuracy.

Nowadays, DFT-based calculations have received tremendous attention among researchers for predicting electronic structure and other molecular and spectral properties, due to computational accuracy when compared to HF and semi-empirical methods. However, computational accuracy highly depends on the functional and basis set used for a particular system. Thus, it is highly recommended to undergo a literature survey to wisely choose a functional and a basis set for a certain system.

1.3.3 Semi-Empirical Methods

Semi-empirical methods are modifications of the ab initio methods, where computationally expensive steps are replaced by approximations and parameters obtained from empirical data. These are usually employed when the system of interest is large, and ab initio methods become too computationally expensive. Some of these approximations include excluding the typical core electrons, minimizing the basis sets, omitting the two-electron integrals, using parameters to estimate the omitted values, etc. The parameterization of semi-empirical methods is based on the experimental or ab initio methods and can be customized for specific properties such as the electronic spectra, NMR shifts, etc. The excluded integrals are often replaced by these parameters.

Semi–empirical calculations are faster compared to the ab initio methods since most of the computationally intensive calculations are skipped. They are quite efficient for moderate-sized organic molecules and have some specific methods designed for inorganic chemistry [12]. Limitations of semi-empirical methods include the unpredictability of results or the predictability of limited properties [12]. The accuracy of this method depends on the similarity of the parameterization set with the molecules under investigation. Also, special cases like cubane and cyclopropane might not yield good results [12].

In this method, it is possible to compute the properties beyond the parameterization set [9]. It does not require zero-point corrections to calculate the heat of formation and is very useful for calculating the electronic excited states through cis-calculations [12]. It is more flexible than molecular mechanical calculations and also successful in organic chemistry, but with limited elements.

1.3.3.1 Hückel

Developed in the 1930s by Erich Hückel, it is one of the earliest and simplest semi-empirical methods and is used to calculate the electrons and shapes of π molecular orbitals, which are crucial for understanding the properties and chemical reactions of organic molecules [25]. It is mainly used for planar, conjugated systems like butadiene, benzene, and other aromatic compounds [25].

This method also uses some assumptions for performing calculations. These assumptions include treating σ and π orbitals separately, considering the overlap between adjacent p_z orbitals to be constant, assuming that only the p_z orbitals contribute to the π-bonding, and parameterization of coulomb and exchange integrals [26]. It also includes the secular determinant used for solving the molecular orbital energies. This method has a few limitations, such as lower accuracy for larger systems, neglect of electron-electron repulsion, and limited application only to planar molecules. Due to these limitations, the Extended Hückel Theory (EHT) was developed, which helped address some limitations of Hückel by including all valence orbitals [27].

1.3.3.2 Pariser-Parr-Pople (PPP)

This method got its name from its developers, Rudolph Pariser, Robert G. Parr, and John A. Pople, who independently developed it in the 1950s [28–30]. This method is used to calculate the electronic properties of π-conjugated systems by improving the Hückel method and including the electron-electron interactions [31]. This method uses the minimal basis set, usually consisting of one p-orbital per atom. This method acts as a bridge between the normal Hückel theory and the modern ab initio methods [31]. One electron integral and two electron repulsions are included in the PPP Hamiltonian.

It normally deals with the parameters such as the resonance integrals (β), one-center repulsion integrals (U), and two-electron repulsion integrals (γ). The important development included in this method over the Hückel method is that it gives better estimations of excitation energies, accounts for correlation effects of electrons, and is more accurate in describing charged and excited states. It uses the SCF approach and CI for excited states.

This method can be utilized for studying electronic transitions in conjugated molecules, calculating UV-visible spectra, and probing charge transfer in organic semiconductors [29]. The limitation of this method is that it depends mostly on empirical parameters, is restricted to π-electron systems, and is less accurate for non-planar molecules. This method still serves as the conceptual foundation for understanding the electronic structure in conjugated systems.

1.3.3.3 Extended Hückel

Extended Hückel Theory (EHT) was developed by Roald Hoffmann in the 1960s based on the work of Hückel [27]. It is a semi-empirical method in quantum chemistry used for calculating molecular orbitals and their energies for a wide range of molecules, which includes non-planar molecules and systems having heteroatoms. This system includes all valence orbitals and not just the π orbitals. It uses a larger set of atomic orbitals and also considers overlap between non-neighboring atoms [27].

This method uses the Roothaan equation to solve the molecular orbital energies and coefficients. The advantages of this method include its computational efficiency, its ability to provide qualitatively correct results for a wide range of molecules, and its ability to manage larger systems than the ab initio methods. This method also has a few limitations, such as neglecting the electron-electron interactions and dependence on empirical parameters. Also, the results provided are less accurate when compared with the results of advanced methods [12].

Even with the availability of many modern methods, EHT is still used in specialized areas, especially inorganic and organometallic chemistry [12]. We also find the use of EHT in the qualitative analysis of electronic structure and initial optimization of geometry. In fact, EHT also serves as an initial point for many advanced calculations.

1.3.3.4 CNDO

Complete Neglect Differential Overlap (CNDO) is an important semi-empirical approach in computational chemistry. It was improved by John Pople and his colleagues in the 1960s as a part of the series of semi-empirical methods [32, 33]. It is used for calculating the molecular electronic structure and properties by providing a balance between computational accuracy and efficiency. It involves the one-electron terms and the simplified form of two-electron interactions. This method includes all valence electrons, neglects the differential overlap in two-electron integrals, and uses a minimal basis set consisting of one *s* and three *p* orbitals per atom. The approximations included in this method neglect three- and four-centered integrals, while two-centered two-electron integrals have zero-differential overlap, and the remaining integrals are parameterized. The parameters considered in this method are the atomic orbital energies, bonding parameters, and electron-electron repulsion parameters.

This method is more comprehensive than the Hückel and extended Hückel methods. It also includes some electron correlation effects while being applicable to a wide range of molecular systems. It is generally used for calculating molecular properties, such as ionization potentials, dipole moments, geometry optimization, and qualitative analysis of electronic structure. This method still has some limitations, which include its lower accuracy compared to the ab initio methods, results being sensitive to the parameter choices, and problems with systems involving strong electron correlation. It has three variations, namely, CNDO/1, which is the original formulation; CNDO/2, which provides better experimental data through improved parameterization; and CNDO/S, which is the formulation specifically for spectroscopic calculations, yielding improved predictions of excitation energies, but may compromise molecular geometry predictions. The CNDO method laid the path for the development of more advanced semi-empirical methods, such as the Intermediate Neglect of Differential Overlap (INDO) and the Neglect of Diatomic Differential Overlap (NDDO).

1.3.3.5 INDO

The Intermediate Neglect of Differential Overlap (INDO) method is a semi-empirical formulation. It was formulated in the 1960s by John Pople and his colleagues to improve the CNDO method [33, 34]. This method is used to calculate molecular properties and electronic structure. This method works by including all valence electrons; it also retains one-center exchange integrals and uses a minimal basis set consisting of one *s* and three *p* orbital atoms. The methodology consists of repeatedly solving the modified Roothaan-Hall equation using the SCF approach. It is this method that bridged the gap between CNDO and more sophisticated approaches.

The parameters considered in this method are the bonding parameters, the atomic orbital energies, and the one-center electron repulsion integrals. Variations of this method include INDO/1, INDO/2, which have different parameterizations for the improvement of performance, and INDO/S, which is specifically used and parameterized for spectroscopic calculations. This method provides better treatment of electron exchange while giving a better description of atomic spectra. INDO calculations are more accurate for open-shell systems and are applicable to a wide range of organic molecules while being computationally efficient. The INDO method also finds use in spectroscopic predictions, qualitative analysis of electronic structure, geometry optimization, and calculating molecular properties such as dipole moments and ionization potentials. However, like all other methods, it also has disadvantages, such as lower accuracy compared to modern methods, variable performance depending on the type of molecule or its properties, and unsuitability for certain systems, such as transition-metal compounds.

1.3.3.6 MINDO

Modified Intermediate Neglect of Differential Overlap (MINDO) was developed in the 1970s by Michael Dewar and his colleagues to improve the INDO method [35]. This method plays a major role in the development of semi-empirical methods by providing the groundwork for more advanced methods such as the Modified Neglect of Diatomic Overlap (MNDO) and the Austin Model 1 (AM1). It is mostly used for calculating the molecular electronic structure, properties, and energies, improving accuracy over earlier semi-empirical methods. It has three versions, namely, MINDO/1, which is the initial formulation; MINDO/2, which includes improved parameterization; and MINDO/3, which is both successful and widely used [36].

This method uses the SCF approach and repetitive solution for the modified Roothaan-Hall equations. It includes atomic and diatomic parameters that are optimized for reproducing experimental molecular properties. This method is computationally efficient, rationally accurate for many organic molecules, and has the ability to handle large systems when compared with the ab initio methods. This method also provides a better core-core repulsion treatment, handles heteroatoms well, and is very accurate in predicting the molecular geometries and heat of formation. The limitations of this method include its unsuitability for transition-metal compounds, lower accuracy compared to modern methods, and the variation of performance based on the type of molecule or the property being studied. Even now, MINDO/3 is still occasionally used for obtaining quick, preliminary results.

1.3.3.7 MNDO

MNDO is an empirical formula developed by Michael Dewar and Walter Thiel in 1977, an advancement in MINDO/3 [37]. This method is useful for calculating molecular electronic structure, energies, and properties, providing accuracy over earlier semi-empirical methods [12]. This method also uses the SCF approach and repetitive solution for the modified Roothaan-Hall equations. This method also includes all valence electrons, uses a minimal basis set of one *s* and three *p* orbital atoms, and works based on the Neglect of Diatomic Differential Overlap (NNDO) approximation. The approximation in this method is that it neglects differential overlap between atomic orbitals of different atoms, while it retains all one-centered integrals and uses parameterized forms for the two-centered integrals.

This method is normally useful for calculating molecular properties, geometry optimization, conformational studies, and the analysis of reaction pathways [38, 39]. The parameters include optimization to reproduce the experimental molecular properties, including diatomic and atomic parameters. It is also typically parameterized for elements up to fluorine in the periodic table. This method provides better treatment of hydrogen bonding and improved descriptions of hypervalent compounds, while providing accurate prediction of molecular geometries and heat of formation. It is computationally efficient and relatively accurate for organic compounds and has good performance for ground state properties. However, this method has limitations, such as lower accuracy compared with advanced computational methods, and its performance varies depending on the property or type of molecule, and is not well-suited for transition-metal compounds. It also sometimes underestimates the electronic excitation energies and activation barriers calculated using this method, which predict higher values [12]. This method made a significant contribution to the improvement of the accuracy of semi-empirical methods while laying the groundwork for many advanced methods, such as AM1 and Parameterization Method 3 (PM3). The MNDO has two variants: MNDO/d, which includes the *d*-orbital for a better description of hypervalent compounds and transition metals; and MNDO/c, which includes core-core repulsion corrections.

1.3.3.8 ZINDO

Zerner's Intermediate Neglect of Differential Overlap (ZINDO) is one of the important semi-empirical computational methods of quantum chemistry. It was developed by Michael Zerner and colleagues in the 1970s and 1980s as an extension of INDO methods [40]. The primary use of this method is in calculating the electronic spectra and properties of large molecules such as organic chromophores and transition-metal complexes. This method has two variants, namely, ZINDO/1, which is parameterized for ground-state properties, and ZINDO/S, which is specifically optimized for electronic spectra [12].

This method works based on the INDO framework, specially parameterized for spectroscopic calculations. It also includes *d*-orbitals, making it suitable for transition-metal compounds. The parameters are initially optimized for reproducing the experimental spectroscopic data and include specific parameters for many transition metals. It usually uses a minimal basis set of *s*, *p*, and *d*-orbitals and involves the INDO/S Hamiltonian with modified parameters. This method is often combined with CI calculations for the excited states. It helps in predicting electronic transitions, excited states, UV-Visible spectra, and photochemical processes, and it also facilitates the study of large organic and organometallic systems [41, 42].

This method is computationally efficient and allows the calculation of large systems. It is good at providing the accuracy of spectroscopic properties and can solve transition-metal complexes. It is also quite useful for qualitative and semi-quantitative interpretations of electronic spectra. Historically, this method is significant as it filled the crucial gap in predicting the electronic spectra of large molecules and complexes. It also showed the potential of tailored semi-empirical methods for specific properties. This method is still used for the spectroscopic calculations of large molecules, especially those with large systems [43]. It is also still used for the initial screening of chromophores and photosensitizers, and often serves as a benchmark for more advanced spectroscopic calculation methods. Nevertheless, this method has its own limitations, such as less accuracy when compared to the high-level ab initio or DFT methods, the variation in performance depending on the system studied, and not being as reliable for ground state properties or geometries.

1.3.3.9 SINDO1

Symmetrically Orthogonalized Intermediate Neglect of Differential Overlap – version 1 (SINDO1) is a semi-empirical quantum chemical approach in computational chemistry. It was introduced in the 1980s as an advancement of the then-existing INDO-based methods by Karl Jug and his colleagues [44]. This method is structured to calculate molecular properties and geometry with improved accuracy compared to earlier semi-empirical methods, particularly for systems containing main-group elements.

This method works based on using the INDO approximation for two-electron integrals. It normally uses a minimal basis set of one *s* and three *p* orbitals per atom, while using symmetrically orthogonalized atomic orbitals, and includes all valence electrons. This method is parameterized to optimize reproduced experimental molecular properties and is applicable to various main-group elements. This method is computationally efficient, provides better accuracy for main-group elements, and can handle larger molecular systems [44]. Nevertheless, this method serves as the reference point for the development of more advanced semi-empirical methods. However, the setbacks arise as its performance can vary depending on the specific property and systems.

1.3.3.10 PRDDO

The Partial Retention of Diatomic Differential Overlap (PRDDO) method is a semi-empirical method for calculating molecular electronic structure and properties, improving earlier semi-empirical methods by accurately computing one-electron integrals over Slater-type atomic orbitals. It was introduced as an intermediate approach between the INDO and NNDO methods. This method is based on the Hartree-Fock-Roothaan formula and involves an SCF approach [45]. This method also consists of a few diatomic differential overlap integrals, which are neglected in INDO. It partially retains some two-center two-electron integrals while including all valence electrons. This method includes a few approximations, such as neglecting some but not all differential overlaps between the atomic orbitals on differential atoms while retaining all one-center integrals. It uses parameterized forms of two-centered integrals. This method is more accurate when calculating a few molecular properties [45]. This method especially helps us with electron-structure analysis and the study of reaction mechanisms, along with geometry optimization and calculations of molecular properties. This method is computationally more accurate than the INDO methods and also computationally more efficient for handling larger molecular systems than present ab initio methods. This method is not widely used or parameterized like some other semi-empirical methods. This method attempts to bridge the gap between the INDO method and advanced methods such as AM1 and NDDO.

1.3.3.11 AM1

AM1 is a semi-empirical quantum chemical approach in computational chemistry. It was developed in 1985 by Michael Dewar and his colleagues as an improvement of the MNDO method [46]. This method works based on the NDDO approximation and uses a minimal basis set that includes all valence electrons. The improvements of this method over MNDO include a modified core-core repulsion function, the addition of Gaussian functions to model the attractive and repulsive interactions, and the better treatment of hydrogen atoms. This method is optimized to produce a wide range of experimental molecular properties, with parameters being available for elements up to Nobelium (atomic number, $Z = 102$). This method facilitates geometry optimization, calculation of various molecular properties, analysis of reaction pathways, conformational studies, drug design, and QSAR studies.

This method, which is computationally efficient, is more accurate than MNDO when calculating many properties. It also has the ability to handle large molecular systems and provides good performance for organic molecules. However, the main drawback of this method is that it sometimes overestimates the hydrogen bond strengths. This method is widely adopted in various computational packages. This method has two variations, namely AM1-D, which includes corrections for empirical dispersions, and AM1-Bond Charge Correction (AM1-BCC), which is a modified version with improved partial-charge calculations in force fields. In modern times, this method is used for performing quick calculations on large systems, employed in initial screening for drug discovery, and behaves as a starting point for more advanced calculations. This method, when integrated with other methods, is often used in Quantum Mechanics

(QM) and Molecular Mechanics (MM) calculations. It can also provide initial structure for higher-level calculations and is extensively tested and compared against high-level theoretical methods and experimental data.

1.3.3.12 PM3

PM3 was designed by James J.P. Stewart in 1989 as a parametrized and improved version of the AM1 method [47]. It is designed to calculate the molecular electronic structure, energies, and properties with improved results. It works based on the NDDO approximation, including all valence electrons while using a minimal basis set, normally containing one *s* and three *p* orbitals per atom. This method has more extensive parameterization with a modified core-core repulsion function. It also has a different approach for parameter optimization. The parameters are optimized using a large dataset of experimental and high-level theoretical data, and they are available for a wide range of elements, including transition metals.

This method finds its use in the calculation of molecular properties, optimizing geometry, analyzing conformations, and in drug design and QSAR studies. It is computationally efficient and more accurate than the AM1 systems. This method can handle large molecular systems and give good results in both organic and inorganic compounds. However, this method is less accurate when compared with the high-level ab initio or DFT methods, and this system sometimes faces problems with including the hydrogen bonds.

This method has two extended variations named PM3-CARB1, which is specifically parameterized for carbohydrates, and PM3-PDDG, which is the pairwise-distance-directed Gaussian modification for improved variation. In modern times, this is still used for quick calculations on large systems and is often employed in the initial screening of drug discovery.

1.3.3.13 PM3(tm)

PM3(tm) is a semi-empirical method used for accurately modelling transition metal (tm)-containing systems. It was developed as an improvised version of PM3 for the better handling of transition metals by J.J.P. Stewart and his colleagues [48]. It is based on the NDDO approximation and includes the *d*-orbitals of the transition metals while retaining the core structure of the PM3 model for transition metals. It employs the SCF approach and uses a modified set of parameters for transition metals.

The parameter is optimized using high-level theoretical calculations and experimental data on transition-metal compounds. Parameters are available for all first-row transition metals and some of the second-row transition metals. This method has enhanced accuracy for transition-metal systems when compared to standard PM3. It is also computationally efficient, allowing for calculations on large organometallic systems. It is also capable of handling different spin multiplicities and oxidation states. When compared with the standard PM3 method, it also provides a better description of metal-ligand bonding, predicts the geometries of organometallic compounds more accurately, and provides improved energetics for reactions involving transition metals. While this method is very useful, it faces problems when dealing with highly ionic metal-ligand bonds.

This method finds applications in the geometry optimization of organometallic compounds, the calculation of molecular properties of transition-metal complexes, studying the reaction mechanism of metal centers, and understanding catalytic processes. In modern times, this method is still used for quick calculations on large organometallic systems and has proven valuable for initial screening in computational catalytic studies. The PM3 modified methods, i.e., PM7 and PM6, are also reported.

1.3.3.14 Fenske-Hall

Developed in the 1970s by Marvin Hall and Robert Fenske, the Fenske-Hall method is a simplified molecular orbital approach. It is a quantum chemical approach used in computational chemistry, especially for studying the electronic structure and bonding properties of transition-metal complexes, organometallic compounds, and other inorganic compounds. It is a semi-empirical method based on Hückel theory that includes the *d*-orbitals of the transition metals [49]. It uses a minimal basis set comprising only valence orbitals and employs the parameterized Coulomb and exchange integrals. The Fenske-Hall method, when compared to other methods, does not use the SCF repetitions. This method calculates molecular orbitals and their energies, and provides information about the charge distribution and bonding. It is computationally efficient, provides a good qualitative description of bonding in metal complexes, and is also very useful in analyzing the trends in series of related compounds. This method finds applications in the analysis of electronic structure in organometallic compounds, studying metal-ligand interactions, investigating spectroscopic properties, and examining reactivity trends. This method focuses on valence orbitals, neglects core electrons, and uses experimentally derived parameters for atomic orbital energies [49]. It is simple compared to other ab initio methods or DFT methods and is more specialized in cases of transition metals than the general semi-empirical methods like MNDO or AM1. It is this method that helped us understand the bonding in organometallic compounds while contributing to the development of molecular orbital theory for inorganic chemistry. But the limitation of this method is that it is not suitable for geometry optimization and is limited to single-point calculations.

1.3.3.15 SAM1

Semi Ab Initio Model 1 (SAM1) was introduced by Michael J.S. Dewar and his colleagues in 1993 to improve the earlier semi-empirical methods like AM1 and PM3 [50]. The main aim of this method is to bridge the gap between fully empirical and ab initio methods to calculate the molecular electronic structure, energies, and properties. This method includes all valence electrons by incorporating some ab initio-like elements and functions based on the NDDO approximation. This method involves elements of the ab initio calculations for deriving the parameters. This method has a more rigorous theoretical foundation, improved treatment of core-core repulsion, and better description of hydrogen bonding when compared to the methods used before it. This method finds its use in geometry optimization, studying the reaction mechanism, calculating molecular properties, and investigating hydrogen-bonded systems. This method is more accurate than the AM1 and PM3 for many properties and provides better theoretical justification for its approximations. The performance of this method can largely vary depending on the type of molecule or property, and is less accurate than the high-level ab initio or DFT calculations. This method is not as widely used as other semi-empirical methods in commercial software packages, but it has influenced the development of more advanced semi-empirical approaches.

1.3.3.16 Gaussian Methods

Gaussian methods are a widely used family of computational techniques in quantum chemistry. This method was developed by John Pople and his colleagues in 1970, with the first version being the Gaussian 70 [51]. It uses the linear combinations of Gaussian-type orbitals (GTOs) to approximate the molecular orbitals and implements various levels of theory ranging from semi-empirical to high-level ab initio methods. It provides a wide range of computational methods, including HF, DFT, MP2, and Coupled Cluster Singles and Doubles (CCSD), and has an extensive basis set library. In the Gaussian basis set, Gaussian functions are used to represent the atomic orbitals. Its basis set has various levels of basis sets available and is more efficient than Slater-type orbitals.

Gaussian methods are capable of performing geometry optimization, frequency calculations, and searching transition states [52–54]. It is designed to provide highly accurate thermochemical data like enthalpies of formation, electron affinities, ionization energies, etc. It aims to attain the accuracy of FCI calculations at a fraction of the computational cost. Thus, it finds applications in excited-state calculations, geometry optimization, studying the reaction pathways, predicting the spectroscopic properties, performing thermochemistry calculations, and vibrational frequency analysis. This method is highly versatile and widely used. It is also well tested and documented. It provides the effective implementation of many quantum-chemical methods. Although this method has various benefits, it is computationally expensive for large systems or high-level methods. A few such examples of such calculations are shown in Figure 1.2.

Gaussian methods may be further categorized into three variations, namely Gaussian-1 (G1), Gaussian-2 (G2), and Gaussian-3 (G3). These are a series of composite quantum chemical approaches developed to achieve high accuracy in predicting molecular properties. They use a series of single-point energy calculations on an optimized geometry. Briefly, Gaussian-1 (G1), introduced in 1989, uses HF/6-31G* level of theory for geometry optimization and includes MP4 calculations with increasingly large basis sets. Also, it applies empirical corrections. The Gaussian-2 (G2) variation

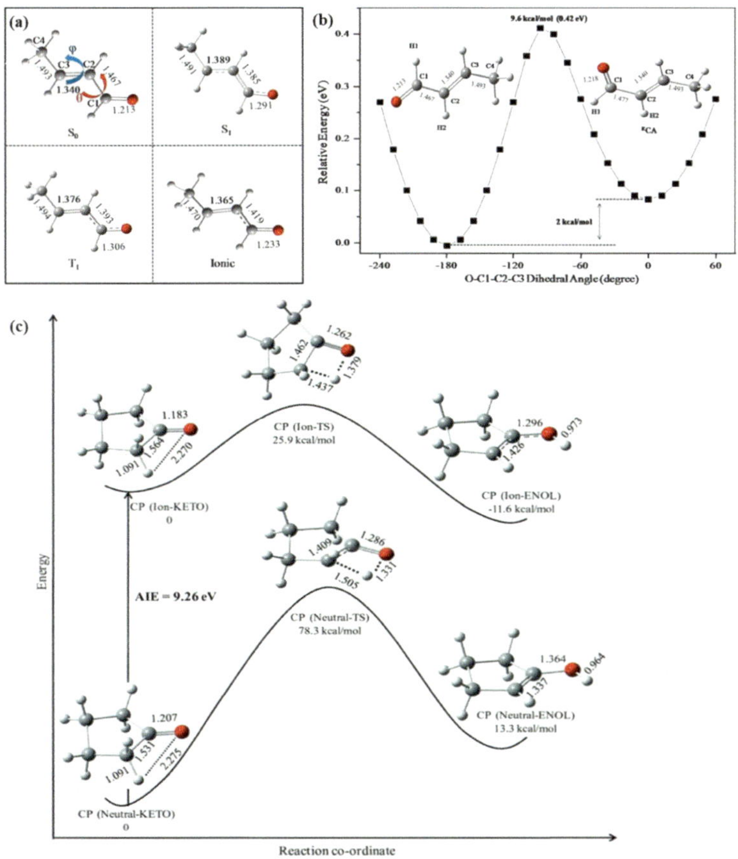

Figure 1.2 (a) Geometries of trans-crotonaldehyde in S_0, S_1, T_1, and ionic states optimized at DFT/B3LYP-6311++G(d,p) and TDDFT/B3LYP-6311++G(d,p) levels for the excited states. (b) Energy profile for internal rotation of crotonaldehyde molecule about the C1–C2 bond in the S_0 state. (c) Potential energy profile for tautomerization of cyclopentanone (CP) monomer in the neutral (bottom) and ionic (top) ground states calculated at DFT/B3LYP/6-311++G(d,p) level of theory. The C–C and C–O bond lengths are mentioned in Å units. Reproduced from ref. [53, 54] with permissions.

was introduced in 1991 as an improvement for G1 and includes the Quadratic Configuration Interaction Calculation (QCISD(T)). Lastly, the Gaussian-3 (G3) version was introduced in 1998 for further refinement of G2 and includes spin-orbit corrections of atoms. Out of all these, G3 is normally the most accurate, and typically the accuracy in terms of mean absolute deviation of experimental values is in the order G3 < G2 < G1. It is still used for high-accuracy calculations on small to medium-sized molecules. It also has two more extensions: G4 theory, which is the more recent extension of the Gaussian-n series, and G3(MP2), which is a cost-effective version of G3 using MP2 instead of MP4 calculations.

1.4 Molecular Mechanics

When it comes to modelling of larger molecules such as biomolecules or macromolecules, ab initio methods do not prove computationally feasible even on the most powerful computers. Also, the semi-empirical methods used for large organic molecules are computationally challenging for most biomolecular systems. Thus, molecular mechanics provides an alternative pathway for treating very large molecules that cannot be effectively handled by semi-empirical treatments by avoiding quantum mechanics completely [12].

The foundation of molecular mechanics is based on classical mechanics instead of quantum mechanics and treats molecules as collections of atoms held together by bonds. The force field is an important part of molecular mechanics and describes the potential energy of a system as a function of atomic positions. It also contains mathematical formulations and parameters. The force fields are parameterized using experimental data and high-level quantum calculations. These parameters are specific to a certain class of molecules. This method consists of two energy components: bonded interactions, such as bond stretching, angle bending, and torsional rotation, and the non-bonded interactions, such as the van der Waals forces and electrostatic interactions. In molecular mechanics, various methods, such as steepest descent and conjugate gradient, are utilized to determine the lowest-energy conformation of a molecule. However, molecular mechanics cannot model the electronic properties or chemical reactions. Also, the accuracy of this method depends on the quality of the force field and its parameters. Commonly used force fields are CHARMM, AMBER, GROMOS, etc., and certain special force fields are also used for specific types of systems.

Molecular mechanics is often used to simulate bulk properties, such as protein folding, drug designing, and material properties. This method allows simulations of large systems for relatively long timescales.

1.4.1 Existing Force Fields

Over the years, many force fields have been developed. Generally, researchers mostly do not parameterize such force fields. Rarely, in some specific cases, researchers may add an additional term. Some commonly used force fields in software packages are discussed below. However, there may be minor differences in the implementation of these force fields across software, which may lead to minor differences in the obtained results.

1.4.1.1 AMBER

The Assisted Model Building with Energy Refinement (AMBER) is a widely used force field in computational chemistry and MD simulations. It was initially developed for proteins and nucleic acids, but has been expanded to cover a wide range of organic molecules since then. Its potential energy function includes fine-bonding and non-bonding terms along with sophisticated electrostatic treatment, and no cross terms are included. Due to its well-parameterized protein and nucleic acid force fields, AMBER is used in biomolecular simulations. It is important to remember that the AMBER has limitations and might not be equally accurate for all molecules under all conditions.

1.4.1.2 CHARMM

Chemistry at Harvard Macromolecule Mechanics (CHARMM) is another important force field. It was developed at Harvard University and is known for its accuracy in modelling various biological and chemical systems, which include proteins, nucleic acids, lipids, and small molecules. This force field also includes the bonded and non-bonded interactions. CHARMM uses additional energy terms, such as Urey-Bradley terms and improper dihedral angles, that help improve the accuracy in certain situations [58]. CHARMM 36 is an improved version of CHARMM that provides improved performance in membrane simulations.

1.4.1.3 CFF

Consistent Force Field (CFF) is widely used in computational chemistry and MD simulations. It has proved useful to accurately predict the molecular structures and properties of a wide range of organic and inorganic compounds [59–60]. It uses the quadratic terms for bond stretching and angle bending, cross-terms to account for coupling between different types of integral coordinates, which are the key features of the CFF force field. It is known for its ability to reproduce experimental vibrational frequencies across different molecular systems [61–65]. It successfully simulates small organic molecules, polymers, and biomolecules. This force field has different variations, such as the Urey-Bradley version (UBCF) and valence version (CVFF).

1.4.1.4 CHEAT

Carbohydrate Hydroxyls represented by External Atoms (CHEAT) is specially used for modelling carbohydrates.

1.4.1.5 DREIDING

It is an all-purpose force field developed. It is famous for its simplicity and broad applications in organic, biological, and main-group inorganic molecules. The unique features of DREIDING include the use of general force constants and geometry parameters based on simple hybridization considerations instead of specific atom types [66–68]. Even with its simple formulation, DREIDING shows good accuracy in predicting structures and energies across a wide range of molecular systems, making it important for initial studies and large-scale simulations in cases where computational efficiency is crucial.

1.4.1.6 ECEPP

The Empirical Conformational Energy Program for Peptides (ECEPP) is the name of both a force field and a computer program. This method was developed by Harold Scheraga and colleagues and is known for its use in fixing bond lengths and bond angles and focusing on torsional angles as the primary degrees of freedom [71]. It also has two improved versions, ECEPP/2 and ECEPP/3, which include improvements in its energy functions and parameters. Thus, the use of ECEPP is quite useful in studying protein folding, predicting peptide conformations, and analyzing the energy conformations of the biomolecules.

1.5 MD and MC Simulations

Statistical mechanics provides a means to determine physical properties not just associated with a single molecule but rather a macroscopic sample, which is a result of properties of different molecules in different conformations, energy states, etc. However, the difficult part is not statistical mechanics but obtaining all the information about the possible energy levels, confirmations, etc. MD and MC simulations are the methods used to obtain this information [12]. These two methods are fundamental computing techniques used for studying the behavior of molecular systems at both the atomic and molecular levels. So, these methods have become irreplaceable tools in the field of computational chemistry, biophysics, and materials science. Both these methods help researchers learn about phenomena such as phase transition, protein folding, and chemical reactions that are difficult or impossible to observe experimentally. The approaches of both these methods are quite different; however, they accompany each other and provide a clear and detailed understanding of the molecular system.

1.5.1 MD

MD simulations solve the time evolution of a system by numerically solving Newton's equations of motion for all particles and giving information about dynamic processes and time-dependent properties [69]. The key aspects of MD simulations include:

- The force field defines the potential energy functions governing the particle interactions.
- The integral algorithms, such as Verlet or leap-frog methods, can update the particle positions and energies.
- The periodic boundary conditions are used for simulating the bulk systems.

- Thermostats and barostats (temperature and pressure controlling algorithms) for maintaining the desired ensemble conditions.
- Time steps, generally in femtosecond scale, for capturing the high-frequency motions.
- And trajectory analysis for extracting the structural, dynamic, and thermodynamic properties.

The main setback of this method is the accuracy of the force fields and the accessible time scales that are typically of nano- and microsecond range for the all-atom simulations.

1.5.2 MC Simulations

MC simulations use random sampling for complex problems in molecular modelling. It explores the configurational space of a system by generating equilibrium properties using statistical averaging [70]. The key aspects of MC simulations include:

- The metropolis algorithms generate a sequence of system configurations based on the acceptance and rejection criteria.
- Sampling for focusing on the most relevant regions of the phase space.
- Markov chain process for ensuring proper sampling of the equilibrium distribution.
- Ensemble distributions, such as canonical and grand canonical, are used for determining the fixed value parameters.
- The move sets, such as particle displacements or rotations, define how the system is perturbed.
- Convergence criteria for determining when the equilibrium is reached.
- Ergodicity ensures that all possible states are reached when provided with enough time.

MC simulations are very good at calculating equilibrium properties and investigating complex energy landscapes. They are particularly useful for studying the adsorption phenomenon, phase transitions, and polymer conformations.

1.6 Summary

In this chapter, a brief history and introduction of the multiple theoretical methods used in the various fields of computational chemistry have been presented. Initially, a discussion on the ab initio methods, such as HF and correlation-based calculations. HF methods are useful for initial geometry optimization of molecular structures. However, HF methods predict higher energies, often leading to errors in activation energies and bond lengths. Nevertheless, these errors are rectified once the initial HF calculations are further refined using correlation methods such as MP, CI, MCSCF, etc. Recently, DFT-based calculations have become very popular for accurate prediction of electronic structures and energies. The TDDFT method has also been useful for predicting excited-state geometries and energies. The computational cost of running DFT calculations is higher compared to HF methods. On the other hand, semi-empirical methods, which are often used for calculating properties of moderately large systems, are computationally less expensive. The accuracy of semi-empirical methods is determined by the similarity of the parameterization set with the molecules. For studying the properties of very large systems such as macromolecules and biomolecules, molecular mechanics provides a good alternative. Molecular mechanics is based on classical mechanics rather than quantum mechanics, and the use of force fields is fundamental to molecular mechanics. The choice of force field determines the accuracy of the prediction. Some commonly used force fields include AMBER, CHARMM, CHEAT, etc. MD and MC simulations are computational tools that aid in studying various phenomena such as phase transitions, polymer conformations, and protein folding. These two methods predict the time evolution of molecular systems with the aid of Newton's motion equation and statistical mechanics. Last but not least, the introduction of artificial intelligence and other machine learning methods in the field of computational chemistry is opening new techniques and methodologies to ease the computational cost involved in theoretical calculations.

References

1. Bisong, E.A., Louis, H., Unimuke, T.O., Odey, J.O., Ubana, E.I., Edim, M.M., Tizhe, F.T., Agwupuye, J.A. and Utsu, P.M., 2020. Vibrational, electronic, spectroscopic properties, and NBO analysis of p-xylene, 3,6-difluoro-p-xylene, 3,6-dichloro-p-xylene, and 3,6-dibromo-pxylene: DFT study. *Heliyon*, *6*(12).
2. Khanum, G., Ali, A., Shabbir, S., Fatima, A., Alsaiari, N., Fatima, Y., Ahmad, M., Siddiqui, N., Javed, S. and Gupta, M., 2022. Vibrational spectroscopy, quantum computational, and molecular docking studies on 2-[(1H-benzimidazol-1-yl)-methyl] benzoic acid. *Crystals*, *12*(3), p.337.
3. Wijaya, C.D., Sumathi, R. and Green, W.H., 2003. Thermodynamic properties and kinetic parameters for cyclic ether formation from hydroperoxyalkyl radicals. *The Journal of Physical Chemistry A*, *107*(24), pp.4908–4920.
4. Schmidt, J., Marques, M.R., Botti, S. and Marques, M.A., 2019. Recent advances and applications of machine learning in solid-state materials science. *NPJ Computational Materials*, *5*(1), p.83.
5. Geng, H., Chen, F., Ye, J. and Jiang, F., 2019. Applications of molecular dynamics simulation in structure prediction of peptides and proteins. *Computational and Structural Biotechnology Journal*, *17*, pp.1162–1170.
6. Dehabadi, M.H., Saidi, H., Zafari, F. and Irani, M., 2024. Assessing the accuracy and efficacy of multiscale computational methods in predicting reaction mechanisms and kinetics of SN2 reactions and Claisen rearrangement. *Scientific Reports*, *14*(1), p.16791.
7. Mahoney, M.S., 2011. *Histories of Computing*. Harvard University Press.
8. Boys, S.F., Cook, G.B., Reeves, C.M. and Shavitt, I., 1956. Automatic fundamental calculations of molecular structure. *Nature*, *178*(4544), 1207–1209.
9. Pople, J.A. and Beveridge, D.L., 1970. *Molecular Orbital Theory. CO., NY.*
10. Ma, X., 2022. Development of computational chemistry and application of computational methods. In *Journal of Physics: Conference Series*, 2386(1), pp.012005, IOP Publishing.
11. Bailer-Jones, D.M., 2009. *Scientific Models in Philosophy of Science*. University of Pittsburgh Press.
12. Young, D.C., 2001. A practical guide for applying techniques to real-world problems. *Computational Chemistry, New York*, *9*, pp.390.

13. McGibbon, R.T., Taube, A.G., Donchev, A.G., Siva, K., Hernández, F., Hargus, C., Law, K.H., Klepeis, J.L. and Shaw, D.E., 2017. Improving the accuracy of Møller-Plesset perturbation theory with neural networks. *The Journal of Chemical Physics, 147*(16).

14. Shavitt, I., 1977. The method of configuration interaction. In: Schaefer, H.F. (eds) *Methods of Electronic Structure Theory: Modern Theoretical Chemistry, 3*. Springer, Boston, MA.

15. Al-Saidi, W.A., Zhang, S. and Krakauer, H., 2007. Bond breaking with auxiliary-field quantum Monte Carlo. *The Journal of Chemical Physics, 127*(14).

16. Helgaker, T., Jorgensen, P. and Olsen, J., 2013. *Molecular Electronic-Structure Theory*. John Wiley & Sons.

17. Sun, Q., Yang, J. and Chan, G.K.L., 2017. A general second-order complete active space self-consistent-field solver for large-scale systems. *Chemical Physics Letters, 683*, pp.291–299.

18. Tran, T., Segarra-Martí, J., Bearpark, M.J. and Robb, M.A., 2019. Molecular vertical excitation energies studied with first-order RASSCF (RAS [1, 1]): Balancing covalent and ionic excited states. *The Journal of Physical Chemistry A, 123*(25), pp.5223–5230.

19. Li, R.R. and Hoffmann, M.R., 2020. On the development and implementation of multi-CPU parallel versions of accurate, general-purpose methods of multireference perturbation theories. In *Advances in Quantum Chemistry, 81*, pp.105–141. Academic Press.

20. Bartlett, R.J., 2012. Coupled-cluster theory and its equation-of-motion extensions. *Wiley Interdisciplinary Reviews: Computational Molecular Science, 2*(1), pp.126–138.

21. Nakatsuji, H., 1997. SAC-CI method: Theoretical aspects and some recent topics. In *Computational Chemistry: Reviews of Current Trends*, pp.62–124.

22. Hirao, K. and Nakatsuji, H., 1981. Cluster expansion of the wavefunction: Symmetry-adapted-cluster (SAC) theory for excited states. *Chemical Physics Letters, 79*(2), pp.292–298.

23. Handy, N.C., Pople, J.A., Head-Gordon, M., Raghavachari, K. and Trucks, G.W., 1989. Size-consistent Brueckner theory limited to double substitutions. *Chemical Physics Letters, 164*(2–3), pp.185–192.

24. Toulouse, J., Assaraf, R. and Umrigar, C.J., 2016. Introduction to the variational and diffusion Monte Carlo methods. In *Advances in Quantum Chemistry, 73*, pp.285–314. Academic Press.

25. Seeman, J.I., 2023. Revolutions in chemistry: Assessment of six 20th-century candidates (The Instrumental Revolution; Hückel Molecular Orbital Theory; Hückel's 4 n+ 2 Rule; the Woodward–Hoffmann Rules; Quantum Chemistry; and Retrosynthetic Analysis). *JACS Au, 3*(9), pp.2378–2401.

26. Yates, K., 2012. *Hückel Molecular Orbital Theory*. Elsevier.

27. Hoffmann, R., 1963. An extended Hückel theory. I. Hydrocarbons. *The Journal of Chemical Physics, 39*(6), pp.1397–1412.

28. Pariser, R. and Parr, R.G., 1953. A semi-empirical theory of the electronic spectra and electronic structure of complex unsaturated molecules. I. *The Journal of Chemical Physics, 21*(3), pp.466–471.

29. Pariser, R. and Parr, R.G., 1953. A semi-empirical theory of the electronic spectra and electronic structure of complex unsaturated molecules. II. *The Journal of Chemical Physics, 21*(5), pp.767–776.

30. Pople, J.A., 1953. Electron interaction in unsaturated hydrocarbons. *Transactions of the Faraday Society, 49*, pp.1375–1385.

31. Kutzelnigg, W., 2007. What I like about Hückel theory. *Journal of Computational Chemistry, 28*(1), pp.25–34.

32. Pople, J.A., Santry, D.P. and Segal, G.A., 1965. Approximate self-consistent molecular orbital theory. I. Invariant procedures. *The Journal of Chemical Physics, 43*(10), pp.S129–S135.

33. Pople, J.A. and Segal, G.A., 1965. Approximate self-consistent molecular orbital theory. II. Calculations with complete neglect of differential overlap. *The Journal of Chemical Physics, 43*(10), pp.S136–S151.

34. Pople, J.A., Beveridge, D.L. and Dobosh, P.A., 1967. Approximate self-consistent molecular-orbital theory. V. Intermediate neglect of differential overlap. *The Journal of Chemical Physics, 47*(6), pp.2026–2033.

35. Baird, N.C. and Dewar, M.J., 1969. Ground states of σ-bonded molecules. IV. The MINDO method and its application to hydrocarbons. *The Journal of Chemical Physics, 50*(3), pp.1262–1274.

36. Bingham, R.C., Dewar, M.J. and Lo, D.H., 1975. Ground states of molecules. XXV. MINDO/3. Improved version of the MINDO semiempirical SCF-MO method. *Journal of the American Chemical Society, 97*(6), pp.1285–1293.

37. Dewar, M.J. and Thiel, W., 1977. Ground states of molecules. 38. The MNDO method. Approximations and parameters. *Journal of the American Chemical Society, 99*(15), pp.4899–4907.

38. Al-Resayes, S., Al-Farhan, K., HI, M.M. and Hasanein, A.A., 1998. MNDO optimized molecular geometries of some pyrimidine and purine derivatives. *Journal of Indian Chemical Society, 75*, pp.506–510.

39. Wierschke, S.G., 1987. Conformational analysis of poly-2,5-benzimidazole (ABPBI), poly-2,5-benzoxazole (ABPBO), and poly-2,6-benzothiazole (ABPBT) dimers by the modified neglect of diatomic overlap (MNDO), and Austin method 1 (AM1) semiempirical molecular orbital methods. Armed Forces Wright Aeronautical Laboratories, AFWL-TR-87-4034. Wright-Patterson Air Force Base, Ohio.

40. Ridley, J. and Zerner, M., 1973. An intermediate neglect of differential overlap technique for spectroscopy: Pyrrole and the azines. *Theoretica Chimica Acta, 32*, pp.111–134.

41. Patalinghug, W.C., Chang, M. and Solis, J., 2007. Predicting the shifts of absorption maxima of azulene derivatives using molecular modeling and ZINDO CI calculations of UV–Vis spectra. *Journal of Chemical Education, 84*(12), pp.1945.

42. Vlček Jr, A. and Záliš, S., 2007. Modeling of charge-transfer transitions and excited states in d6 transition metal complexes by DFT techniques. *Coordination Chemistry Reviews, 251*(3–4), pp.258–287.

43. Zemer, M.C., 1997. On calculating the electronic spectroscopy of very large molecules. *Problem Solving in Computational Molecular Science: Molecules in Different Environments*, pp.249–289.

44. Nanda, D.N. and Jug, K., 1980. SINDO1: A semiempirical SCF MO method for molecular binding energy and geometry I. Approximations and parametrization. *Theoretica Chimica Acta, 57*, pp.95–106.

45. Halgren, T.A. and Lipscomb, W.N., 1973. Self-consistent-field wavefunctions for complex molecules: The approximation of partial retention of diatomic differential overlap. *The Journal of Chemical Physics, 58*(4), pp.1569–1591.

46. Dewar, M.J., Zoebisch, E.G., Healy, E.F. and Stewart, J.J., 1985. Development and use of quantum mechanical molecular models. 76. AM1: A new general-purpose quantum mechanical molecular model. *Journal of the American Chemical Society, 107*(13), pp.3902–3909.

47. Stewart, J.J., 1989. Optimization of parameters for semiempirical methods II. Applications. *Journal of Computational Chemistry, 10*(2), pp.221–264.

48. Stewart, J.J., 1991. Optimization of parameters for semiempirical methods. III Extension of PM3 to Be, Mg, Zn, Ga, Ge, As, Se, Cd, In, Sn, Sb, Te, Hg, Tl, Pb, and Bi. *Journal of Computational Chemistry, 12*(3), pp.320–341.

49. Hall, M.B. and Fenske, R.F., 1972. Electronic structure and bonding in methyl-and perfluoromethyl (pentacarbonyl) manganese. *Inorganic Chemistry*, *11*(4), pp.768–775.
50. Dewar, M.J., Jie, C. and Yu, J., 1993. SAM1: The first of a new series of general-purpose quantum mechanical molecular models. *Tetrahedron*, *49*(23), pp.5003–5038.
51. Hehre, W.J., Lathan, W.A., Ditchfield, R., Newton, M.D. and Pople, J.A., 1970. *Gaussian 70* (Quantum Chemistry Program Exchange, Program No. 237).
52. Ghosh, A.K., Chattopadhyay, A., Mukhopadhyay, A. and Chakraborty, T., 2013. Isomeric effects on fragmentations of crotonaldehyde and methacrolein in low-energy electron–molecule collisions. *Chemical Physics Letters*, *561*, pp.24–30.
53. Ghosh, A.K., Datta, S., Mukhopadhyay, A. and Chakraborty, T., 2013. Signatures of isomerization in photodissociation of trans-crotonaldehyde probed by multiphoton ionization mass spectrometry. *The Journal of Physical Chemistry A*, *117*(36), pp.8710–8717.
54. Ghosh, A.K., Chatterjee, P. and Chakraborty, T., 2014. Keto-enol tautomerization and intermolecular proton transfer in photoionized cyclopentanone dimer in the gas phase. *The Journal of Chemical Physics*, *141*(4), pp.044303.
55. Kohn, W. and Sham, L.J., 1965. Self-consistent equations including exchange and correlation effects. *Physical Review*, *140*(4A), pp.A1133.
56. Greengard, L. and Rokhlin, V., 1987. A fast algorithm for particle simulations. *Journal of Computational Physics*, *73*(2), pp.325–348.
57. White, C.A., Johnson, B.G., Gill, P.M. and Head-Gordon, M., 1994. The continuous fast multipole method. *Chemical Physics Letters*, *230*(1–2), pp.8–16.
58. Brooks, B.R., Brooks III, C.L., Mackerell Jr, A.D., Nilsson, L., Petrella, R.J., Roux, B., Won, Y., Archontis, G., Bartels, C., Boresch, S. and Caflisch, A., 2009. CHARMM: The biomolecular simulation program. *Journal of Computational Chemistry*, *30*(10), pp.1545–1614.
59. Warshel, A., Levitt, M. and Lifson, S., 1970. Consistent force field for calculation of vibrational spectra and conformations of some amides and lactam rings. *Journal of Molecular Spectroscopy*, *33*(1), pp.84–99.
60. Rasmussen, K.J., Engelsen, S.B., Fabricius, J. and Rasmussen, B., 1993. The consistent force field: Development of potential energy functions for conformational analysis. In: Fausto, R. (eds). *Recent Experimental and Computational Advances in Molecular Spectroscopy. NATO ASI Series, 406*. Springer, Dordrecht.

61. Hagler, A.T., Huler, E. and Lifson, S., 1974. Energy functions for peptides and proteins. I. Derivation of a consistent force field including the hydrogen bond from amide crystals. *Journal of the American Chemical Society*, *96*(17), pp.5319–5327.

62. Hagler, A.T., Lifson, S. and Dauber, P., 1979. Consistent force field studies of intermolecular forces in hydrogen-bonded crystals. 2. A benchmark for the objective comparison of alternative force fields. *Journal of the American Chemical Society*, *101*(18), pp.5122–5130.

63. Hagler, A.T., Dauber, P. and Lifson, S., 1979. Consistent force-field studies of inter-molecular forces in hydrogen-bonded crystals. 3. C=O... HO hydrogen-bond and the analysis of the energetics and packing of carboxylic acids. *Journal of the American Chemical Society*, *101*(18), pp.5131–5141.

64. Michalska, D., Schaad, L.J., Carsky, P., Andes Hess Jr, B. and Ewig, C.S., 1988. Basis set effects and the choice of reference geometry in ab initio calculations of vibrational spectra. *Journal of Computational Chemistry*, *9*(5), pp.495–504.

65. Hwang, M.J., Stockfisch, T.P. and Hagler, A.T., 1994. Derivation of class II force fields. 2. Derivation and characterization of a class II force field, CFF93, for the alkyl functional group and alkane molecules. *Journal of the American Chemical Society*, *116*(6), pp.2515–2525.

66. Hwang, M.J., Stockfisch, T.P. and Hagler, A.T., 1994. Derivation of class II force fields. 2. Derivation and characterization of a class II force field, CFF93, for the alkyl functional group and alkane molecules. *Journal of the American Chemical Society*, *116*(6), pp.2515–2525.

67. Colón, Y.J. and Snurr, R.Q., 2014. High-throughput computational screening of metal–organic frameworks. *Chemical Society Reviews*, *43*(16), pp.5735–5749.

68. Dokur, D. and Keskin, S., 2018. Effects of force field selection on the computational ranking of MOFs for CO_2 separations. *Industrial & Engineering Chemistry Research*, *57*(6), pp.2298–2309.

69. Hollingsworth, S.A. and Dror, R.O., 2018. Molecular dynamics simulation for all. *Neuron*, *99*(6), pp.1129–1143.

70. Paquet, E. and Viktor, H.L., 2015. Molecular dynamics, Monte Carlo simulations, and Langevin dynamics: A computational review. *BioMed Research International*, *2015*(1), pp.183918.

71. Nicosia, G. and Stracquadanio, G., 2008. Generalized pattern search algorithm for peptide structure prediction. *Biophysical Journal*, *95*(10), pp.4988–4999.

Chapter 2

A Molecular Modelling Approach to Condensed Phase Chemical Dynamics Simulations

Manju Siyaram Yadav,[a] Ankita Agarwal,[a,b] Sk. Samir Ahamed,[a,c] and Amit Kumar Paul[d]

[a]*Department of Chemical and Biological Sciences, National Institute of Technology Meghalaya, Saitsohpen, Sohra, Meghalaya, India*

[b]*Department of Chemistry and Sustainability, School of Basic and Applied Sciences, Galgotias University, Greater Noida, Uttar Pradesh, India*

[c]*Department of Chemical and Biological Sciences, S. N. Bose National Centre for Basic Sciences, Salt Lake, Bidhannagar, Kolkata, West Bengal, India*

[d]*Department of Chemical Sciences, Bose Institute, Salt Lake, Kolkata, West Bengal, India*

amit.paul@jcbose.ac.in

This book chapter presents molecular modelling approaches for studying chemical dynamics simulations in the condensed phase. It emphasizes mainly two models, viz., molecular mechanics/molecular mechanics (MM/MM) and quantum mechanics/molecular mechanics (QM/MM). The simulation results are used to perform a systematic

Computational Methods for Molecular Modelling and Docking
Edited by Arup Kumar Ghosh and Suban Kumar Sahoo

ISBN 978-981-5352-25-2 (Hardcover), 978-1-042-00204-7 (eBook)
www.jennystanford.com

DOI: 10.1201/9781042002047-2

investigation, focusing on understanding the roles of environmental molecules in chemical reactions and intermolecular energy transfer processes. Under the MM/MM model, two types of condensed-phase models are explored: single and mixed-bath models. In the single-bath model, a bath of N_2 molecules is used to study energy transfer and association-dissociation dynamics of molecular systems. In the mixed-bath model, the bath consists of N_2 molecules along with other species such as NO or benzene, allowing a deeper understanding of how bath composition influences intermolecular energy transfer (IET). The QM/MM model is employed to study chemically reactive systems, where an accurate quantum-mechanical treatment of the system is essential. In this study, simulations are conducted in a single bath of N_2 molecules at varying densities and excitation temperatures to investigate post-transition-state dynamics, product distribution channels, and the role of solvent-mediated energy dissipation. By integrating these simulation methodologies and bath models, the chapter offers a comprehensive view of how condensed-phase dynamics differ from gas-phase dynamics and how the nature of the surrounding environment governs molecular reactivity and energy flow.

2.1 Introduction

Molecular modelling is rapidly advancing in science, encompassing activities from constructing molecular structures to visualizing them [1]. A key component of this field is molecular mechanics (MM), which uses the principles of classical or Newtonian mechanics to explain the forces and interactions in these models. In MM, atoms are typically represented as point masses with charges, and the bonds among them are modelled as springs. Additional forces, such as van der Waals interactions, describe how nearby atoms influence one another. This method is one of the simplest and quickest ways to analyze molecular systems. The core idea of MM is to calculate the potential energy of a molecular system based on the positions of its atoms. It provides a static view of molecular structure, where the system's energy is minimized to find the most stable configuration. This technique is fast and can handle larger systems, but it does not provide information on the dynamics or time-dependent behavior of molecules. Molecular dynamics (MD) builds upon the principles of MM but extends the analysis to simulate the time evolution of molecular systems. MD uses the equation(s) of motion to describe how atoms move and interact over a period of time. The force on each atom is calculated from the MM force field, and MD then solves the equations of motion to determine how the positions and velocities of atoms change at each small time step. MM provides the foundation for MD. The forces used to describe interactions in MD come from the MM force field. Without MM, MD would not have the rules to describe how atoms and molecules interact. MD is dynamic and MM is static. This is because MM focuses on finding low-energy configurations, while MD simulates the movement and interactions of molecules over time, enabling the study of kinetic properties.

In conclusion, molecular dynamics bridges the gap between the two computational chemistry approaches by simulating the time-dependent behavior of molecules using the potential energy surface produced by MM. MD, an essential computational approach, goes further by simulating the structural evolution and dynamic behavior of molecules, providing valuable insights into their physical properties and functions. MD helps understand the behavior of systems over time, making it a crucial tool across fields such as biophysics, materials science, and drug discovery.

MD simulations can be applied to various phases of matter, including liquids, solids, and gases, and typically aim to study phenomena such as protein folding, diffusion, molecular collisions, chemical reactions (on a classical scale), heat transfer, and energy redistribution in a molecular system. MD simulations can be performed in different environments (e.g., gas, liquid, solid, or vacuum), using periodic boundary conditions and various thermodynamic ensembles (NVE, NVT, NPT). Gas-phase dynamics focuses specifically on the behavior and interaction of molecules in the gas phase, where molecules are relatively free to move, with minimal interactions apart from occasional collisions. In gas-phase dynamics, molecular interactions and collisions are studied over short time scales (typically femtoseconds to picoseconds), and the systems are often simpler than condensed-phase environments. A key focus of gas-phase dynamics is on molecular collisions and how energy is transferred during these events. MD simulations are often used to investigate molecular collision dynamics, including how energy is distributed among the translational, rotational, and vibrational modes during a collision. MD can be simplified in the gas phase since the interactions among molecules are less frequent, and fewer forces need to be considered compared to condensed-phase simulations. Gas-phase dynamics are also often used to investigate chemical reactions, such as those in combustion or atmospheric chemistry [1–7]. MD can simulate these reactions by tracking how molecular trajectories evolve and how reactants transform into products over time.

While gas-phase dynamics are simpler, focusing on individual molecules and their infrequent interactions, condensed-phase dynamics are far more complex due to many-body collisions at every instant. The principles of molecular motion are the same, but the scale and complexity of interactions change dramatically when moving from gas-phase to condensed-phase systems. This connection is essential in understanding how molecular behavior in isolated conditions (gas phase) can influence or predict behavior in more complex, real-world environments (condensed phase). Condensed-phase dynamics gives the real scenario of how reactions like atmospheric and combustion are carried out in the real world and what the effect of surrounding molecules is on the solute system [8–13]. The surrounding molecules around the solute system can be a single type or multiple types. The latter is commonly termed as 'mixed-bath systems' [10, 12].

The rest of this chapter is arranged into three sections: Section 2.2 includes methodologies used for the dynamics, Section 2.3 reports all the results with related discussion, and Section 2.4 summarizes the whole content.

2.2 Methodology

In computational chemistry and molecular modelling [14], MM [4, 5, 7, 9, 10, 12, 15–17] and Quantum Mechanics (QM) [8, 18] are two approaches used to simulate the behavior of molecular systems. Each has its strengths and limitations, making them suitable for different types of simulations. In MM, molecules are modelled as collections of atoms connected by bonds, where atoms behave like particles. The system's potential energy is defined using classical physics rather than QM. In MM, empirical force fields are used to model interactions among atoms, including bonded interactions like bond stretching, angle bending, and torsion, as well as non-bonded interactions such as van der Waals forces and electrostatic interactions. MM is computationally efficient, making it possible to simulate large-scale systems, like proteins, lipids [19–23], or biomolecular systems [24–28], over long periods. It cannot model electronic structure changes, such as bond formation or breaking. It is best suited for systems where complicated chemical reactions do not take place. Assisted Model Building with Energy Refinement (AMBER) [29], Chemistry at Harvard Macromolecular Mechanics (CHARMM) [30–35], GROningen MOlecular Simulation (GROMOS) [36–42], and Optimized Potentials for Liquid Simulations (OPLS) [43–48] force fields are widely utilized in MM simulations.

QM simulations solve the Schrödinger equation to model the electron behavior within molecules. This approach captures the electronic structure and can account for bond formation, breaking, and charge distribution. QM is highly accurate and can describe complex chemical processes, such as reactions involving bond making/breaking [49], excited states [50], and charge transfer [51]. Common QM methods include Hartree-Fock (HF) [52, 53], Density Functional Theory (DFT) [54], and post-HF methods (e.g., MP2, CCSD) [55, 56]. QM simulations are computationally expensive, especially for larger systems. As the electronic size of the system grows, the computational expense rises substantially. QM is used for small systems where high accuracy is needed, such as chemical reactions, spectroscopy, and catalytic processes.

Hybrid QM/MM [8, 18] methods are commonly utilized to leverage the strengths of both approaches. These methods divide the system into two regions: a smaller, high-accuracy QM region and a larger, computationally efficient molecular MM region. QM/MM methods are especially valuable for investigating enzymatic reactions [57] as they allow the active site to be modeled using QM, while the surrounding environment is described using MM. This combined approach allows for the modeling of systems where both large-scale interactions (like those in proteins or solvents) and detailed chemical events (like reactions or bond changes) occur.

A comprehensive discussion on simulations utilizing the MM and QM/MM models is presented in Section 2.3.

2.2.1 MM Model

The model of MM is discussed in this article for the simulation performed for the polyaromatic system [9, 10, 12]. To perform chemical dynamics simulations for studying the behavior of polyaromatic hydrocarbons in the atmosphere, it is essential to determine the complete potential energy surface of the system, incorporating both intramolecular and intermolecular interactions. In a single molecule, intramolecular potential governs the interactions among atoms within the molecule itself, like bond elongation, angle deformation, and torsional forces, while intermolecular potential can also exist within the molecule when different parts of the same molecule interact through non-covalent forces, like hydrogen bonding or van der Waals forces, especially in huge or flexible molecules where distant regions may come into contact. Thus, both intramolecular and intermolecular potentials can simultaneously influence the molecule's structure and behavior.

Three simulations, including the MM model, are discussed in the chapter. Firstly, the association process of the molecule following the subsequent dissociation dynamics of the benzene (Bz) dimer molecule [9], and secondly, the dynamics of energy transfer from a vibrationally excited hexafluorobenzene (HFB*) to the bath of Nitric oxide/Nitrogen (NO/N_2) molecules [12] and lastly, the dynamics of energy transfer from a vibrationally excited benzene (Bz*) to a mixed bath of Bz/N_2 bath [10]. For all the simulations mentioned above, we need the intramolecular potential of each molecule and the intermolecular potential between the solute and solvent molecules for the particular simulation. The intramolecular potential energy Bz and HFB, utilized as solute molecules in the study, comprises Morse potential terms for in-plane C–F, C–H, and C–C stretching modes. Additionally, it includes harmonic bending terms for C–C–F, C–C–H, and C–C–C angles, as well as harmonic wagging modes for out-of-plane C–C–F and C–C–H movements, and torsional interactions. The intramolecular parameters for Bz and HFB are shown in Table 2.1 [4].

Table 2.1 The parameters for the intramolecular potential energy of the Bz molecule and the HFB molecule

Forcefield	Equilibrium parameters and force constants	Units	Bz	HFB
CH/CF stretching	r_e	Å	1.084	1.327
	f_s	mdyn $Å^{-1}$	5.076	6.640
CC stretching	r_e	Å	1.397	1.394
	f_s	mdyn $Å^{-1}$	6.640	5.800
CCH/CCF bending	θ_e	deg	120.0	120.0
	f_θ	mdyn Å rad^{-2}	0.521	0.800
CCC bending	θ_e	deg	120.0	120.0
	f_θ	mdyn Å rad^{-2}	1.065	1.550
CCH/CCF wagging	f_α	mdyn Å rad^{-2}	0.297	0.297
Torsional barrier	V_0	kcal mol^{-1}	24.00	24.00

The intramolecular potential energy of solvent molecules used in the study, such as N_2 and NO, is described by Morse stretches, which are given by Eq. (2.1).

$$V(r) = D_e\left(1 - e^{-\beta_e(r - r_e)}\right)^2 \tag{2.1}$$

The values of r_e, D_e, and β_e for the N_2 molecule are 1.09 Å, 228.3 kcal mol^{-1}, and 2.70 $Å^{-1}$, respectively [58]. Similarly, for NO, these parameters are 1.16 Å, 152.8 kcal mol^{-1}, and 2.83 $Å^{-1}$, respectively [59]. The computed stretching harmonic frequencies of NO and N_2 are 1966 and 2367 cm^{-1}, respectively. The corresponding experimental values are 1876 and 2330 cm^{-1} [60], respectively. Condensed-phase simulations require potential energy functions to account for all intramolecular and intermolecular interactions, including solute-solute, solute-solvent, and solvent-solvent interactions. Two-body interactions are described by analytical functions such as the Lennard-Jones and Buckingham functions. The general expression for all the intermolecular interactions is given by Eq. (2.2) [11].

$$V = \sum_i \sum_j V_{ij} = \sum_i \sum_j A_{ij} \exp\left(-B_{ij}\left(r_{ij} - b_{ij}\right)\right) + \frac{C_{ij}}{\left(r_{ij} - c_{ij}\right)^n} + \frac{D_{ij}}{\left(r_{ij} - d_{ij}\right)^m} + \frac{E_{ij}}{\left(r_{ij} - e_{ij}\right)^l} \tag{2.2}$$

Here, V represents the total intermolecular potential energy, while V_{ij} denotes the potential energy for a specific dual-body interaction among the ith and j^{th} atoms. The intermolecular potential parameters for interactions, including Bz–N_2, HFB–N_2, HFB–NO, N_2–NO, NO–NO, N_2–N_2, and Bz–Bz, are summarized in Table 2.2 [2, 11, 12].

The VENUS chemical dynamics program [61] is utilized to perform chemical dynamics simulations based on the system's potential energy. The simulations are performed while keeping the temperatures of solute molecules at lower and higher temperatures. The temperatures are converted to the corresponding quantum energies using standard statistical equations, and then, those energies are randomly added to the molecules using a quasiclassical microcanonical sampling technique [62], which produces a vibrationally excited molecule. The quantum vibrational energies, associated with vibrational modes, are calculated based on the principles of the harmonic oscillator model [63]. Using a standard algorithm, these energies are transformed into Cartesian coordinates and momenta, which serve as the initial conditions for the solute molecule. Additionally, the initial center of mass translational and rotational energy of a solute molecule is drawn from the Boltzmann distribution at a specified temperature.

In the condensed-phase implementation [61, 64], an extra algorithm is incorporated into the gas-phase code. The newly implemented algorithm enables the preparation of a bath system to investigate the influence of surrounding molecules. The solute molecule is positioned at the center of the box, maintaining its initial conditions. The solvent molecules are arranged within the simulation box, placed column by column from one side to the other, with random orientations [11]. If the solvent molecules are of more than one type, then each type of solvent molecule is placed alternately, considering the desired number for each type of solvent molecule [12]. The input parameters specify the spacing among solvent molecules in a column or layer, while the spacing among layers is modified to attain the target bath density. The radius of a sphere surrounding each solvent molecule is set to approximately twice the molecular bond length, ensuring free rotation and preventing overlap. The algorithm ensures that there is no overlap between the solute and solvent molecules. After the algorithm determines the position of a solvent molecule, the center-of-mass

Table 2.2 Parameters[1] for intermolecular potential energy Bz–N_2, HFB–N_2, HFB–NO, N_2–NO, NO–NO, N_2–N_2, and Bz–Bz interactions

Interactions	A	B	b	C	c	n	D	d	m	E	e	l
					Bz–N_2							
C–N	21476.800	3.2292490	–	–1411.4470	–	7	13298.140	–	10	0	0	0
H–N	18694.060	3.7056870	–	–2591.5260	–	8	4584.9460	–	11	0	0	0
					HFB–N_2							
C–N	99569.690	4.4959950	-0.22395410	–51703.280	2.6963240	7	99087.810	0.86939990	10	0	0	0
F–N	854509.80	4.2121510	0.54178930	–19432.070	3.9312110	6	78636.530	3.0781810	9	0	0	0
					HFB–NO							
F–N	90024.82	3.956846	–	-4037.780	–	8	12.21000	–	5	–	–	–
C–N	98997.84	3.549908	–	-15999.00	–	8	433.2978	–	10	–	–	–
F–O	98789.92	4.281916	–	-3538.850	–	8	8042.770	–	10	–	–	–
C–O	71349.20	3.598613	–	-8477.620	–	8	18.97000	–	5	–	–	–
					N_2–NO							
N–O	37970.52	3.442187	–	2728.720	–	7	10.00000	–	10	–	–	–
N–N	98999.99	3.907758	–	2421.660	–	8	13326.06	–	11	–	–	–

(*Continued*)

Table 2.2 *(Continued)*

Interactions	A	B	b	C	c	n	D	d	m	E	e	l
					NO–NO							
N–O	–	–	–	−0.2754268	–	1	16098.57	–	9	−625.3578	–	6
N–N	–	–	–	0.2754268	–	1	14480.38	–	9	−604.4070	–	6
O–O	–	–	–	0.2754268	–	1	18193.86	–	9	−658.8519	–	6
					N_2–N_2							
N–N	13454.280	4.5147010	0.023064110	−214.87920	−0.56835820	6	11305.300	−0.29055750	9	0	0	0
					Bz–Bz							
C–C	–	–	–	4.395186	–	1	1121755.6	–	12	−560.4387	–	6
C–H	–	–	–	−4.395186	–	1	73693.969	–	12	−116.2253	–	6
H–H	–	–	–	4.395186	–	1	4841.3406	–	12	−24.10313	–	6

[1]The units for the variables are as follows: A, B, and E are in kcal mol^{-1}; b, c, and e are in Å; C is in kcal $Å^n$/mol; D is in kcal $Å^m$/mol; and d is in kcal $Å^l$/mol.

spacing between the solvent and solute molecules is computed. If this distance falls below a specified threshold, the solvent molecule is prohibited from occupying that position, and an alternate location is attempted. The threshold is set to be slightly greater than the sum of the solute and solvent molecule lengths. After the bath is prepared with the required number of solvent molecules and desired density, an MD simulation is conducted to equilibrate it. The starting velocities of the solvent molecules are selected based on a Boltzmann distribution that matches the temperature of the bath. Equilibrating the bath is essential to shuffle the solvent molecules, which were initially arranged column by column during the preparation process. The equation of motion for the solute molecules is not solved in the equilibration process. During this process, solvent molecules sense the appearance of solute molecules and prevent them from overlapping, because the solute-solvent interaction potential is predefined. The dimensions of the box are established based on the bath density and the total number of bath molecules within the simulation box. Equilibration occurs in multiple stages, each consisting of a phase for sampling and a phase for monitoring. During the sampling, the velocity of each atom of the bath is resampled every five steps of integration. In monitoring, resampling is not performed, and conservation of energy is monitored. Equilibration is halted once the bath reaches the desired temperature, independent of the solute molecule's temperature. At this point, the translational, rotational, and vibrational degrees of freedom (DOF) of the solvent molecules acquire energy in accordance with the equipartition principle. Various configurations of the solvent molecules are then backed up to initiate the trajectory, during which the equation of motion for both solute and solvent molecules is solved. Once the trajectory gets started, the interaction between solute and solvent molecules starts. A 6th-order symplectic integrator [65] is employed to solve Hamilton's equations of motion with the specified timestep.

2.2.2 QM/MM Model

The QM/MM model is used for the simulation performed for the Catechol-ozone (cat-O_3) post transition state (TS) dynamics discussed in this section. The simulation is performed using the QM/MM model, a hybrid method in which the solute part of the system is solved by employing QM, whereas the solvent system is treated by MM. The detailed discussion about the simulation performed using the QM/MM model is on-fly-dynamics simulations of post-TS of ozonolysis of catechol reaction in an N_2 bath. The potential energy of the system is represented by Eq. (2.3).

$$V = V_{C_6H_6O_5} + \sum_{i=1}^{n} V_{N_2} + \sum_{i}^{n}\sum_{j\neq 1}^{n} V_{N_2-N_2} + \sum_{i=1}^{n} V_{C_6H_6O_3-N_2} \tag{2.3}$$

Here n represents the total number of N_2 molecules in the bath with i and j being indices of the N_2 molecules, $V_{C_6H_6O_5}$ and V_{N_2} are the intramolecular potential energies of cat-O_3 endo TS and N_2, respectively, while $V_{C_6H_6O_3-N_2}$ and $V_{N_2-N_2}$ denote the intermolecular potential energies among N_2 and N_2 and among $C_6H_6O_5$ and N_2, respectively. The intramolecular potential energy of the cat-O3 endo TS is derived from the semiempirical PM7 calculation using the MOPAC-2012 QM package. The intramolecular potential energy of N_2 is defined by an analytic function, as shown in Eq. (2.1). The values of D_e, β_e, and r_e for N_2, which are provided in Section 2.2.1, are taken from the literature. The ab initio curves developed for cat-O_3–N_2 interaction are by KS, B3LYP/6-311G+(2df,2p) theory [8], and for N_2–N_2 interaction is by MP2/aug-cc-pVTZ level of theory [66]. The ab initio curves of cat-O_3–N_2 interactions are fitted to the analytic function given by Eq. (2.4) considering four different orientations of cat-O_3–N_2 interactions and parameters A, B, C, n, D, and m are 10787.124 kcal mol^{-1}, 3.780 Å^{-1}, −228.549 kcal Å^{n}/mol, 6, 0.00 kcal Å^{m}/mol, and 0, respectively, for H–N interaction, 15470.024 kcal mol^{-1}, 3.190 Å^{-1}, −1411.447 kcal Å^{n}/mol, 7, 10298.139 kcal Å^{m}/mol, and 10, respectively, for C–N interaction, and 54138.126 kcal mol^{-1}, 3.634 Å^{-1}, −7718.414 kcal Å^{n}/mol, 9, 0.00 kcal Å^{m}/mol, and 11, respectively, for O–N interactions. The potential energies for the N_2–N_2 interaction are already mentioned in Section 2.2.1. Optimization of endo TS was previously carried out using CCSD(T)/extrapolated, MN15/6-311G+G(2df,2p), B3LYP/6-311G+(2df,2p) along with several semiempirical methods [18].

$$V(R) = A\exp(-Br) + \frac{C}{r^n} + \frac{D}{r^m} \tag{2.4}$$

The B3LYP/6-311G+(2df,2p) method was found to be both accurate and cost-effective. The semiempirical PM7 method yields potential energies that are nearly identical to those obtained using the B3LYP/6-311G+(2df,2p) method [18]. Thus, the QM portion of the dynamics was carried out using the PM7 method within the VENUS-MOPAC integrated condensed-phase code. The VENUS-MOPAC integrated condensed-phase code is the modified version of the VENUS chemical dynamics package, enhanced with condensed-phase algorithms [3], interfaced with a semiempirical electronic structure code MOPAC-2012 [67, 68]. This interface was developed to enable QM/MM simulations. The simulations are carried out at two different densities, i.e., 20 kg m^{-3} and 324 kg m^{-3}, with solute temperature as 800, 1000, and 1500 K at each density. The potential energy and its gradient for the solute system are derived from MOPAC-2012 (QM package), while the remaining calculations are performed using MM methods. The potential energy and its gradient from MOPAC-2012 for the reactive system are transferred to the VENUS code, which then solves Hamilton's equations of motion to update the coordinates of the reactive system during the simulation. In a Born-Oppenheimer (BO) direct dynamics [69, 70] simulation, the Hessian, along with the potential and its gradient, is directly calculated from the time-independent Schrödinger equation at each integration step. A sixth-order integrator is used for integration with a step size of 0.1 fs.

The formation and equilibration of the simulation box are the same as the MM model provided in Section 2.2.1.

2.3 Results and Discussion

2.3.1 MM/MM Simulations

2.3.1.1 Association followed by ensuing dissociation dynamics of Bz + Bz system in an N_2 bath

The chemical dynamics simulation of the association ensuing dissociation reaction of Bz–Bz dimer was performed to better understand the kinetics of association and dissociation dynamics of aromatic complexes. The association ensuing dissociation of Bz–Bz dimer in gas and condensed phases is performed and compared.

2.3.1.1.1 *Association dynamics of Bz + Bz*

In these simulations, the condensed-phase calculations were performed at excitation temperatures of each Bz of 1000, 1500, and 2000 K, while the gas-phase counterpart was studied at 1000, 1200, 1500, 1800, and 2000 K. For gas-phase simulations, impact parameters ranged from 0 to 6 Å, and for condensed-phase simulations, from 0 to 12 Å at each temperature. The criteria for forming a Bz–Bz complex include: (i) the center-of-mass distance among two Bz molecules must be ≤ 6 Å, (ii) the complex must have a lifetime of at least 1.2 ps, and (iii) the complex's lifetime is measured when the center-of-mass separation spans from 7.5 to 14 Å, with at least one center-of-mass oscillation within this duration. A trajectory is counted as a complex forming one if it meets all three aforementioned conditions.

For each temperature, 1000 trajectories were computed in the gas phase, while only 300 were integrated in the condensed phase due to the higher computational demands of the latter. The association probability of the Bz–Bz dimer is calculated as the ratio of complex-forming trajectories to the total trajectories. Table 2.3 lists the number of associated complexes per 1000 trajectories in the gas phase and per 300 trajectories in the condensed phase, at each temperature and impact parameter, along with the complex lifetime ranges for gas-phase simulations.

The percentage of associated complexes decreases as the simulation temperature rises in both the gas and condensed phases. In gas-phase simulations, the highest number of associated complexes occurs at an impact parameter of b = 6 Å for 1000 K, at b = 2 Å for 1500 K, and at b = 4 Å for 2000 K. The minimum energy geometry of the Bz dimer complex has a center-of-mass distance of 4.93 Å, while the orientation-averaged potential energy calculation shows a center-of-mass distance of 6.58 Å. As the temperature increases, the maximum association shifts to lower impact parameters. Table 2.4 lists the association probabilities across impact parameters from 0 to 12 Å at each temperature.

Table 2.3 The details of the trajectories forming complex: complex lifetime, the dissociated complexes in 3, 4, and 5 ps at the simulation temperatures of 1000, 1500, and 2000 K with an impact parameter range of 0–6 Å (for gas phase) and of 0–12 Å (for condensed phase) at each temperature

T (K)	Impact parameter, *b* (Å)	Complexes formed[2]	Complex time (τ) range (ps)	Number of complexes dissociated (%)[3] 3 ps	4 ps	5 ps
Gas Phase						
1000	0	133	2.03–18.53	31(23.3%)	60(45.1%)	80(60.2%)
	2	180	1.98–30.30	46(25.6%)	83(46.1%)	129(71.7%)
	4	199	1.93–13.28	98(49.2%)	150(75.4%)	166(83.4%)
	6	242	1.73–7.20	195(80.6%)	228(94.2%)	237(97.9%)
1500	0	70	2.05–12.4	15(21.4%)	31(44.3%)	48(68.6%)
	2	81	2.00–11.8	23(28.4%)	50(61.7%)	60(74.1%)
	4	52	1.15–8.78	24(46.2%)	34(65.4%)	41(78.9%)
	6	45	1.93–3.63	43(95.6%)	45(100.0%)	-
2000	0	27	1.73–17.6	13(48.2%)	18(66.7%)	19(70.4%)
	2	26	1.65–7.70	9(34.6%)	12(46.2%)	20(76.9%)
	4	33	0.20–12.3	17(51.5%)	25(75.8%)	30(90.9%)
	6	5	1.23–2.38	5(100.0%)	-	-

(*Continued*)

Table 2.3 *(Continued)*

T (K)	Impact parameter, b (Å)	Complexes formed[2]	Complex time (τ) range (ps)	Number of complexes dissociated (%)[3]		
				3 ps	4 ps	5 ps
Condensed Phase						
1000	0	57	1.92–15.50	9(15.8%)	24(42.1%)	37(64.9%)
	2	92	2.16–19.80	20(21.7%)	39(42.4%)	56(60.9%)
	4	84	2.06–18.60	30(35.7%)	50(59.5%)	63(75.0%)
	6	72	1.90–16.18	40(55.5%)	48(66.7%)	53(73.6%)
	8	36	2.06–20.14	12(33.3%)	20(55.6%)	26(72.2%)
	10	19	2.38–15.20	5(26.3%)	7(36.8%)	11(57.9%)
	12	3	3.34–10.54	1(33.3%)	-	-
1500	0	31	2.00–11.74	9(29.03%)	14(45.2%)	22(71.0%)
	2	29	2.48–21.78	6(20.7%)	12(41.4%)	17(58.6%)
	4	45	2.16–17.20	13(28.9%)	28(62.2%)	34(75.6%)
	6	29	2.08–7.44	14(48.3%)	23(79.3%)	25(86.2%)
	8	8	2.04–4.22	7(87.5%)	7(87.5%)	8(100.0%)
	10	5	2.00–3.96	2(40.0%)	5(100.0%)	-
2000	0	17	1.08–14.76	6(35.3%)	8(47.1%)	10(58.8%)
	2	24	0.20–20.68	4(16.7%)	10(41.7%)	19(79.2%)
	4	17	1.28–9.98	8(47.1%)	12(70.6%)	13(76.5%)
	6	16	2.38–11.84	9(56.2%)	12(75.0%)	12(75.0%)
	8	7	0.18–6.34	4(57.1%)	5(71.4%)	6(85.7%)

[2]The numbers are obtained from 1000 trajectories in the gas phase and 300 trajectories in the condensed phase.
[3]The percentage is calculated with respect to the total number of complexes formed.

Table 2.4 Association probabilities for impact parameters ranging from 0 to 12 Å at each temperature

T (K)	Impact parameter, *b* (Å)	Association probabilities (gas phase)	Association probabilities (condensed phase)
1000	0	0.133 ± 0.0107	0.190 ± 0.0226
	2	0.180 ± 0.0121	0.307 ± 0.0266
	4	0.199 ± 0.0126	0.280 ± 0.0259
	6	0.242 ± 0.0135	0.240 ± 0.0246
	8	0	0.120 ± 0.0187
	10	0	0.0630 ± 0.0140
	12	–	0.0100 ± 0.00570
1500	0	0.0700 ± 0.00807	0.103 ± 0.0175
	2	0.0810 ± 0.00863	0.0960 ± 0.0170
	4	0.0520 ± 0.00702	0.150 ± 0.0206
	6	0.0450 ± 0.00656	0.0967 ± 0.0170
	8	–	0.0267 ± 0.00930
	10	–	0.0167 ± 0.00730
2000	0	0.0270 ± 0.00513	0.0567 ± 0.0133
	2	0.0260 ± 0.00503	0.0800 ± 0.0156
	4	0.0330 ± 0.00565	0.0567 ± 0.0133
	6	0.00500 ± 0.00622	0.0533 ± 0.0129
	8	–	0.0233 ± 0.00870

In the gas phase at 1000 K, the association probability increases from b = 0 Å to b = 6 Å, then sharply decreases at b = 8 Å. At 1500 K, the probability rises up to b = 2 Å and starts to decrease from b = 4 Å, while the highest association probability at 2000 K occurs at b = 4 Å. In the condensed phase at 1000 K, association probability increases from b = 0 Å to b = 2 Å, then declines starting at b = 4 Å. At 1500 K, the probability increases up to b = 4 Å before decreasing from b = 6 Å, with the peak association probability at b = 2 Å for 2000 K. Figure 2.1 shows the association probabilities for both gas and condensed phases across different impact parameters.

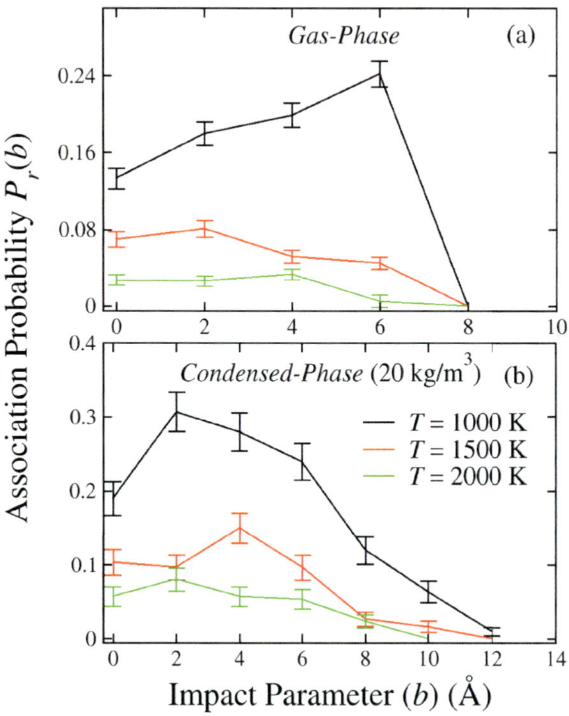

Figure 2.1 Association probabilities as a function of impact parameters at 1000, 1500, and 2000 K simulations. (a) for gas-phase simulation (b) for condensed-phase simulation at a density of 20 kg m^{-3} (Taken from ref. [9]).

To calculate the association cross-section (σ), the association probability is used. σ is obtained by integrating $\int P_r(b)2\pi bdb$, where $P_r(b)$ is the association probability obtained at the impact parameter (b). The σ values for the gas-phase simulations at 1000 K are 32.77 ± 1.960 Å^2, at 1500 K and 2000 K, the values are 8.038 ± 1.063 Å^2 and 2.689 ± 0.8792 Å^2, respectively. For condensed-phase simulation, values are 61.37 ± 8.324 Å^2, 23.32 ± 4.597 Å^2, and 11.13 ± 2.908 Å^2 at 1000, 1500, and 2000 K, respectively. As σ depends on association probability, it decreases with the increase in temperature in both phases.

The temperature-dependent association rate constant $k_{asso}(T)$ is calculated from Eq. (2.5), for which we required the value of relative velocity $v(E_{rel})$ calculated from Eq. (2.6), where E_{rel} is the relative translational energy, μ is the reduced mass.

$$k_{asso}(T) = \sigma(\bar{E}_{rel})\,v(\bar{E}_{rel}) \tag{2.5}$$

$$v(E_{rel}) = \sqrt{\frac{2E_{rel}}{\mu}} \times 0.2045 \times 10^4 \tag{2.6}$$

The $v(E_{rel})$ is calculated using Eq. (2.6) at 2.98, 4.47, and 5.96 kcal mol^{-1} of relative translational energy at 1000, 1500, and 2000 K. The $v(E_{rel})$ at 1000, 1500, and 2000 K is obtained as 799.50, 979.18, and 1130.7 m/s. As $v(E_{rel})$ does not depend on association probability and only depends on temperature, these three values are the same for both gas and condensed phase at the three respective temperatures. The $k_{asso}(T)$ at 1000, 1500, and 2000 K for gas and condensed phase is mentioned in Table 2.5.

Table 2.5 The $k_{asso}(T)$ at 1000, 1500, and 2000 K for gas and condensed phase

T (K)	$k_{asso}(T)$ for gas phase (cm^3 $molecule^{-1}$ s^{-1})	$k_{asso}(T)$ for condensed phase (20 kg m^{-3}) (cm^3 $molecule^{-1}$ s^{-1})
1000	$(2.621 \pm 0.1568) \times 10^{-10}$	$(4.907 \pm 0.6655) \times 10^{-10}$
1500	$(0.7872 \pm 0.1041) \times 10^{-10}$	$(2.283 \pm 0.4501) \times 10^{-10}$
2000	$(0.3041 \pm 0.09944) \times 10^{-10}$	$(1.259 \pm 0.3287) \times 10^{-10}$

The $k_{asso}(T)$ decreases with an increase in temperature and increases when moving from the gas phase to the condensed phase. As the temperature changes from 1000 to 1500 and 1500 to 2000 K, $k_{asso}(T)$ for a 20 kg m^{-3} density simulation is almost double. $k_{asso}(T)$ at 1500 K is 1/3rd of 1000 K and double that of 2000 K.

The activation energy (E_a) as well as frequency factor (A) for the association dynamics are obtained by fitting $k_{asso}(T)$ at 1000, 1500, and 2000 K versus $1/T$, considering the Arrhenius equation. The fitted E_a and A values for the gas phase are –8.513 kcal mol^{-1} and 0.0370×10^{-10} cm^3 $molecule^{-1}$ s^{-1}, and for the condensed phase (20 kg/m^3) are –5.597 kcal mol^{-1} and 0.351×10^{-10} cm^3 $molecule^{-1}$ s^{-1}. The activation energy barrier is higher for the gas phase as compared to the condensed phase, which concludes that association is more favoured in the condensed phase. In the condensed phase, there is an additional phenomenon called collisional intermolecular energy transfer (CIET), which transfers energy from the complex to the bath, causing more association.

A simulation at a density of 324 kg m^{-3} is also performed to see the effect of higher density. As we move from a density of 20 kg m^{-3} to 324 kg m^{-3}, association probability increases at lower impact parameters. The average total energy of the Bz–Bz dimer complex is calculated over the simulation time for both 20 kg m^{-3} and 324 kg m^{-3} and plotted in Figure 2.2.

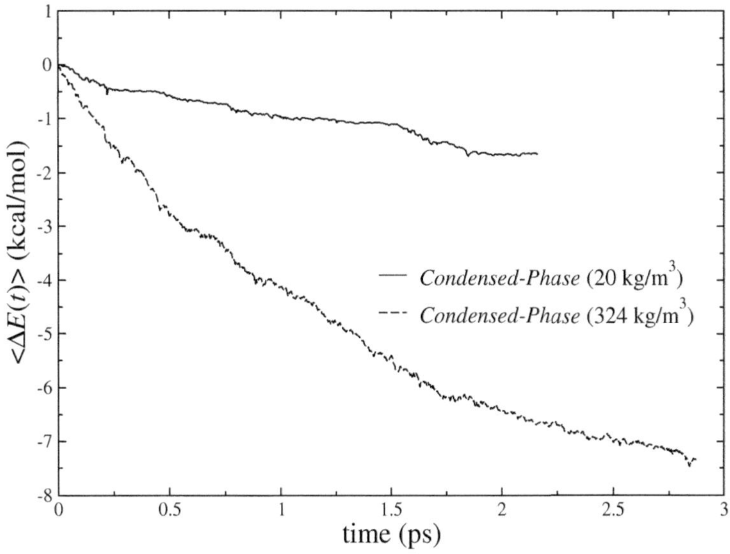

Figure 2.2 The change in average energy (total) of two Bz* to its initial energy versus time is presented at 20 kg m^{-3} and 324 kg m^{-3} bath densities. The results are at 1500 K and at a *b* value of 4 Å, and for the time before the formation of the complex (the data for this figure is taken from ref. [9]).

The total average energy of the excited Bz dimer is transferred more to the bath at a density of 324 kg m^{-3}, which causes more time for the association process at a density of 324 kg m^{-3} as compared to a density of 20 kg m^{-3}.

2.3.1.1.2 *Ensuing dissociation dynamics of Bz dimer*

The lifetime of the complex is counted from when it is formed, i.e., when the center of mass separation reaches 7.5 Å, till it reaches 14 Å. The ensuing dissociation data gives an idea of the complex's lifetime and provides the percentage of direct dissociation trajectories at a particular simulation condition. $N(t)/N(0)$ is calculated for each temperature at each impact parameter. $N(t)$ is the number of undissociated trajectories, and $N(0)$ is the number of total trajectories. Since there is no complex formation in the gas phase at b = 8–10 Å and there is no significant number of complex formation at b = 8–12 Å in the condensed phase, therefore, plot of $N(t)/N(0)$ at fixed impact parameter by varying the temperature is plotted in Figure 2.3 for gas phase and condensed phase for impact parameter b = 0 and 4 Å.

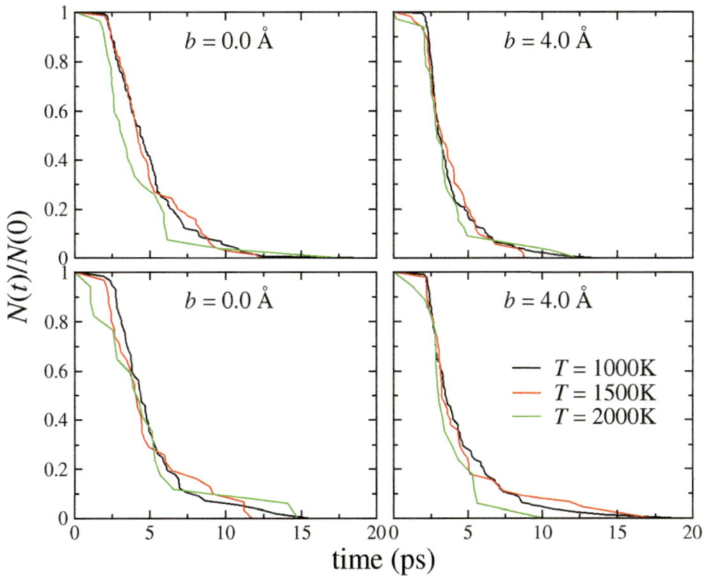

Figure 2.3 The $N(t)/N(0)$ versus time is plotted at a fixed b value for gas-phase simulations in the top row (top left at b = 0 Å and top right at b = 4 Å) and for the condensed phase in the bottom row (bottom left at b = 0 Å and bottom right at b = 4 Å). The black, red, and green curves represent 1000, 1500, and 2000 K simulation results, respectively (the full result is available in ref. [9]).

The dissociation at a lower time in ensuing dissociation is less than the unimolecular dissociation. In unimolecular dissociation dynamics, the initial excitation energy is randomly distributed, but in ensuing dissociation, the energy distribution in all modes is non-random. The time lag (t_0) before ensuing dissociation varies with temperature and impact parameter, i.e., decreases with an increase in temperature and impact parameter. The dissociation rate is calculated from the exponential equation (Eq. (2.7)).

$$\frac{N(t-t_0)}{N(t_0)} = \exp(-k(t-t_0)) \tag{2.7}$$

where, $N(t - t_0)$ is the number of undissociated trajectories at time $t - t_0$ and $N(t_0)$ is the total number of trajectories at t_0. t_0 is the minimum time that the complex spent before dissociation, and k is the dissociation rate constant. By fitting the dissociation curves at different temperatures and impact parameters, t_0 and k values are obtained for gas-phase and condensed-phase simulations, which are mentioned in Table 2.6.

Table 2.6 t_0 and k values at 1000, 1500, and 2000 K temperature with the range of impact parameter b = 0 to 6 Å

		Gas phase		**Condensed phase (20 kg m^{-3})**	
***T* (K)**	**Impact parameter, *b* (Å)**	**t_0 (ps)**	**k (ps^{-1})**	**t_0 (ps)**	**k (ps^{-1})**
1000	0	2.30	0.350	2.45	0.380
	2	2.20	0.390	2.26	0.350
	4	2.10	0.640	2.15	0.430
	6	2.10	1.650	1.95	0.475
1500	0	2.30	0.360	1.90	0.340
	2	2.10	0.460	2.28	0.310
	4	2.00	0.560	2.05	0.430
	6	1.95	3.000	2.05	0.730
2000	0	1.60	0.400	0.850	0.280
	2	1.70	0.390	2.50	0.550
	4	1.90	0.690	2.00	0.620
	6	–	–	2.25	0.770

k values in the gas phase increase with an increase in impact parameters at particular temperatures, and are similar or higher at higher temperatures. The variation in t_0 and k values at 2000 K is not similar to that at 1000 and 1500 K, which may be due to the lower number of trajectories associated with this temperature and, therefore, fewer trajectories for ensuing dissociation. In the condensed phase, the dissociation rate increases with an increase in temperature, and it is lower than in the gas phase due to energy transfer to bath molecules; i.e., energy is not enough in the dissociating vibrational mode so that it can dissociate. As shown in Table 2.3, the dissociation percentage is higher at b = 4 and 6 Å, which concludes that faster association favors faster dissociation.

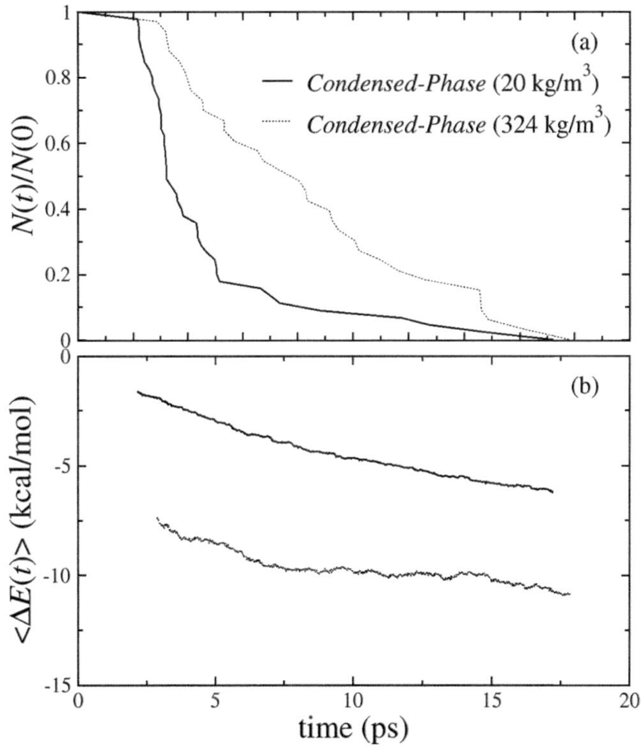

Figure 2.4 A comparative study of the ensuing dissociation (panel a) and change in average total energy (panel b) of the complex is presented at bath densities of 20 and 324 kg m^{-3}. Both results are obtained at a simulation temperature of 1500 K and an impact parameter b = 4 Å (the data for this figure is taken from ref. [9]).

The dissociation rate is higher at 20 kg m^{-3} than at 324 kg m^{-3} and is shown in Figure 2.4. From the figure, it is clear that the change in average total energy of the complex at a bath density of 20 kg m^{-3} is lower than 324 kg m^{-3} because at a higher bath density, the energy is transferred from the complex to the bath. The transfer of energy is mainly from the vibrational mode of the complex to the translational and rotational modes of N_2 bath molecules.

The association of the complex is favored more at a higher density of bath due to the energy transfer by a higher collision frequency among the molecules. Energy is transferred from the dimer to the bath molecules. The subsequent dissociation of the complex is more favored at lower bath density because the complex has a higher amount of energy due to less transfer to the bath molecules.

2.3.1.2 Energy transfer study from vibrationally excited C_6F_6 to NO/N_2 mixed bath

In this simulation using the MM model, the energy transferred from the reactive system HFB* to the bath molecules N_2 and NO is monitored and quantified. Two bath systems, small and large, are used: a small bath with 100 NO and 100 N_2 molecules each and a large bath with 500 NO and 500 N_2 molecules each. HFB* and bath molecules' average energies are monitored versus time, and a detailed description is mentioned below. The effect of the bath size is also studied to determine and explain further. Energy transferred per collision is calculated and compared with the previous simulation of HFB* in a single bath of N_2 molecules.

HFB* average energies are transferred with time, which was fitted nicely to a bi-exponential function given by Eq. (2.8).

$$\langle E(t)\rangle = (\langle E_0\rangle - E_\infty)\times\left[f_1\exp(-k_1t)+f_2\exp(-k_2t)\right]+E_\infty \qquad (2.8)$$

where $f_1 + f_2 = 1$ and k_1, k_2 are rate constants of the energy transfer process. Optimize parameters f_1, f_2, k_1, and k_2 are obtained from Eq. (2.8). $<E_0>$ denotes HFB* average total initial energy and E_∞ denotes HFB* thermal energy at 300K. HFB* total energy transfer for small and large mixed baths is shown in Figure 2.5. HFB* total energy transfer for only a single bath of N_2 molecules is also shown in the same figure [3]. The energy transferred for the single bath is observed to be highest, and for a large mixed bath, it is similar but somewhat lower than the single bath, and for a small mixed bath, it is lowest.

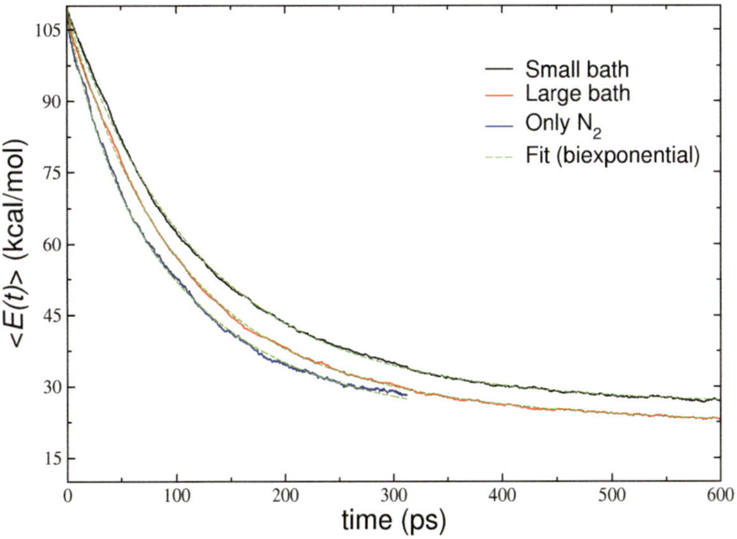

Figure 2.5 HFB* Averaged total energy plotted versus time. The black curve represents the smaller bath, and the red curve represents the larger bath used in the current study. The blue (only N_2) curve represents a simulation of the previous study [3]. Green dashed lines represent the fit to Eq. (2.8) (Figure taken from ref. [12]).

The optimized parameters for the small as well as large bath are given in Table 2.7 using Eq. (2.8).

Table 2.7 Energy transfer optimizes parameters fit to Eq. (2.8)

Number of NO molecules in the bath	Number of N_2 molecules in the bath	E_∞ kcal mol^{-1}	f_1	f_2	k_1 ps^{-1}	k_2 ps^{-1}
100	100	25.50	0.8000	0.2000	0.009283	0.004268
500	500	20.92	0.7700	0.2300	0.01113	0.003751
0	1000[4]	22.20	0.2560	0.7440	0.02770	0.008100

[4]The fitting parameters for a single bath of 1000 N_2 molecules [3].

The center of mass translational, rotational, and vibrational energy of HFB* is also examined with respect to time, which is represented in Figure 2.6, which clearly states that the translational and rotational energy are not much affected, but the vibrational energy of HFB* is much decreased from the initial time.

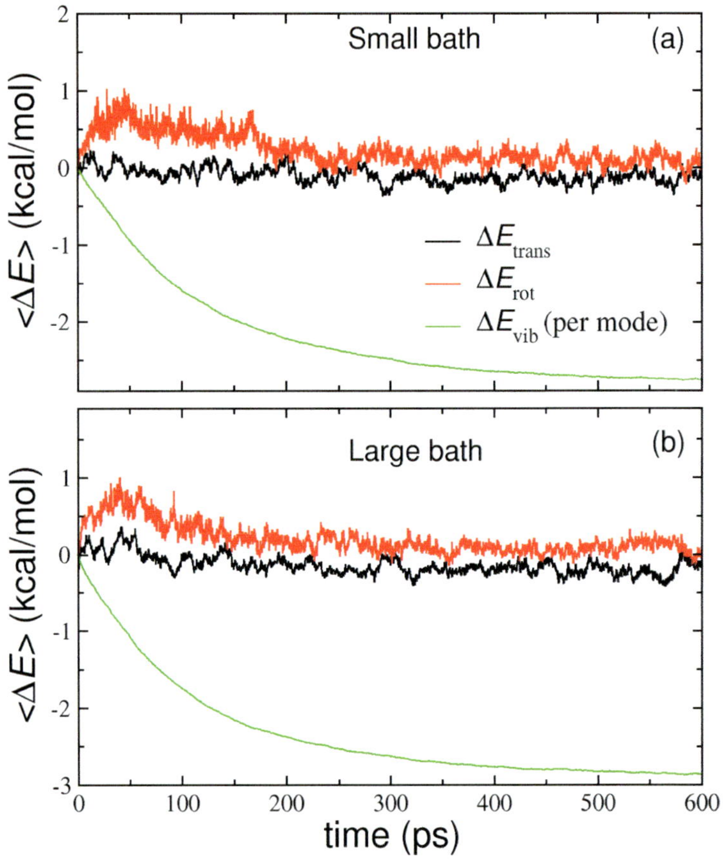

Figure 2.6 The variation in average center-of-mass translational, rotational, and vibrational energies with respect to the initial energy values of the HFB* is shown by black, red, and green curves, respectively. Panel (a) at the top represents the small bath, while panel (b) at the bottom illustrates the large bath (Figure taken from ref. [12]).

There is a loss of ~2.8 kcal mol^{-1} of energy in each vibrational mode during the last part of the simulation. All 30 vibrational modes of HFB* lose a total amount of ~84 kcal mol^{-1} in a large bath, whereas in a small bath, the respective values are ~2.7 and 81 kcal mol^{-1}. The mode vibrational energy of HFB* initially is 3.57±0.06, and finally it is 0.82±0.02 in a small bath. The same values in the large bath are 3.55±0.06 and 0.70±0.02. The rotational mode received energy of about 1.0 kcal mol^{-1} above during the simulation at 40 ps, and the same mode finishes with a 0 kcal mol^{-1} value of <ΔE> at the last time of the simulation period. There is no change in the translational mode energy, and it remains almost constant, and <Δ*E*> is 0. Hence, it is seen that transfer of energy from vibrational to rotational modes but no energy transfers from vibrational to translational modes.

The center of mass translational, rotational, and vibrational energies of the bath molecules are monitored during the simulation time, which is shown in Figure 2.7.

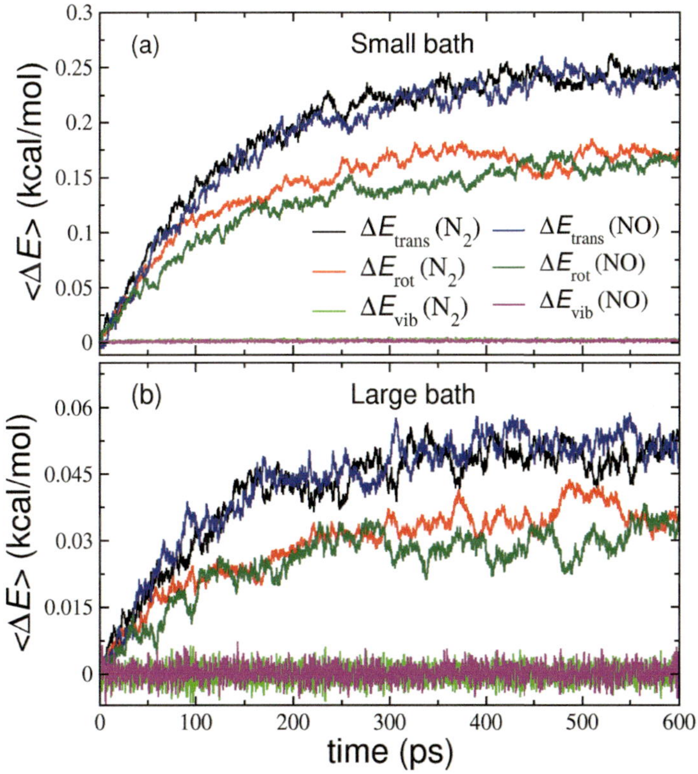

Figure 2.7 The variation in average center of mass translational, rotational, and vibrational energies with respect to the initial energy values of the N_2 molecules is shown by black, red, and light green colors. The same is shown by the blue, deep green, and magenta colors of NO molecules. The top panel (a) and bottom panel (b) are for the small bath and large bath, respectively (Figure taken from ref. [12]).

Energy for each DOF is calculated for the N_2 bath and NO bath separately. Momentum coordinates and those of N_2 and NO molecules are separated out, and then the energy components of each DOF are calculated. From Figure 2.7, no change is observed in the vibrational energy DOF of N_2 and NO molecules in small as well as large bath, but some energy is gained in the center-of-mass translational and rotational DOF of N_2 and NO from HFB*. The ratio of the increase in translational and rotational energy modes of a small bath to a large bath is almost four. It can also be observed that the rotational DOF of NO molecules is receiving less energy as compared to N_2 molecules. A similar phenomenon is observed for the translational DOF of a small bath. Therefore, due to the presence of NO molecules in the bath, the bath becomes less efficient at Intermolecular Energy Transfer (IET), as reflected in the HFB* average energy-time profile. It is also observed from Figure 2.7 that the results of the larger mixed bath are much closer to the only-N_2 molecule bath results [3]. The ratio between $k(\Delta E)$, which is calculated as $\frac{1}{(\langle E_0\rangle - E_\infty)}\frac{d\langle E(t)\rangle}{dt}$ at $t = 0$ from Eq. (2.8) and the data given in Table 2.8, of only-N_2 and large bath (500 N_2 and 500 NO) is 1.4.

Equilibration of smaller and larger baths is performed at 300 K. At every integration step during the simulation, the temperature is calculated using Eq. (2.9).

$$\frac{3}{2}nk_BT = \sum_{i=1}^{n}\frac{1}{2}m_iv_i^2 \tag{2.9}$$

where n is the number of atoms present in the bath, k_B is the Boltzmann constant, and m_i and v_i are the mass and the velocity of the ith atom in the bath. The sum in the RHS of the above equation is for all the atoms. The number of bath molecules is also a deciding factor in the temperature of the bath. Energy transferred from hot HFB* due to the collision with the bath molecules increases the temperature of the bath. Thus, if the bath density is fixed, the fewer the number of bath molecules, the less will be the bath heating. For the smaller bath where 200 molecules are present, the relative density of HFB* in the bath is larger than in the larger bath, where 1000 molecules are present. The temperature rise can be estimated at a complete bath re-equilibration. At $t = 0$ and initial temperature (T_i) is 300 K, the system's total energy is the summation of center-of-mass translational, rotational, and vibrational energies of HFB* and 200/1000 bath molecules, i.e., $3/2\ k_BT_i + 3/2\ k_BT_i + E_{vib}$ (HFB*) + $n \times (3/2\ k_BT_i + k_BT_i + k_BT_i)$, with n = 200 or 1000. For the smaller bath, the total energy is ~526 kcal mol^{-1}, whereas it is 2195 kcal mol^{-1} for the large bath. When the bath is completely equilibrated, these energies should be equal to $3/2\ k_BT_f + 3/2\ k_BT_f + 30\ k_BT_f + n \times (3/2\ k_BT_f + k_BT_f + k_BT_f)$. For the smaller bath, the final temperature Tf is calculated to be 361 K with a ΔT of 61 K, whereas for the larger bath, ΔT is 13 K. In Figure 2.8, the rotational temperature of N_2 molecules as a function of time is shown for both small and large baths. The rotational ΔT for the smaller and larger bath is ~85 K and ~20 K at 600 ps, respectively. The differences between these ΔT and those mentioned above are due to the non-statistical behavior of the IET dynamics, in which the translational/rotational modes of NO/N2 absorb more energy, while the vibrational modes remain inactive. This type of bath heating affects the process of energy transfer. In the current study, the smaller bath is also less effective at energy transfer than the larger bath, as discussed previously. The energy transfer dynamics are purely non-statistical as seen before [71].

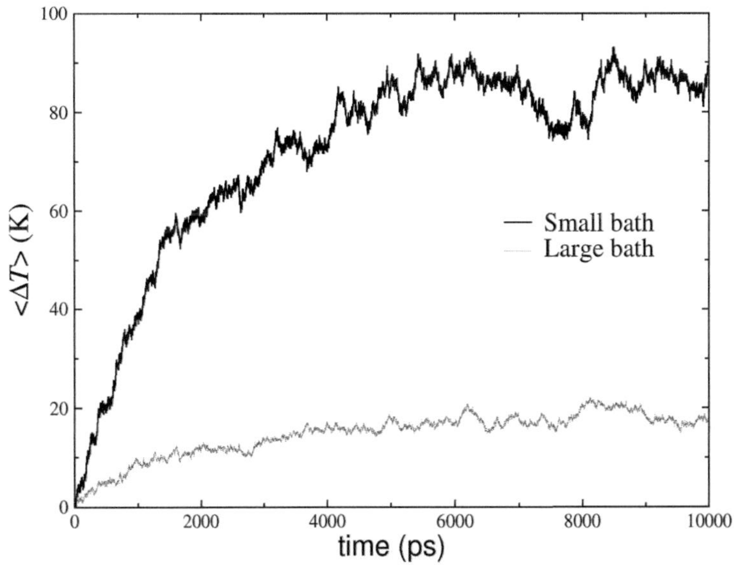

Figure 2.8 Solid line and dashed line represent the change in averaged N_2 rotational temperature for the small and the large bath to its initial temperature, respectively (The data for this figure is taken from ref. [12]).

2.3.1.2.1 *Energy transferred per collision and comparison with C_6F_6* + N_2 simulations*

The average energy transferred per collision, $<\Delta E_c>$, is calculated for the present simulations. A comprehensive explanation for calculating $<\Delta E_c>$ is provided in the reference [3]. The differentiation of average energy of HFB* with respect to time, i.e., $[d<E(t)>/dt]$, provides average energy transferred per unit of time, which is divided by the number of collisions per unit of time, ω, which gives $<\Delta E_c>$, i.e., $<\Delta E_c> = [d<E(t)>/dt]/\omega$. The collision frequency ω is a sum of the collision frequency among HFB* with N_2 and HFB* with NO [71], i.e., Eq. (2.10)

$$\omega = \omega\left(\mathrm{HFB}^* - N_2\right) + \omega\left(\mathrm{HFB}^* - \mathrm{NO}\right) \tag{2.10}$$

ω is calculated by Eq. (2.11).

$$\omega = 4.415 \times 10^7 \sigma^2 p (T\mu)^{-\frac{1}{2}} \Omega_{2,2}^* \tag{2.11}$$

where σ is the collision diameter in Å, p is the pressure (Torr), T is the temperature (K) of the bath, μ is the reduced mass of HFB and N_2 or HFB and NO in atomic mass units, and $\Omega_{2,2}{}^*$ is the collision integral represented by Eq. (2.12)

$$\Omega_{2,2}^* = \frac{A}{\left(T^*\right)^B} + \frac{C}{e^{DT^*}} + \frac{E}{e^{FT^*}} \tag{2.12}$$

$T^* = k_B T/\varepsilon$, where k_B is the Boltzmann constant, ε is the intermolecular well depth, and the constant values are A = 1.16145, B = 0.14874, C = 0.52487, D = 0.77320, E = 2.16178, and F = 2.43787, which are obtained from the literature. The Lennard-Jones (LJ) parameter values of σ and ε are used as 5.12 Å and 250 K for HFB*–N_2, and 5.30 Å and 137 K for HFB*–NO. The parameters for HFB*–N_2 are taken from reference [3]. However, a different set of these values is also obtained from literature [72] and used to calculate $<\Delta E_c>$. The corresponding values are 4.97 Å and 163 K for HFB*–N_2 and 4.84 Å and 194 K for HFB*–NO, respectively. These values are obtained from the relations $\sigma = \sigma\left(\sigma_{\mathrm{HFB}} + \sigma_{\mathrm{N_2/NO}}\right)/2$ and $\varepsilon = \sqrt{\varepsilon_{\mathrm{HFB}}\varepsilon_{\mathrm{N_2/NO}}}$. The corresponding values of σ_{HFB}, $\sigma_{\mathrm{N_2}}$, and σ_{NO} are 6.19 Å [3], 3.74 Å [72], and 3.49 Å [72], respectively, whereas the values of $\varepsilon_{\mathrm{HFB}}$, $\varepsilon_{\mathrm{N_2}}$, and $\varepsilon_{\mathrm{NO}}$ are 323, 82, and 117 K, respectively. In Figure 2.9, the <ΔE_c> is presented versus <$E(t)$> of HFB* for the LJ parameters obtained from the potential energy used in the simulations. Also presented in the figure is the <ΔE_c> obtained from a 1000 N_2 bath. As one can see, the <ΔE_c> profiles between 1000 N_2 and 500 N_2 + 500 NO are almost similar, but deviate in the case of the smaller bath, i.e., 100 N_2 + 100 NO.

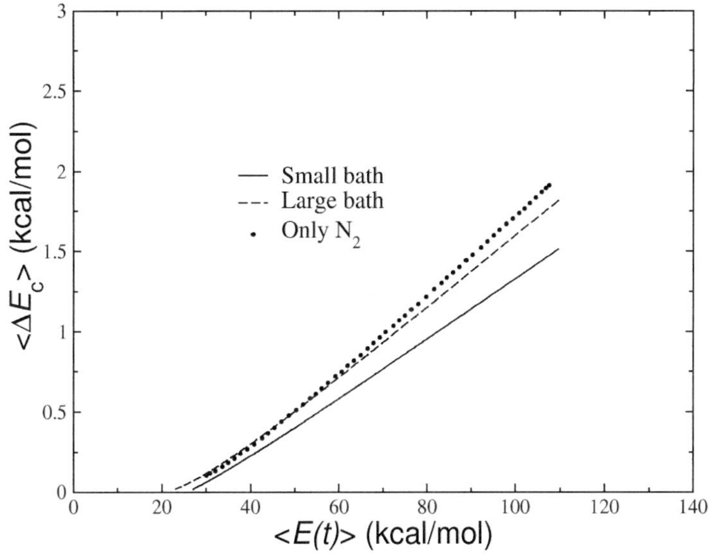

Figure 2.9 Solid, dashed, and dotted lines represent the averaged energy transferred per collision versus average HFB* energy for small, large, and N_2-only baths [3], respectively (The figure is adopted from ref. [12].

In conclusion, energy is transferred from the HFB* to the bath molecules NO and N_2, mainly to the N_2 molecules. The energy is transferred from the vibrational modes of HFB* to the translational and rotational modes of NO and N_2 molecules, and the rotational modes are somewhat inactive in this process. The smaller bath has a higher rise in temperature than the larger bath because it contains fewer bath molecules. Results of energy transferred per collision in a larger bath are similar to the results of a simulation previously performed with only N_2 bath molecules.

2.3.1.3 IET from vibrationally excited benzene (Bz*) to N_2–Bz mixed baths

The energy transfer from vibrationally excited Bz* molecule to the bath of N_2 molecules, where some Bz molecules are also included in the bath. How those Bz molecules included in the bath affect the energy transfer process is discussed in detail below. Complex formation of Bz, such as its dimer, trimer, and tetramer possibilities, is monitored during the simulation. The changes in energy of Bz* and bath molecules are checked.

2.3.1.3.1 *Possibilities of complex formation of Bz molecules*

The formation of dimers, trimers, and tetramers of Bz molecules due to collisions is monitored during the simulation at 140 K. The complex formation of Bz molecules is determined by calculating the center-of-mass distances between them. One excited Bz and three unexcited are present in the simulation box. The distances between the centers of mass of the Bz molecules are plotted as a function of time in Figure 2.10 for a trajectory with the greatest potential for complex formation.

For four Bz molecules, there are six center-of-mass distances calculated among them. Two simultaneous criteria must be fulfilled for complex formation: firstly, the center-of-mass distances among two Bz molecules should be less than 10 Å, where the equilibrium center of mass distance among Bz molecules in the Bz dimer is 4.93 Å [2], and a distance of less than 10 Å should be maintained for at least 10 ps.

The previously reported orientation-averaged potential energy curve [2 showed −0.039 kcal mol−1 of potential energy at a center-of-mass distance of 10 Å between two Bz molecules. The selection of the second criterion of complex formation is explained in detail in the reference [10]. Out of the total simulation time, dimer complex formation time percentage was 17.8%, whereas trimer and tetramer complex formation time percentages were very less as 2.25% and 0.34%, respectively. Total complex formation, including Bz*is 7.90%, dimer formation percentage is 6.72%, trimer formation percentage is 1.01%, and tetramer formation percentage is 0.17%. For the sample trajectory depicted in Figure 2.10, the total percentage of complex formation was 37.1%. Out of this, 13.8% was a dimer, 11.7% was a trimer, and 11.6% was a tetramer.

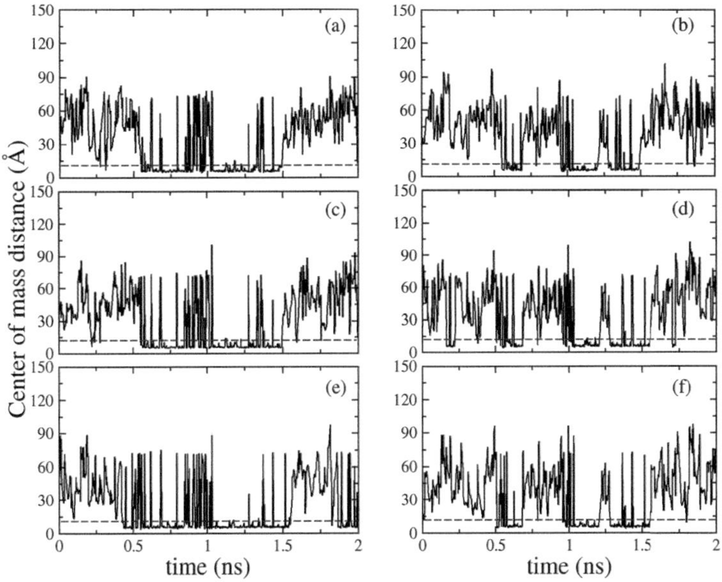

Figure 2.10 The center of-mass distance among two Bz molecules is plotted as function of time at 140 K for the trajectory where the possibility of complex formation is very less: (a) Bz_1–Bz_2; (b) Bz_1–Bz_3; (c) Bz_1–Bz_4; (d) Bz_2–Bz_3; (e) Bz_2 Bz_4; (f) Bz_3–Bz_4. Bz_1 is Bz^*, and Bz_2, Bz_3, and Bz_4 are unexcited three Bz molecules in the bath. 10 Å center-of-mass distance is considered the threshold for complex formation, represented by dotted lines (the data for this figure is taken from ref. [10]).

2.3.1.3.2 *Bz* and bath molecules N_2 and Bz average energies with time*

Bz* average total energy is monitored with time and plotted in Figure 2.11. N_2 and Bz bath molecules' translational, rotational, and vibrational energies at 140 K are plotted in Figure 2.12.

It was shown that the IET efficiency of different Bz vibrational modes is different [73]. However, this method of calculating the per-mode vibrational energy yields an average energy and allows comparison with the individual translational and rotational energies of the bath. Results are shown for 39 trajectories. For a N_2 bath molecule, the initial average translational, rotational, and vibrational energies, i.e., $\langle E_{trans}\rangle$, $\langle E_{rot}\rangle$, and $\langle E_{vib}\rangle$, are 0.416 ± 0.003 kcal mol^{-1}, 0.278 ± 0.002 kcal mol^{-1}, and 0.276 ± 0.002 kcal mol^{-1}, respectively, and at termination of trajectories nearly 3 ns, these numbers are 0.470 ± 0.003 kcal mol^{-1}, 0.313 ± 0.003 kcal mol^{-1}, and 0.277 ± 0.002 kcal mol^{-1}. Initial average translational, rotational, and vibrational temperatures are at 140 K each, but at the termination of the trajectories, these temperatures are 158 K ± 1 K, 157 K ± 2 K, and 140 K ± 1 K, respectively which clearly shows that heating up of translational and rotational modes of N_2 molecules take place and is similar but there in no change in temperature of vibrational mode of N_2 bath molecules which is similar previous simulation where 298 K was the initial bath temperature [3].

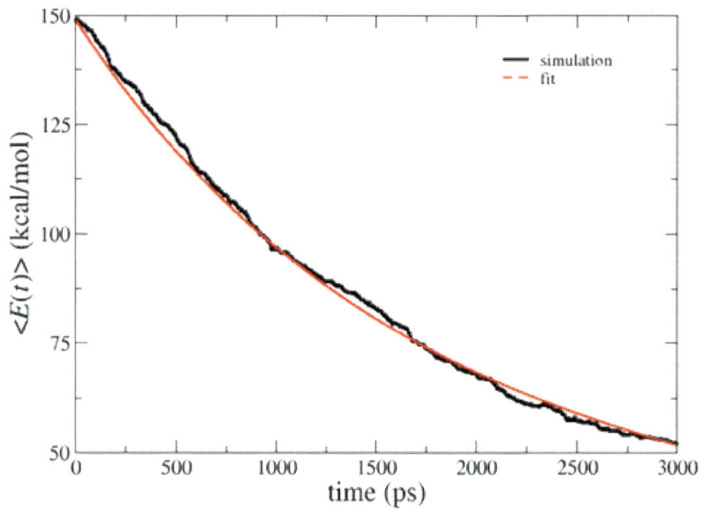

Figure 2.11 Total energy of the Bz* in Bz/N_2 bath at 140 K is plotted as a function of time (Figure taken from ref. [10]).

For the thermalized Bz in bath, the initial $\langle E_{trans} \rangle$, $\langle E_{rot} \rangle$, and $\langle E_{vib} \rangle$ are 0.445 ± 0.03 kcal mol^{-1}, 0.433 ± 0.04 kcal mol^{-1}, and 0.270 ± 0.01 kcal mol^{-1}, respectively. At the 3 ns, these numbers are 0.450 ± 0.04 kcal mol^{-1}, 0.560 ± 0.04 kcal mol^{-1}, and 1.08 ± 0.04 kcal mol^{-1}. Initial average translational, rotational, and vibrational temperatures are 140 K each, and at 3 ns, these temperatures are 150 K ± 13 K, 188 K ± 13 K, and 543 K ± 20 K. Here energy transfer from vibrational modes of Bz* to vibrational modes of Bz in bath molecules is more efficient than translational and rotational modes which was also observed in previous simulation with bath temperature 300 K [71]. Energy transfer from Bz* vibrational modes to rotational modes of Bz bath molecules is more efficient than to translational modes, which is also observed in the previous simulation [71]. $\langle E_{trans} \rangle$, $\langle E_{rot} \rangle$, and $\langle E_{vib} \rangle$ per mode of Bz* in Bz/N_2 bath as a function of time are plotted in Figure 2.13.

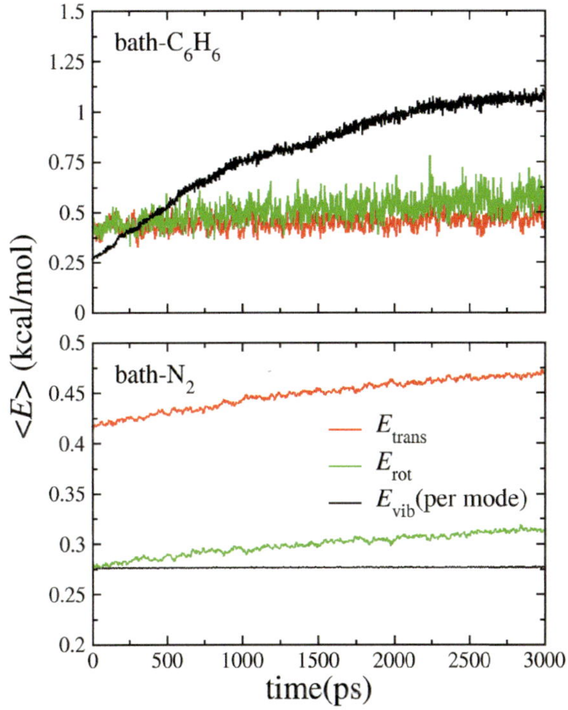

Figure 2.12 Red, green, and black colors represent average translational, rotational, and vibrational energies of Bz bath molecules in the top panel and N_2 bath molecules in the bottom panel as a function of time, respectively, at 140 K bath temperature (Figure taken from ref. [10]).

The initial $\langle E_{trans} \rangle$, $\langle E_{rot} \rangle$, and $\langle E_{vib} \rangle$ of Bz* are 0.45 ± 0.05 kcal mol^{-1} and 0.48 ± 0.07 kcal mol^{-1}, and 4.93 ± 0.40 kcal mol^{-1}, respectively. At a simulation time of 3 ns, these energy values are 0.49 ± 0.06 kcal mol^{-1}, 0.67 ± 0.14 kcal mol^{-1}, and 1.70 ± 0.14 kcal mol^{-1}, respectively. $\langle E_{trans} \rangle$ does not change with time, and $\langle E_{rot} \rangle$ increases rapidly and then decreases, but the value is higher than that of $\langle E_{trans} \rangle$. At 3 ns simulation time, the Bz* translational, rotational, and vibrational temperatures are 164 K ± 20 K, 225 K ± 47 K, and 855 K ± 70 K. The translational, rotational, and vibrational temperatures for the N_2 and Bz bath molecules and Bz* at 3 ns simulation time are summarized in Table 2.8.

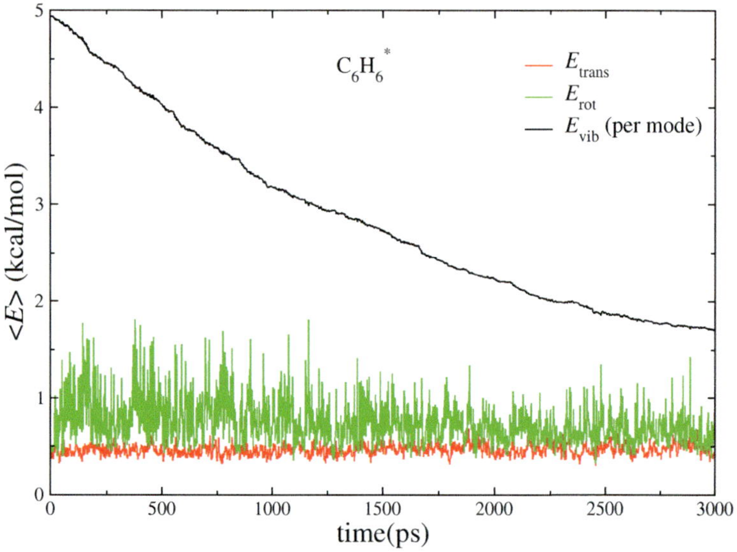

Figure 2.13 Red, green, and black colors represent Bz* average translational, rotational, and vibrational energies in Bz/N_2 bath at 140 K as a function of time (Figure taken from ref. [10]).

Table 2.8 Molecular temperatures at 3 ns simulation time[5]

Molecule	Bz*	Bz	N_2
Translational temperature	164	150	158
Rotational temperature	225	188	157
Vibrational temperature	855	543	140

[5]These are the temperatures (K) for the molecular systems present in the bath. The initial vibrational temperature of Bz* is 2484 K, and the Bz* translational/rotational temperature is 300 K.

It should be noted that a complete re-equilibration of the Bz* vibration has not been achieved at 3 ns, i.e., at the termination of the trajectories. The final Bz* per mode vibrational energy at 3 ns is

1.70 ± 0.14 kcal mol^{-1}, which is about one third of its initial vibrational energy. At 3 ns, energy of Bz bath molecules for the same mode is 1.08 ± 0.04 kcal mol^{-1}. The average energies of Bz* with time, for the simulations with the N_2 bath at 140 K and 300 K, are plotted in Figure 2.14. These results are obtained from simulations in which there is no unexcited Bz molecule in the bath. One can see from this figure that the energy transfer process at 300 K is more efficient than that at 140 K. The translational, rotational, and vibrational temperatures from these simulations at both 140 K and 300 K are obtained and compared with those from mixed bath simulations mentioned above.

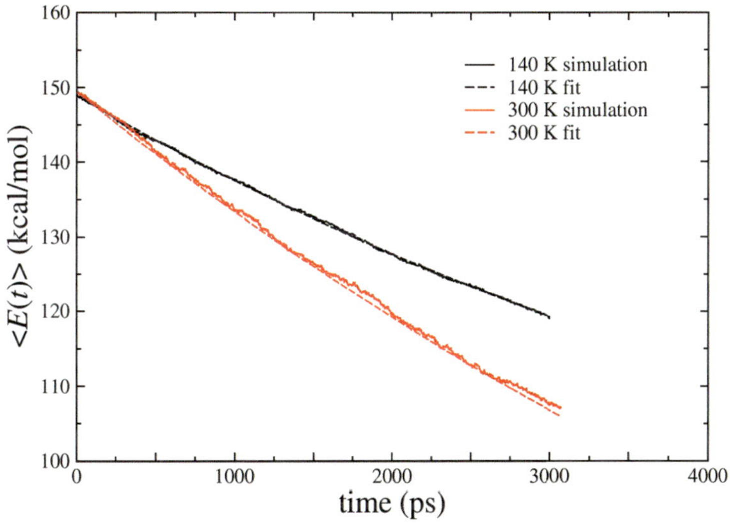

Figure 2.14 Solid lines (black for 140 K simulation and red for 300 K simulation) represent the total energy of the Bz* in N_2 bath as a function of time, and dashed lines are the fit to simulation curves (The data for this figure is taken from ref. [10]).

At 3 ns of the 140 K bath-temperature simulation, the translational and rotational temperatures of the N2 bath molecule are 155 K ± 1 K and 157 K ± 1 K, respectively. The translational, rotational, and vibrational temperatures of Bz* are 137 K ± 16 K, 319 K ± 74 K, and 1974 K ± 25 K, respectively. For the 300 K bath temperature simulation of the N_2 bath, translational and rotational temperatures are 316 K ± 2 K and 317 K ± 2 K of N_2, and translational, rotational, and vibrational temperatures of Bz* are 310 K ± 40 K, 362 K ± 52 K, and 1749 K ± 52 K, respectively. These temperatures, except for the vibrational temperature, are the same as those found for the previous Bz/N_2 bath simulation at 300 K [71]. However, for both the 140 K and 300 K N_2 bath simulations, there is some indication that the rotational temperature for Bz* is a bit higher than that of translational and rotational of N_2 and translational of Bz*. However, better statistics are required to determine if this is indeed the case.

2.3.1.3.3 *Energy transfer per collision*

The average energy transferred per collision, $\langle \Delta E_c \rangle$, from vibrationally Bz* is determined from the simulation and $\langle E(t) \rangle$ given by the equation mentioned in Section 2.3.1.2.3. As found for previous simulations [2, 74–78], the $\langle E(t) \rangle$ are well fit by the bi-exponential Eq. (2.8), and the description is mentioned in Section 2.3.1.2.3. The fits are shown in Figure 2.14, and the optimized parameters are stated in Table 2.9. The optimized parameters of the previous Bz/N_2 bath simulation at 300 K are also included [71].

Table 2.9 Bath conditions and optimized parameters for four simulations of $\langle E(t) \rangle$

Bath[6]	***T* (K)**	**300**[8]	**140**	**300**	**140**
	P	32.50	16.20	35.00	16.30
	N_2	190	396	400	399
	Bz	9	4	0	0
Fit parameters[7]	$E(0)$	149.90	149.00	149.50	149.00
	$E(\infty)$	23.40	12.20	19.80	51.90
	f_1	0.9010	0.6000	0.9980	0.6000
	f_2	0.0890	0.4000	0.0020	0.4000
	k_1	1.330×10^{-3}	7.240×10^{-4}	1.330×10^{-4}	1.740×10^{-4}
	k_2	7.990×10^{-4}	1.980×10^{-4}	1.300×10^{-4}	5.600×10^{-5}

[6]Bath temperature (K), bath pressure (atm), and the number of N_2 and Bz bath molecules.

[7]$E(0)$ and $E(\infty)$ are in kcal mol^{-1}, $f_1 + f_2 = 1$, and k_1 and k_2 are in ps^{-1}.

[8]These parameters are collected from the reference [71].

$\langle \Delta E_c \rangle$ in the equation mentioned in Section 2.3.1.2.3 includes all up and down collisions, which transfer energy to and from Bz*, respectively. The collision frequency, for the simulation with the Bz/N_2 bath, is the sum of the collision frequencies for Bz* – N_2 and Bz* – Bz collisions, i.e., ω (overall) = ω (Bz*–N_2) + ω (Bz*–Bz). The component-wise collision frequency of the bath can be expressed as $\omega = \omega_P \times P$, where P is the partial pressure of the bath gas. The ω_P values are the same as the values to mimic the experiments [79], with which the simulations are compared. All the values for N_2 and Bz bath are summarized in Table 2.10 for both 140 K and 300 K bath calculations.

Table 2.10 Partial pressures and collision frequency parameters for two mixed bath simulations at 140 K and 300 K

Bath[9]	T	300	140
	$P(N_2)$	31.00	16.10
	P(Bz)	1.500	0.120
Frequency parameters[10]	$\omega_P(N_2)$	1.140×10^{10}	2.410×10^{10}
	ω_P(Bz)	1.740×10^{10}	3.680×10^{10}
	ω(Bz*–Bz)	2.580×10^{10}	4.490×10^{9}
	ω(Bz*–N_2)	3.530×10^{11}	3.870×10^{11}

[9]The bath temperatures and partial pressures are given in K and atm, respectively.
[10]ω_P values are written in $atm^{-1}\ s^{-1}$, and ω values are in s^{-1}.

When the simulation started, the pressures for the bath were utilized to calculate the collision frequencies. Table 2.10 also displays the N_2 partial pressures for the 140 K and 300 K mixed bath simulations. At 300 K and 140 K, the pure N_2 bath has a pressure of 35.0 atm and 16.3 atm, respectively. It is interesting to note that the percentage of Bz*–Bz collisions at 140 K is lower than that of the Bz*–N_2 and Bz*–Bz collisions in the 2.4 ns trajectory calculation at 300 K, which are approximately 847 and 62, respectively, compared to the 3 ns trajectory integration at 140 K, which is approximately 1161 and 13. $\langle \Delta E_c \rangle$ for Bz* energy transfer to the N_2 and Bz bath molecules, i.e., $\langle \Delta E_c \rangle$ N_2 and $\langle \Delta E_c \rangle$ Bz are of interest from the study of $\langle E(t) \rangle$, as described by the equation mentioned in Section 2.3.1.2. Simulation plots of $\langle E(t) \rangle$ in Figure 2.14 and the optimized parameters in

Table 2.9 were used to calculate values of $\langle \Delta E_c \rangle$ N_2 for collisions of N_2 with Bz* at 140 K and 300K. Figure 2.15 plots the resulting values of $\langle \Delta E_c \rangle$ N_2 vs. $\langle E(t) \rangle$, showing that $\langle \Delta E_c \rangle$ N_2 is substantially smaller at 140 K than at 300 K.

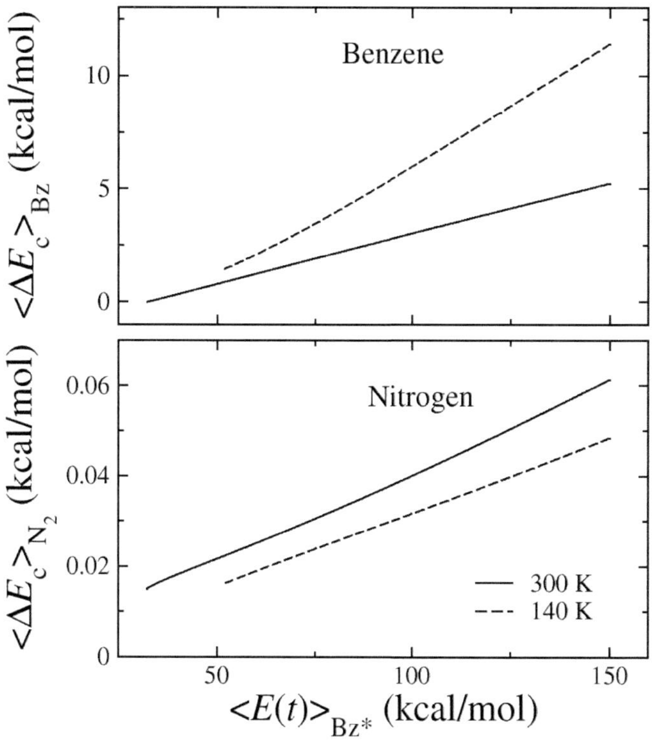

Figure 2.15 Average energies transferred to the N_2 and Bz bath molecules, i.e., $\langle \Delta E_c \rangle$Bz (top panel) and $\langle \Delta E_c \rangle N_2$ (bottom panel), from vibrationally Bz*, for the simulations at 140 K and 300 K. Results are given vs the average vibrational energy of Bz* (Data for this figure is taken from ref. [10]).

Using the fitting parameters in Table 2.9, the same analysis was performed on $\langle E(t) \rangle$ vs time in Figure 2.14 to calculate $\langle \Delta E_c \rangle$ vs $\langle E(t) \rangle$ for Bz* collisions with the Bz/N_2 bath at 140 K. This data was previously provided for the 300 K simulations. Eq. (2.13) was used to calculate the values of $\langle \Delta E_c \rangle$ for Bz* collisions with the Bz/N_2 bath and $\langle \Delta E_c \rangle N_2$, as well as $\langle \Delta E_c \rangle$Bz for Bz* collisions with the Bz bath molecules.

$$\langle \Delta E_c \rangle_{\mathrm{Bz}} = \frac{\omega}{\omega\left(\mathrm{Bz}^* - \mathrm{Bz}\right)} \langle \Delta E_c \rangle_{\mathrm{Bz}^*} - \frac{\omega\left(\mathrm{Bz}^* - \mathrm{N}_2\right)}{\omega\left(\mathrm{Bz}^* - \mathrm{Bz}\right)} \langle \Delta E_c \rangle_{\mathrm{N}_2} \quad (2.13)$$

Figure 2.15 displays the values and indicates that the Bz* + Bz energy transfer is significantly more efficient at 140 K than at 300 K. At 140 K, the $\langle \Delta E_c \rangle$Bz is around 11 kcal mol^{-1}, while at 300 K, with an excitation energy of 148.1 kcal mol^{-1}, it is approximately 5 kcal mol^{-1}. It should be noted that the $\langle \Delta E_c \rangle$Bz was determined to be around 2.6 kcal mol^{-1} at a Bz* excitation energy of 116.4 kcal mol^{-1} and a Bz temperature of 300 K in a prior classical trajectory study [73] on Bz* + Bz single-collision energy transfer. Nevertheless, it was also shown that implementing the rotationally frozen condition in classical trajectory research increased the likelihood of forming collisional complexes with longer lifetimes, thereby raising the $\langle \Delta E_c \rangle$Bz value. The $\langle \Delta E_c \rangle$Bz from our simulations in Figure 2.15 is ~3.7 kcal mol^{-1} at the same excitation energy, $\langle E(t) \rangle$Bz* = 116.4 kcal mol^{-1}. This could be because the current simulation model forms more collisional complexes. At 140 K, however, $\langle \Delta E_c \rangle$Bz increases by nearly two times at $\langle E(t) \rangle$Bz* = 116.4 kcal mol^{-1} compared to 300 K. The lowest Bz temperature simulated in the previously described classical trajectory study [80] was 200 K, where $\langle \Delta E_c \rangle$Bz was determined to be approximately 4% higher than the one at 300 K and was approximately 2.7 kcal mol^{-1} for the down Bz*–Bz collisions. As a result, it may be presumed that when the temperature drops from 200 K to 140 K, the value of $\langle \Delta E_c \rangle$Bz increases significantly. A later section discusses the potential cause of this behavior.

2.3.1.3.4 *Comparison between simulation and experiment*

For meaningful comparisons with experimental data, it is crucial to determine simulation values for the average energy transfer per collision, $\langle \Delta E_c \rangle$Bz*, from vibrationally Bz* as a function of bath temperature. The atomistic simulation approach used here provides detailed insights into IET dynamics, including the effects of intermolecular potential energy parameters, the energy distribution within the excited molecule, and the energy partitioning into translational, rotational, and vibrational modes among the collision partners. This model also identifies specific modes that act as gateways for energy transfer. The simulations yield significantly more IET parameters than can be experimentally measured; for instance, experiments have only measured the transient rotational/translational temperature of N_2 in the bath, using a 1% concentration of NO as a proxy.

In prior work [71], a simulation was conducted for Bz* relaxation in a Bz/N_2 bath at 300 K. For the present study, a similar simulation was performed at 140 K. However, a direct comparison of $\langle \Delta E_c \rangle$Bz* vs. $\langle E(t) \rangle$Bz* between the two simulations is challenging because the Bz/N_2 bath composition was different. The 300 K simulation included 1 Bz*, 9 Bz, and 190 N_2 molecules, modeling a bath with 5% Bz and 10% vibrationally Bz*, as described in the reference [71]. In contrast, the 140 K simulation used a bath of 1 Bz*, 3 Bz, and 396 N_2 molecules, representing a 1% Bz composition with 25% vibrationally Bz*. This 140 K simulation model corresponds to experimental conditions on which Bz accounts for 1.25% of the Bz/N_2 bath, with 16%–30% vibrationally Bz*.

While $\langle \Delta E_c \rangle$Bz* values from the two simulations cannot be directly compared due to bath composition differences, the energy transfer rates to the N_2 and Bz bath molecules, $\langle \Delta E_c \rangle N_2$ and $\langle \Delta E_c \rangle$Bz, respectively, allow us to investigate the temperature dependence of $\langle \Delta E_c \rangle$Bz*. In the earlier 300 K simulation, $\langle \Delta E_c \rangle$Bz* was found to be 0.44 kcal mol^{-1} for Bz* with an average energy $\langle E(t) \rangle$Bz* of 150 kcal mol^{-1}. To align simulated $\langle \Delta E_c \rangle$Bz* values at 140 K and 300 K for Bz* at $\langle E(t) \rangle$Bz* = 150 kcal mol^{-1}, the 140 K bath composition (396 N_2 + 3 Bz) was converted to match the 300 K composition (190 N_2 + 9 Bz) by applying $\langle \Delta E_c \rangle N_2$ and $\langle \Delta E_c \rangle$Bz values from Figure 2.15 in Eq. (2.13). This yielded an $\langle \Delta E_c \rangle$Bz* of 0.83 kcal mol^{-1}, approximately twice the 300 K value. A similar comparison for the 140 K simulation ($\langle \Delta E_c \rangle$Bz* = 0.18 kcal mol^{-1} at $\langle E(t) \rangle$Bz* = 150 kcal mol^{-1}) produced an adjusted $\langle \Delta E_c \rangle$Bz* of 0.12 kcal mol^{-1} when the bath composition was converted to 140 K using 300 K $\langle \Delta E_c \rangle N_2$ and $\langle \Delta E_c \rangle$Bz values, about 1.5 times lower than the 140 K value. Figure 2.16 presents $\langle \Delta E_c \rangle$Bz* at 140 K and 300 K for both bath compositions as a function of $\langle E(t) \rangle$Bz*.

In the study, $\langle \Delta E_c \rangle$Bz* at 140 K is approximately 1.5–2 times higher than ⟨E(t)⟩Bz* across all baths. Experimental procedures have been previously detailed [81, 82], and prior experiments at 300 K are summarized in a comparison of both simulated and experimental studies of Bz* relaxation in a Bz/N_2 bath initially at

300 K [71]. In these experiments, Bz* was generated through 193 nm photoexcitation followed by rapid internal conversion, after which Bz* underwent collisional relaxation in Bz/N_2 baths with initial temperatures of either 300 K or 140 K. A transient increase in the rotational translational temperature of N_2 (97.75%) was tracked via laser-induced fluorescence (LIF) measurements of seeded NO (1% of the bath), chosen for its rapid RT-RT equilibration with N_2, its similar heat capacity to N_2 (preserving Laval flow fields at low temperatures), and its strong signal-to-noise ratio in LIF.

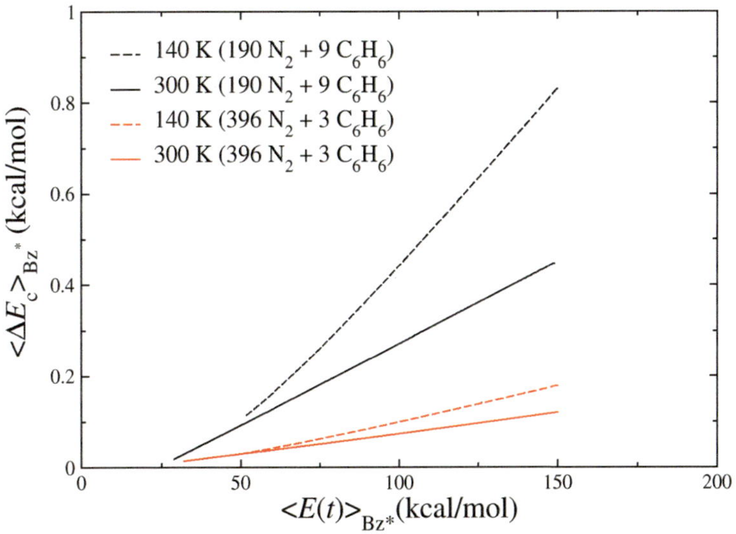

Figure 2.16 Values of $\langle \Delta E_c \rangle$ vs $\langle E \rangle$ of Bz* for Bz/N_2 at 140 K and 300 K are shown. Results for two bath mixes are presented. $\langle \Delta E_c \rangle$ at 140 K (dashed line) and 300 K (solid line) for a bath containing 9 Bz and 190 N_2 molecules (black); $\langle \Delta E_c \rangle$ at 140 K (dashed line) and 300 K (solid line) for a bath containing 3 Bz and 396 N_2 molecules (red) (The data taken from ref. [10]).

Temperature-dependent collision frequencies among Bz*–N_2 and Bz*–Bz were estimated using Lennard-Jones parameters for Bz and N_2. These collision frequencies were consistently applied in both experimental and simulation analyses to ensure comparability. Previous studies of the 300 K data found that $\langle \Delta E_c \rangle$Bz* values from experiments and simulations were in agreement and highlighted Bz*–Bz vibrational energy transfer as critical in the collisional relaxation of Bz*. For the present 140 K experiments, the Bz/N_2 bath contained 1.25% Bz, and it was estimated that about 16%–30% of the Bz molecules were vibrationally excited. The bath's temperature rose to around 95% of its maximum value (approximately 225 K) within 10 μs, estimated to occur over roughly 400 Bz*–N_2 and 8 Bz*–Bz collisions. Following this period, the NO rotational temperature showed no further notable increase, suggesting that either Bz* collisional relaxation had completed or any remaining relaxation occurred on a timescale too slow for detection in the Laval apparatus.

The experimental $\langle \Delta E_c \rangle$Bz* is observed to be about 14 times greater than the simulated $\langle \Delta E_c \rangle$Bz* at an $\langle E(t) \rangle$Bz* of 150 kcal mol^{-1}. Additionally, a discrepancy is noted in the final N_2 rotational temperature, where the experimental ΔT exceeds the simulated ΔT by approximately a factor of five. Further sections address potential reasons for these differences in $\langle \Delta E_c \rangle$Bz*. Comparisons with earlier studies [79] suggest that the average energy transferred per collision rises as temperature decreases below 300 K, contrasting with behavior above 300 K. The IET efficiency at 140 K, which saw the temperature increase from 140 K to 225 K, is approximately double that for a temperature rise from 300 K to 610 K, indicating that the rate of IET accelerates more rapidly per kelvin below 300 K than above. Explanations for the more efficient energy transfer observed at 140 K are explored in a later section.

2.3.1.3.5 *Bz*–Bz transient dimers for vibrational relaxation of Bz* at low temperature*

In the 140 K experiments, the concentration of Bz molecules was up to 5.6 times lower than in the 300 K experiments. At lower temperatures, it is worth considering whether the longer lifetimes of Bz*–Bz and Bz–Bz collisions, due to extended "chattering" collisions, as well as the potential formation of collisionally stabilized dimers, might facilitate more efficient Bz*–Bz IET. Based on an orientation-averaged calculation done [2], the Bz–Bz potential energy parameters used in the simulations yield a minimum potential energy of −0.54 kcal mol^{-1} at a Bz–Bz center-of-mass distance of 6.7 Å. However, the global minimum was found to be −2.32 kcal mol^{-1} for a tilted-T configuration. This tendency to form long-lived complexes aligns with previous simulations [80], which showed that "chattering" collisions lasting around 7 ps formed at 200 K, in contrast to shorter ~30 fs "hit-and-run" collisions. At 140 K, these complexes would have longer lifetimes due to slower molecular velocities, leading to increased IET efficiency as energy could be transferred multiple times within each "chattering" collision complex.

The simulations also explored the potential formation of Bz*–N_2 complexes at low temperatures. Trajectories were analyzed by tracking the center-of-mass distances among Bz and selected N_2 molecules from the bath, but no evidence of Bz*–N_2 complexes was found. The Bz–N_2 minimum energy was −0.24 kcal mol^{-1}, significantly higher than that of Bz–Bz, and the global minimum for Bz–N_2 was only −0.75 kcal mol^{-1}, much weaker than the −2.32 kcal mol^{-1} seen in Bz dimers. This weaker interaction is likely why no Bz–N_2 complexes formed in the simulations.

To avoid collisionally stabilized Bz dimer formation in the experiments, conditions were chosen to prevent significant dimer formation over the 140 K experiment's timescale, according to rate coefficients [83]. A lower Bz concentration was used at 140 K than in the 300 K experiments, similar to the conditions in prior NO quenching studies [10]. Given Bz's melting point of 278.68 K, it would become solid at 140 K over extended timescales, so measurements were taken quickly and at a low pressure to minimize stabilized dimer formation. Collisions with other bath species occurred approximately every 100 ns during the 10 μs experiment, so if dimers had formed, they might explain the notably higher IET efficiency seen at 140 K. Recent studies of Bz–Bz* IET dynamics in an N_2 bath also indicated increased IET efficiency for aromatic complexes [11], with $\langle \Delta E_c \rangle$ values of approximately 2.7 kcal mol^{-1} and 4.8 kcal mol^{-1} at excitation energies of 67 kcal mol^{-1} and 125 kcal mol^{-1}, respectively, in a bath density of 20 kg m^{-3}. Comparatively, Bz* at 104 kcal mol^{-1} excitation gave a $\langle \Delta E_c \rangle$ of about 2.7 kcal mol^{-1} under similar conditions [1, 3]. If collisionally stabilized dimers were present, Bz*–Bz IET could occur without requiring individual Bz* and Bz collisions, increasing the likelihood of excited Bz dimers and oligomers, both contributing to higher IET rates.

Based on Hamon's rate coefficient measured at a lower pressure, the proportion of Bz–Bz complexes in the 140 K simulation (lasting 3 ns) would be less than 1%. At this timescale, long-lived dimers were rare, with about 2% of simulation time involving Bz dimers persisting for more than 50 ps. These findings align reasonably well with Hamon's results. Although the IET dynamics in the simulation are non-statistical, the final bath temperature would hypothetically reach 186 K if given infinite time, resulting in a ΔT of 46 K. The N_2 rotational ΔT after 3 ns in the 140 K simulation is 17 K, showing that the simulation is about 37% toward a fully equilibrated state. By contrast, the experimental N_2 rotational ΔT was 85 K, approximately double that of the simulation, if complete re-equilibration of translational and rotational modes is assumed.

Several factors may explain the differences in N_2 temperature rise and IET efficiency between experiment and simulation. Possible experimental errors, such as inaccuracies in initial Bz concentration, Bz* generation at 140 K, Bz multiphoton absorption at 193 nm, or enhanced absorption by Bz dimers, may account for some discrepancies. Other potential differences could arise from whether NO adequately represents N_2 behavior at low temperatures or if NO quenches electronically Bz* more efficiently at these temperatures. Simulation-related factors, like intermolecular potential energy parameters or trajectory initial conditions, could also contribute, but these are unlikely to fully explain the observed differences.

2.3.2 QM/MM Simulations

2.3.2.1 Ozonolysis reaction of catechol in N_2 bath

The dynamics performed are to explore the different product channels formed from the cat-O_3 TS, and therefore, the starting reactive species of the dynamics is cat-O_3 TS. There are two TS geometries available, namely, exo and endo, of which the endo TS is more stable and is taken as the starting species for the dynamics, which is why the dynamics are named as post-TS dynamics. Similar product channels, as previously noticed, are found in the gas-phase and the condensed-phase dynamics at 800, 1000, and 1500 K [18]. Following the formation of the POZ (primarily ozonide), the cat-O_3 endo-TS decomposes into a peroxy acid (POA), which in turn generates the five major channels: the CO, CO_2, H_2O, O_2, and SCA channels. Although it is one of the most noticeable channels at higher temperatures, the O_2 channel was not taken into consideration because it was insignificant at lower temperature simulations. With the exception of two channels, every channel's energy profile was previously shown [18].

2.3.2.1.1 *Reaction probabilities*

All trajectories that result in product formation are classified as reactive channels. The trajectories that remain in the POA state by the end of the simulation time are termed as POA channels, while non-reactive channels include trajectories that do not lead to any product formation and do not settle into the POA state. These non-reactive trajectories tend to form POZ and fluctuate among endo and exo POZ configurations throughout the simulation. Reaction probabilities are determined by calculating the proportion of trajectories forming a specific channel relative to the total number of trajectories in each simulation.

The probability of reactive channels in the gas phase over a 100 ps run at 800 K is 53.0%. The percentages at 1000 and 1500 K are 59.3% and 92.0%, respectively. POA formation channel probabilities at 800, 1000, and 1500 K are 46.3%, 40.6%, and 8.0%, respectively. Together, these two channels account for unimolecular reaction pathways, with their combined probabilities reaching 99.3%, 99.9%, and 100% at 800, 1000, and 1500 K.

In condensed-phase simulations at a density of 20 kg m^{-3}, the reactive channel probabilities are 28.1%, 40.9%, and 64.7% at 800, 1000, and 1500 K, respectively, while the corresponding POA formation channel probabilities are 64.8%, 56.6%, and 34.7%. At a higher density of 324 kg m^{-3}, the reactive channel probabilities at those temperatures are 12.4%, 26.2%, and 47.14%, with POA formation channel probabilities at 47.6%, 55.2%, and 34.3%. Figure 2.17 presents the reaction probabilities in a bar diagram.

Table 2.11 Reactive, POA, and unimolecular dissociation channel percentage of gas and condensed phase

	Gas-phase simulation 20 kg m^{-3}			Condensed-phase simulation 324 kg m^{-3}			
T (K)	Reactive channel	POA channel	Unimolecular channel	Reactive channel	POA channel	Reactive channel	POA channel
800	53.0%	46.3%	99.3%	28.1%	64.8%	12.4%	47.6%
1000	59.3%	40.6%	99.9%	40.9%	56.6%	26.2%	55.2%
1500	92.0%	8.0%	100.0%	64.7%	34.7%	47.14%	34.3%

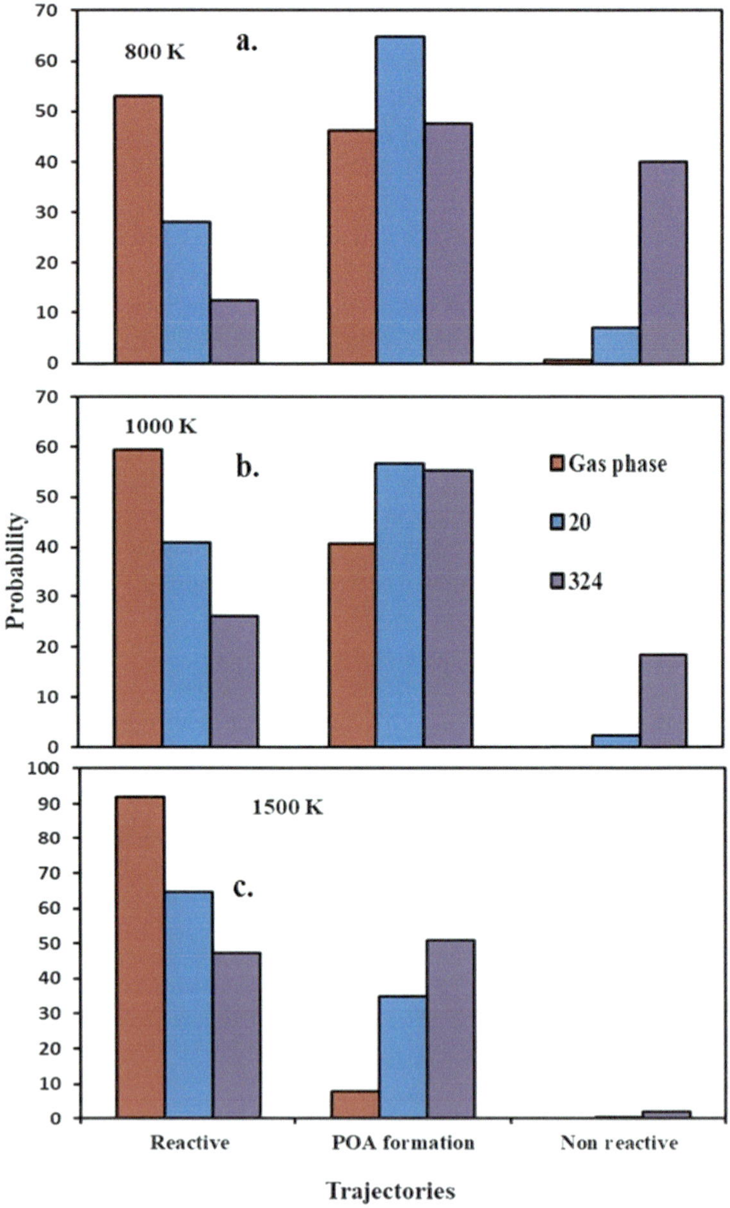

Figure 2.17 Bar chart illustrates the probability variation at different temperatures: (a) 800 K, (b) 1000 K, and (c) 1500 K, comparing results for the gas phase and for densities of 20 kg m^{-3} and 324 kg m^{-3} (Figure taken from ref. [8]).

As the temperature increases, the overall number of reactive trajectories also increases for both densities. However, at any given temperature, the proportion of reactive trajectories is lower at 324 kg m^{-3}. This decrease is likely due to enhanced energy transfer between the bath and the solute, reducing the likelihood of the excited cat-O_3 system forming products at higher densities. In contrast, in gas-phase simulations, where energy transfer with bath molecules does not occur, the probability of reactive channels remains highest.

Previous research indicates that energy transfer from reactive species to bath molecules occurs primarily through low-frequency vibrational modes of the reactive species to translational and rotational modes of N_2 bath molecules [3, 9–13, 66, 71, 84, 85]. Product channel probabilities were also calculated, as shown in Figure 2.18, with the most probable products being CO, CO_2, H_2O, small carboxylic acids (SCA), and O_2. Notably, the O_2 channel, which is prominent at higher temperatures, was minimal at 400 K and 500 K in prior studies [18].

Figure 2.18 shows that, in the gas phase, the most probable product channels are the H_2O channel at 800 and 1500 K and the CO channel at 1000 K. The H_2O and CO channels are kinetically similar, but thermodynamically, CO is more probable due to the energy of the final products. The probabilities of these channels in the simulations are nearly identical within statistical uncertainties due to their close energy values. In the condensed phase, H_2O channels have slightly higher probabilities than CO channels, except at 324 kg m^{-3} and 800 K. For the O_2 channel, its probability increases significantly in the condensed phase compared to the gas phase. CO, H_2O, and SCA channels follow similar reaction paths as seen at lower temperatures (400 and 500 K), but the CO_2 channel at higher temperatures takes a high-energy pathway that was not favorable at lower temperatures. In both gas and condensed phases, SCA channels are less probable than other channels.

Generally, the probabilities of CO, CO_2, H_2O, and SCA channels decrease as density increases from the gas phase to 20 kg m^{-3} and 324 kg m^{-3}, mainly due to reduced reactivity and energy dissipation at higher densities. However, the opposite trend is observed for the O_2 channel, where its probability rises with increased bath density. Being kinetically controlled, this fast reaction stabilizes the excited products via collisional vibrational energy exchange, making it more favorable in the condensed phase and at higher densities.

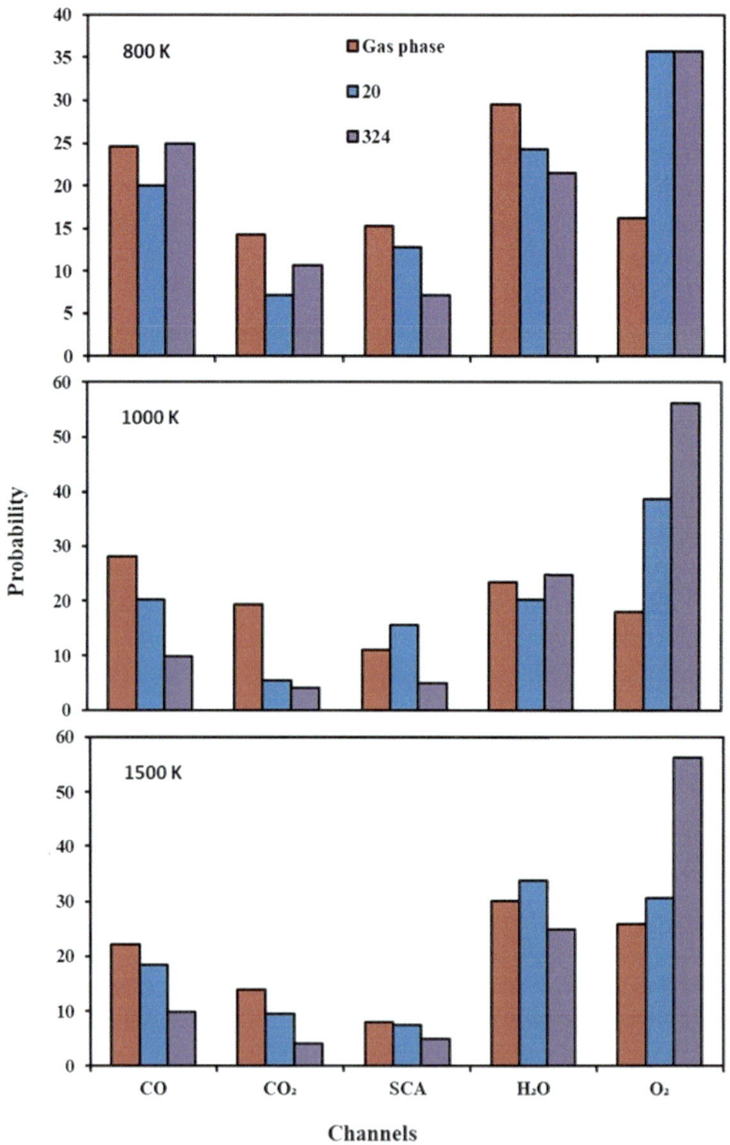

Figure 2.18 Bar chart displays the probability distribution across product channels at temperatures (a) 800 K, (b) 1000 K, and (c) 1500 K, with comparisons for the gas phase and densities of 20 kg m^{-3} and 324 kg m^{-3} (Figure taken from ref. [8]).

2.3.2.1.2 *Unimolecular dissociation*

The creation of POA from the cat-O_3 TS is the unimolecular reaction examined in this work, taking into account both reactive and POA-forming pathways at all temperatures and densities. Figure 2.19 calculates and displays the behavior of the ratio $N(t)/N(0)$ over time for the gas phase (a), 20 kg m^{-3} (b), and 324 kg m^{-3} (c) at all temperatures. $N(0)$ is the initial number of trajectories at each temperature, and $N(t)$ is the number of undissociated trajectories at time t.

It is worth noting that there is an early cut-off of all $N(t)/N(0)$ data before 100 ps, even though the $N(t)/N(0)$ value does not decay to zero. This is because, beyond the cut-off time, no trajectories report a unimolecular reaction within the 100-ps simulation window. For example, if 180 out of 300 trajectories react, and the last reactive trajectory reaches product formation at 40 ps, the $N(t)/N(0)$ value would stop at (300–180)/300 at 40 ps.

Unimolecular rate constants are then derived by fitting those curves to analytical functions, following the basic methods in kinetic theory [4, 86–89]. For unimolecular dissociation reaction in gas and the condensed phase at a density of 20 kg m^{-3} at all temperatures, the survival probabilities fit well to a bi-exponential function of the form:

$$P(t)=\frac{N(t)}{N(0)}=f_1*\exp(-k_1t)+f_2*\exp(-k_2t)+C \tag{2.14}$$

Here, $f_1+f_2=1$. k_1 and k_2 are the rate constants, and C represents the final $N(t)/N(0)$ value at the end of the simulation time.

In contrast, for the time profiles at a density of 324 kg m^{-3} and temperatures of 800, 1000, and 1500 K, the data are better represented by a stretched exponential function of the form:

$$P(t)=\frac{N(t)}{N(0)}=f_1*\exp(-k_1t)+f_2*\exp\left(-k_2t^{\beta(T)}\right)+C \tag{2.15}$$

where $f_1+f_2=1$, $\beta(T)$ is the stretching exponent $(0<\beta(T)<1)$, and C is the final $N(t)/N(0)$ value at the end time.

It is important to note that in condensed-phase calculations, two processes are occurring simultaneously: unimolecular dissociation and collisional energy transfer. Equations (2.14 and 2.15) specifically address the unimolecular dissociation reaction only.

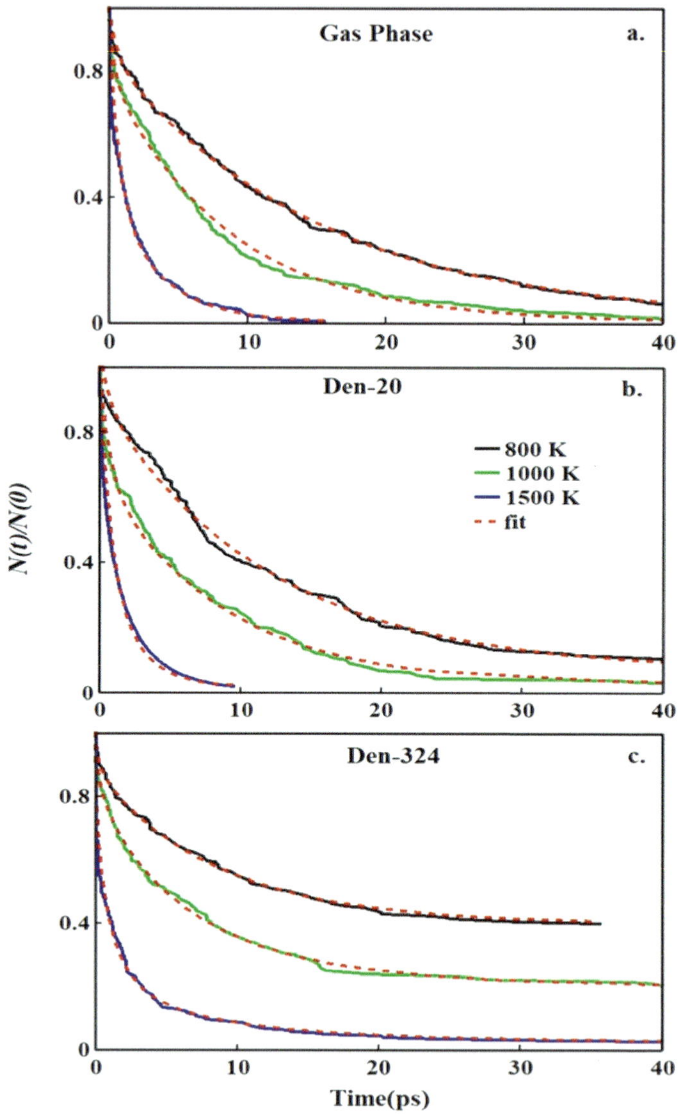

Figure 2.19 The ratio $N(t)/N(0)$ over time was analyzed for the unimolecular dissociation of endo-TS in three settings: (a) in the gas phase, (b) at an N_2 bath density of 20 kg m^{-3}, and (c) at an N_2 bath density of 324 kg m^{-3}. Each scenario was examined at temperatures of 800, 1000, and 1500 K (Figure taken from ref. [8]).

The parameters for the fits are f_1, f_2, k_1, k_2, $\beta(T)$, and C for the unimolecular reaction in the gas phase at densities of 20 and 324 kg m^{-3}, are listed in Table 2.12. This table also includes the microcanonical rate constants, k_m. It is calculated from $d\left(\frac{N(t)}{N(0)}\right)/dt$ at $t = 0$. In Eq. (2.14), k_m is transformed into $f_1k_1 + f_2k_2$, whereas in Eq. (2.15), k_m becomes f_1k_1.

Table 2.12 The fit parameters for $N(t)/N(0)$ for the unimolecular reaction in both gas and condensed phases at temperatures of 800, 1000, and 1500 K

Density	Gas-phase	20	324	Gas-phase	20	324	Gas-phase	20	324
(kg m^{-3})	800 K			1000 K			1500 K		
f_1	0.158	0.159	0.525	0.212	0.344	0.544	0.550	0.550	0.246
f_2	0.842	0.841	0.475	0.788	0.656	0.456	0.450	0.450	0.754
k_1	1.700	1.562	3.340	2.480	1.500	1.180	1.500	1.300	1.000
k_2	0.067	0.085	0.116	0.117	0.120	0.139	0.298	0.470	0.890
$\beta(T)$	-	-	0.300	-	-	0.280	-	-	0.440
C	0.010	0.067	0.400	0.0034	0.029	0.186	0.0034	0.0184	0.0190
SD[11]	0.015	0.032	0.036	0.025	0.040	0.023	0.074	0.105	0.072
k_m (ps^{-1})[12]	0.325	0.320	0.055	0.618	0.595	0.063	0.959	0.926	0.246

[11]SD is the standard deviation of the fit.

[12]k_m is the microcanonical rate constant at $t = 0$.

The bi-exponential function indicates two distinct timescales in the dynamics, whereas a stretched exponential function suggests multi-exponential behavior, likely due to dynamics across multiple timescales. In Eq. (2.15), an exponential term with probability f_1 is combined with a stretched exponential term. For a purely stretched exponential function, the rate constant at time zero (microcanonical rate constant) would theoretically be zero, which is impractical. The current approach calculates this rate constant as f_1k_1.

2.3.2.1.3 *Comparison between gas-phase and condensed-phase results*

In previous work [18], the energy profiles of product channels indicated that POA formation involved an energy barrier of approximately 26 kcal mol^{-1}, which resulted in an initial delay in the reaction. However, at higher temperatures, this delay is significantly reduced. If all trajectories lead to POA formation, or if the number of trajectories remains constant, the total time required for the unimolecular reaction decreases with increasing temperature. In gas-phase simulations, all trajectories resulted in POA formation at 1000 and 1500 K, whereas at 800 K, only 2 out of 300 trajectories failed to form POA. Therefore, in these gas-phase simulations, the overall time for unimolecular reaction decreases with increasing temperature.

In contrast, in condensed-phase simulations, a greater number of non-reactive trajectories was observed, particularly at a density of 324 kg m^{-3}, although the number decreases as the temperature increases. For instance, at 800 K, 84 trajectories were non-reactive, while this number dropped to 39 at 1000 K and 4 at 1500 K. Consequently, reactivity increases with temperature. Since the number of trajectories is not fixed in these simulations, the total time for unimolecular dissociation can either increase or decrease with rising temperature. At this density, the total time for dissociation increases at 1000 K but decreases at 1500 K, which suggests that some trajectories at 1000 K take longer to dissociate. However, if the number of trajectories is held constant, the total dissociation time reduces as the temperature increases. For instance, 126 trajectories dissociate within 36 ps at 800 K, but the same number of trajectories takes only 8.2 ps at 1000 K and 1.2 ps at 1500 K. This indicates that, at a fixed density, higher temperatures accelerate unimolecular dissociation, leading to a higher rate constant. Additionally, when the temperature is constant, and the density of the bath is varied, the energy transfer from the TS to the N_2 bath changes, influencing the unimolecular dissociation rate.

Table 2.13 A comparison of gas-phase and condensed-phase dynamics observations

	Unimolecular dissociation within 20 ps			k_m (ps^{-1})			k_m ratio with respect to the gas phase	
T (K)	Gas phase	20 kg m^{-3}	324 kg m^{-3}	Gas phase	20 kg m^{-3}	324 kg m^{-3}	20 kg m^{-3}	324 kg m^{-3}
800	78.0%	80.0%	57.2%	0.223	0.075	0.0474	2.90	4.70
1000	91.7%	93.3%	76.2%	0.330	0.441	0.0627	0.74	5.62
1500	100%	99.5%	96.2%	0.385	0.106	0.336	1.10	3.60

As summarized in Table 2.13, at 800, 1000, and 1500 K in the gas phase, 78.0%, 91.7%, and 100% of the trajectories undergo unimolecular dissociation within 20 ps. At a density of 20 kg m^{-3}, these values are 80.0%, 93.3%, and 99.5%, indicating that the dissociation probabilities in the gas phase and at 20 kg m^{-3} are quite similar. For a higher density of 324 kg m^{-3}, 57.2%, 76.2%, and 96.2% of trajectories dissociate within 20 ps at 800, 1000, and 1500 K, respectively. The calculated microcanonical rate constants, k_m, at 800, 1000, and 1500 K in the gas phase are 0.223, 0.330, and 0.385 ps^{-1}, while at 20 kg m^{-3}, they are 0.075, 0.441, and 0.106 ps^{-1}, and at 324 kg m^{-3}, they are 0.0474, 0.0627, and 0.336 ps^{-1}. The ratios of these rate constants between the gas and condensed phases are 2.9 and 4.7 at 800 K, 0.74 and 5.62 at 1000 K, and 1.1 and 3.6 at 1500 K for densities of 20 and 324 kg m^{-3}, respectively.

2.3.2.1.4 *The bath effect*

As previously discussed, the probability of the unimolecular reaction decreases as density increases, with a notably distinct behavior observed at 324 kg m^{-3}. At this higher density, the number of non-reactive trajectories is significantly greater at all simulation temperatures compared to those at 20 kg m^{-3}. This effect may be attributed to increased energy transfer from the reactant to the bath at 324 kg m^{-3}. It is important to note that while the bath is maintained at 300 K, the TS excitation temperatures are 800, 1000, and 1500 K, respectively. Consequently, there is ample opportunity for the excited molecule to lose energy through collisions with the bath.

To examine this energy transfer, the average total energy of the bath was tracked at the initial stage of all condensed-phase simulations, as shown in Figure 2.20. The results indicate a significantly larger energy transfer at 324 kg m^{-3} than at 20 kg m^{-3}. The number of N_2 molecules within a 15 Å radius from the reactant's center-of-mass was calculated to better understand this energy transfer behavior, with values of 4 and 49 for 20 and 324 kg m^{-3}, respectively. For an 18 Å radius, these values increase to 6 and 85. Therefore, the density of N_2 molecules near the solute is much higher at 324 kg m^{-3}, leading to more frequent multi-body collisions.

The timescale of energy transfer versus the reaction timescale is also relevant here. At 20 kg m^{-3}, energy transfer occurs at a slower rate compared to that at 324 kg m^{-3}, while the reaction timescale remains relatively consistent across both densities. Consequently, reactivity is reduced in simulations conducted at the 324 kg m^{-3} bath density.

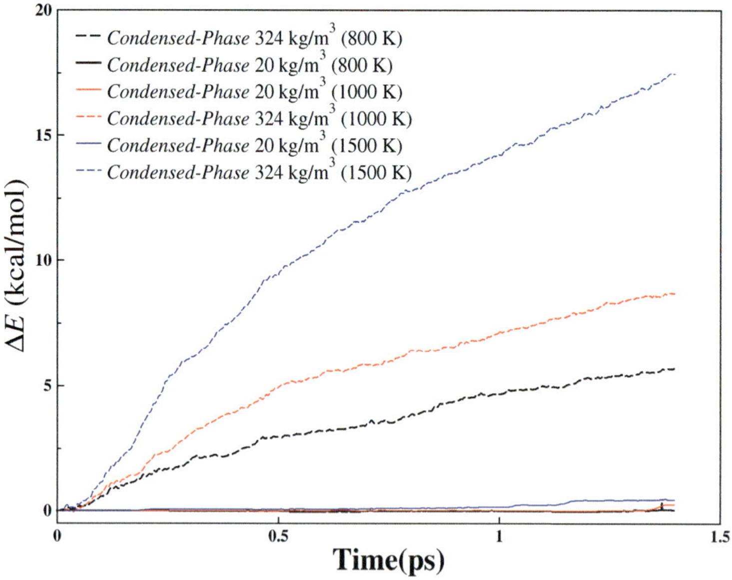

Figure 2.20 The average total energy of the bath versus time for the simulations conducted at 20 and 324 kg m^{-3} with solute's excitation temperatures of 800, 1000, and 1500 K for each case (The data for this figure is taken from ref. [8]).

2.4 Conclusions

In condensed-phase simulations, both MM and QM/MM models offer unique advantages and limitations that can shape the accuracy and efficiency of chemical dynamics studies. The results of simulations performed using the MM model and QM/MM model are explained in detail in the above section.

The MM model is computationally efficient, allowing for large-scale simulations with many atoms over extended timeframes. This is particularly valuable for studying bulk properties, diffusion, and non-reactive interactions in condensed phases, such as dimer association and dissociation reactions, as mentioned in Section 2.3.1.1. However, MM methods mainly rely on empirical force fields and cannot fully capture electronic effects or bond-breaking and bond-forming processes essential to chemical reactions, which limits their applicability for reactive simulations in condensed phases.

In contrast, the QM/MM model combines the strengths of QM, which accurately describes electronic structure changes and reactive events, with the efficiency of MM for non-reactive regions. This hybrid approach enables more realistic modeling of condensed-phase reactions by allowing key reactive sites to be treated with QM precision while the surrounding environment is modeled with MM, as in the simulation described in Section 2.3.2.1. As a result, QM/MM is especially powerful for simulations where electronic interactions within a localized reactive region are critical, such as enzymatic reactions, catalysis, and complex solvent effects.

In summary, MM models are valuable for non-reactive studies at scale, while QM/MM models offer a balanced approach for simulating reactive events in condensed phases with high accuracy and at a feasible computational cost. Together, these methods expand our ability to model and understand condensed-phase dynamics by leveraging the strengths of both classical and quantum mechanical approaches.

References

1. A. K. Paul, S. Kolakkandy, and W. L. Hase, "Dynamics of Na+(Benzene) + benzene association and ensuing Na+(Benzene)$_2$* dissociation," *J. Phys. Chem. A*, vol. 119, no. 28, pp. 7894–7904, 2015, doi: 10.1021/acs.jpca.5b01922.
2. X. Ma, A. K. Paul, and W. L. Hase, "Chemical dynamics simulations of benzene dimer dissociation," *J. Phys. Chem. A*, vol. 119, no. 25, pp. 6631–6640, 2015, doi: 10.1021/acs.jpca.5b03897.
3. A. K. Paul, S. C. Kohale, S. Pratihar, R. Sun, S. W. North, and W. L. Hase, "A unified model for simulating liquid and gas phase, intermolecular energy transfer: N_2 + C_6F_6 collisions," *J. Chem. Phys.*, vol. 140, pp. 194103, 2014, doi: 10.1063/1.4875516.
4. H. Mahanta, D. Baishya, S. S. Ahamed, and A. K. Paul, "A better understanding of the unimolecular dissociation dynamics of weakly bound aromatic compounds at high temperature: A study on C_6H_6-C_6F_6 and comparison with C_6H_6 dimer," *J. Phys. Chem. A*, vol. 123, no. 13, pp. 2517–2526, 2019, doi: 10.1021/acs.jpca.8b12188.
5. H. Mahanta, D. Baishya, S. S. Ahamed, and A. K. Paul, "Chemical dynamics simulations on association and ensuing dissociation of a benzene-hexafluorobenzene molecular system," *J. Phys. Chem. A*, vol. 123, no. 24, pp. 5019–5026, 2019, doi: 10.1021/acs.jpca.9b02332.
6. H. Mahanta and A. K. Paul, "Dynamical behavior of aromatic trimer complexes in unimolecular dissociation reaction at high temperatures: Case studies on C_6H_6–C_6F_6–C_6H_6 and C_6H_6 trimer complexes," *J. Phys. Chem. A*, vol. 126, no. 2, pp. 259–271, 2022, doi: 10.1021/acs.jpca.1c09073.
7. B. Deb, H. Mahanta, N. P. Baruah, M. Khardewsaw, and A. K. Paul, "On the intramolecular vibrational energy redistribution dynamics of aromatic complexes: A comparative study on C_6H_6-C_6H_5Cl, C_6H_6-$C_6H_3Cl_3$, C_6H_6-C_6Cl_6 and C_6H_6-C_6H_5F, C_6H_6-$C_6H_3F_3$, C_6H_6-C_6F_6," *J. Chem. Phys.*, vol. 160, pp. 024307, 2024, doi: 10.1063/5.0174748.
8. A. Agarwal, P. J. Boruah, S. S. Ahamed, S. Baruah, and A. K. Paul, "Post-transition state direct dynamics simulations on the ozonolysis of catechol in an N_2 bath and comparison with gas-phase results," *J. Phys. Chem. A*, vol. 127, no. 32, pp. 6804–6815, 2023, doi: 10.1021/acs.jpca.3c03326.
9. S. S. Ahamed, H. Mahanta, and A. K. Paul, "An advanced bath model to simulate association followed by ensuing dissociation dynamics of benzene + benzene system: A comparative study of gas and condensed

phase results," *Phys. Chem. Chem. Phys.*, vol. 24, no. 38, pp. 23825–23839, 2022, doi: 10.1039/d2cp02483g.

10. S. S. Ahamed et al., "Comparison of intermolecular energy transfer from vibrationally excited benzene in mixed nitrogen-benzene baths at 140 K and 300 K," *J. Chem. Phys.*, vol. 153, pp. 144116, 2020, doi: 10.1063/5.0021293.
11. S. S. Ahamed, H. Mahanta, and A. K. Paul, "A competition between dissociation pathway and energy transfer pathway: Unimolecular dissociation of a benzene-hexafluorobenzene complex in nitrogen bath," *J. Phys. Chem. A*, vol. 123, no. 50, pp. 10663–10675, 2019, doi: 10.1021/acs.jpca.9b07258.
12. S. S. Ahamed, P. Kumar, H. Kalita, and A. K. Paul, "Mode-to-mode collision energy transfer from vibrationally excited C_6F_6 to NO/N_2 mixed bath with the development of new potential energy functions," *ChemistrySelect*, vol. 5, no. 34, pp. 10475–10487, 2020, doi: 10.1002/slct.202002600.
13. A. K. Paul, D. Donzis, and W. L. Hase, "Collisional intermolecular energy transfer from a N_2 bath at room temperature to a vibrationlly 'cold' C_6F_6 molecule using chemical dynamics simulations," *J. Phys. Chem. A*, vol. 121, no. 21, pp. 4049–4057, 2017, doi: 10.1021/acs.jpca.7b00948.
14. M. J. Forster, "Molecular modelling in structural biology," *Micron*, vol. 33, no. 4, pp. 365–384, 2002, doi: 10.1016/S0968-4328(01)00035-X.
15. S. S. Ahamed, H. Mahanta, and A. K. Paul, "Unimolecular dissociation of C_6H_6-C_6F_6 complex in N_2 bath and comparison with gas phase dynamics," *Chem. Phys. Lett.*, vol. 730, pp. 630–633, 2019, doi: 10.1016/j.cplett.2019.06.063.
16. H. Mahanta, S. N, R. Misra, and A. K. Paul, "Unimolecular dissociation dynamics of C_6H_6–C_6Cl_6 complex and the effect of anharmonicity," *Int. J. Mass Spectrom.*, vol. 456, pp. 116392, 2020, doi: 10.1016/j.ijms.2020.116392.
17. B. Deb, S. R. N. Anal, H. Mahanta, Yogita, and A. K. Paul, "Unimolecular dissociation of C_6H_6-C_6H_5Cl, C_6H_6-$C_6H_3Cl_3$, and C_6H_6-C_6Cl_6 complexes using machine learning approach," *J. Chem. Phys.*, vol. 158, pp. 194104, 2023, doi: 10.1063/5.0139864.
18. A. Agarwal, P. J. Boruah, B. Sarkar, and A. K. Paul, "Post-transition-state direct dynamics simulations on the ozonolysis of catechol," *J. Phys. Chem. A*, vol. 126, no. 32, pp. 5314–5327, 2022, doi: 10.1021/acs.jpca.2c04028.

19. R. M. Venable, F. L. H. Brown, and R. W. Pastor, "Mechanical properties of lipid bilayers from molecular dynamics simulation," *Chem. Phys. Lipids*, vol. 192, pp. 60–74, 2015, doi: 10.1016/j.chemphyslip.2015.07.014.

20. S. E. Feller, "Molecular dynamics simulations of lipid bilayers," *Curr. Opin. Colloid Interface Sci.*, vol. 5, no. 3–4, pp. 217–223, 2000, doi: 10.1016/S1359-0294(00)00058-3.

21. H. Leontiadou, A. E. Mark, and S. J. Marrink, "Molecular dynamics simulations of hydrophilic pores in lipid bilayers," *Biophys. J.*, vol. 86, no. 4, pp. 2156–2164, 2004, doi: 10.1016/S0006-3495(04)74275-7.

22. S. Moradi, A. Nowroozi, and M. Shahlaei, "Shedding light on the structural properties of lipid bilayers using molecular dynamics simulation: A review study," *RSC Adv.*, vol. 9, no. 8, pp. 4644–4658, 2019, doi: 10.1039/C8RA08441F.

23. M. Doktorova, G. Khelashvili, R. Ashkar, and M. F. Brown, "Molecular simulations and NMR reveal how lipid fluctuations affect membrane mechanics," *Biophys. J.*, vol. 122, no. 6, pp. 984–1002, 2023, doi: 10.1016/j.bpj.2022.12.007.

24. G. Monard and K. M. Merz, "Combined quantum mechanical/molecular mechanical methodologies applied to biomolecular systems," *Acc. Chem. Res.*, vol. 32, no. 10, pp. 904–911, 1999, doi: 10.1021/ar970218z.

25. T. E. Cheatham, J. L. Miller, T. Fox, T. A. Darden, and P. A. Kollman, "Molecular dynamics simulations on solvated biomolecular systems: The particle mesh ewald method leads to stable trajectories of DNA, RNA, and proteins," *J. Am. Chem. Soc.*, vol. 117, no. 14, pp. 4193–4194, 1995, doi: 10.1021/ja00119a045.

26. D. J. Huggins et al., "Biomolecular simulations: From dynamics and mechanisms to computational assays of biological activity," *Wiley Interdiscip. Rev. Comput. Mol. Sci.*, vol. 9, no. 3, pp. 1–23, 2019, doi: 10.1002/wcms.1393.

27. J. Marelius, K. Kolmodin, I. Feierberg, and J. Åqvist, "Q: A molecular dynamics program for free energy calculations and empirical valence bond simulations in biomolecular systems," *J. Mol. Graph. Model.*, vol. 16, no. 4–6, pp. 213–225, 1998, doi: 10.1016/S1093-3263(99)00012-1.

28. S. Takada, R. Kanada, C. Tan, T. Terakawa, W. Li, and H. Kenzaki, "Modeling structural dynamics of biomolecular complexes by coarse-grained molecular simulations," *Acc. Chem. Res.*, vol. 48, no. 12, pp. 3026–3035, 2015, doi: 10.1021/acs.accounts.5b00338.

29. J. Wang, R. M. Wolf, J. W. Caldwell, P. A. Kollman, and D. A. Case, "Development and testing of a general amber force field" *J. Comput. Chem.*, vol. 25, no. 9, pp. 1157–1174, 2004, doi: 10.1002/jcc.20035.

30. A. Allouche, "Software news and updates gabedit — A graphical user interface for computational chemistry software," *J. Comput. Chem.*, vol. 32, no. 1, pp. 174–182, 2011, doi: 10.1002/jcc.

31. X. Zhu, P. E. M. Lopes, and A. D. Mackerell, "Recent developments and applications of the CHARMM force fields," *Wiley Interdiscip. Rev. Comput. Mol. Sci.*, vol. 2, no. 1, pp. 167–185, 2012, doi: 10.1002/wcms.74.

32. R. W. Pastor and A. D. MacKerell, "Development of the CHARMM force field for lipids," *J. Phys. Chem. Lett.*, vol. 2, no. 13, pp. 1526–1532, 2011, doi: 10.1021/jz200167q.

33. E. Lindahl, P. Bjelkmar, P. Larsson, M. A. Cuendet, and B. Hess, "Implementation of the charmm force field in GROMACS: Analysis of protein stability effects from correction maps, virtual interaction sites, and water models," *J. Chem. Theory Comput.*, vol. 6, no. 2, pp. 459–466, 2010, doi: 10.1021/ct900549r.

34. J. B. Klauda, V. Monje, T. Kim, and W. Im, "Improving the CHARMM force field for polyunsaturated fatty acid chains," *J. Phys. Chem. B*, vol. 116, no. 31, pp. 9424–9431, 2012, doi: 10.1021/jp304056p.

35. K. Vanommeslaeghe and A. D. Mackerell, "CHARMM additive and polarizable force fields for biophysics and computer-aided drug design," *Biochim. Biophys. Acta - Gen. Subj.*, vol. 1850, no. 5, pp. 861–871, 2015, doi: 10.1016/j.bbagen.2014.08.004.

36. N. Schmid et al., "Definition and testing of the GROMOS force-field versions 54A7 and 54B7," *Eur. Biophys. J.*, vol. 40, no. 7, pp. 843–856, 2011, doi: 10.1007/s00249-011-0700-9.

37. C. Oostenbrink, A. Villa, A. E. Mark, and W. F. Van Gunsteren, "A biomolecular force field based on the free enthalpy of hydration and solvation: The GROMOS force-field parameter sets 53A5 and 53A6," *J. Comput. Chem.*, vol. 25, no. 13, pp. 1656–1676, 2004, doi: 10.1002/jcc.20090.

38. C. Oostenbrink, T. A. Soares, N. F. A. Van Der Vegt, and W. F. Van Gunsteren, "Validation of the 53A6 GROMOS force field," *Eur. Biophys. J.*, vol. 34, no. 4, pp. 273–284, 2005, doi: 10.1007/s00249-004-0448-6.

39. R. D. Lins and P. H. Hünenberger, "A new GROMOS force field for hexopyranose-based carbohydrates," *J. Comput. Chem.*, vol. 26, no. 13, pp. 1400–1412, 2005, doi: 10.1002/jcc.20275.

40. T. A. Soares et al., "An improved nucleic acid parameter set for the GROMOS force field," *J. Comput. Chem.*, vol. 26, no. 7, pp. 725–737, 2005, doi: 10.1002/jcc.20193.

41. W. Huang, Z. Lin, and W. F. Van Gunsteren, "Validation of the GROMOS 54A7 force field with respect to β-peptide folding," *J. Chem. Theory Comput.*, vol. 7, no. 5, pp. 1237–1243, 2011, doi: 10.1021/ct100747y.

42. K. Nester, K. Gaweda, and W. Plazinski, "A GROMOS force field for furanose-based carbohydrates," *J. Chem. Theory Comput.*, vol. 15, no. 2, pp. 1168–1186, 2019, doi: 10.1021/acs.jctc.8b00838.

43. S. W. I. Siu, K. Pluhackova, and R. A. Böckmann, "Optimization of the OPLS-AA force field for long hydrocarbons," *J. Chem. Theory Comput.*, vol. 8, no. 4, pp. 1459–1470, 2012, doi: 10.1021/ct200908r.

44. D. Shivakumar, E. Harder, W. Damm, R. A. Friesner, and W. Sherman, "Improving the prediction of absolute solvation free energies using the next generation opls force field," *J. Chem. Theory Comput.*, vol. 8, no. 8, pp. 2553–2558, 2012, doi: 10.1021/ct300203w.

45. D. Kony, W. Damm, S. Stoll, and W. F. Van Gunsteren, "An improved OPLS-AA force field for carbohydrates," *J. Comput. Chem.*, vol. 23, no. 15, pp. 1416–1429, 2002, doi: 10.1002/jcc.10139.

46. S. V. Sambasivarao and O. Acevedo, "Development of OPLS-AA force field parameters for 68 unique ionic liquids," *J. Chem. Theory Comput.*, vol. 5, no. 4, pp. 1038–1050, 2009, doi: 10.1021/ct900009a.

47. B. Doherty and O. Acevedo, "OPLS force field for choline chloride-based deep eutectic solvents," *J. Phys. Chem. B*, vol. 122, no. 43, pp. 9982–9993, 2018, doi: 10.1021/acs.jpcb.8b06647.

48. B. Doherty, X. Zhong, and O. Acevedo, "Virtual site OPLS force field for imidazolium-based ionic liquids," *J. Phys. Chem. B*, vol. 122, no. 11, pp. 2962–2974, 2018, doi: 10.1021/acs.jpcb.7b11996.

49. B. Fu, X. Shan, D. H. Zhang, and D. C. Clary, "Recent advances in quantum scattering calculations on polyatomic bimolecular reactions," *Chem. Soc. Rev.*, vol. 46, no. 24, pp. 7625–7649, 2017, doi: 10.1039/c7cs00526a.

50. M. D. Newton, "The $Co(NH_3)_6^{2+/3+}$ exchange reaction: Ground-state versus thermally excited pathways," *J. Phys. Chem.*, vol. 95, no. 1, pp. 30–38, 1991, doi: 10.1021/j100154a010.

51. P. J. Boruah, V. N, A. Samanta, and A. K. Paul, "A detailed theoretical investigation on intramolecular charge transfer mechanism of primary, secondary, and tertiary p-amino substituted benzaldehyde," *Chem. Phys. Impact*, vol. 8, pp. 100538, 2024, doi: 10.1016/j.chphi.2024.100538.

52. J. C. Slater, "A simplification of the Hartree-Fock method," *Phys. Rev.*, vol. 81, no. 3, pp. 385–390, 1951, doi: 10.1103/PhysRev.81.385.

53. J. Bardeen et al., "Generalized Hartree-Fock method," *Phys. Rev.*, vol. 122, no. 8, pp. 1012–1020, 1961, doi: 10.1103/PhysRev.122.1012.

54. D. M. Brink, "Density functional theory," *Int. J. Phytoremediation*, vol. 12, no. 4, pp. 27–32, 2002, doi: 10.1080/10506890208232107.

55. E. R. Johnson and A. D. Becke, "A post-Hartree-Fock model of intermolecular interactions: Inclusion of higher-order corrections," *J. Chem. Phys.*, vol. 124, pp. 174104, 2006, doi: 10.1063/1.2190220.

56. E. R. Johnson and A. D. Becke, "A post-Hartree-Fock model of intermolecular interactions," *J. Chem. Phys.*, vol. 123, pp. 024101, 2005, doi: 10.1063/1.1949201.

57. H. M. Senn and W. Thiel, "QM/MM studies of enzymes," *Curr. Opin. Chem. Biol.*, vol. 11, no. 2, pp. 182–187, 2007, doi: 10.1016/j.cbpa.2007.01.684.

58. V. Laporta, R. Celiberto, and J. M. Wadehra, "Theoretical vibrational-excitation cross sections and rate coefficients for electron-impact resonant collisions involving rovibrationally excited N_2 and NO molecules," *Plasma Sources Sci. Technol.*, vol. 21, no. 5, 2012, doi: 10.1088/0963-0252/21/5/055018.

59. D. D. Konowalow and J. O. Hirschfelder, "Morse potential parameters for O-O, N-N, and N-O interactions," *Phys. Fluids*, vol. 4, no. 5, pp. 637–642, 1961, doi: 10.1063/1.1706374.

60. K. A. Peterson, D. E. Woon, and T. H. Dunning, "Benchmark calculations with correlated molecular wave functions. IV. The classical barrier height of the $H+H_2 \rightarrow H_2+H$ reaction," *J. Chem. Phys.*, vol. 100, no. 10, pp. 7410–7415, 1994, doi: 10.1063/1.466884.

61. X. Hu, W. L. Hase, and T. Pirraglia, "Vectorization of the general Monte Carlo classical trajectory program VENUS," *J. Comput. Chem.*, vol. 12, no. 8, pp. 1014–1024, 1991, doi: 10.1002/jcc.540120814.

62. K. Park, J. Engelkemier, M. Persico, P. Manikandan, and W. L. Hase, "Algorithms for sampling a quantum microcanonical ensemble of harmonic oscillators at potential minima and conical intersections," *J. Phys. Chem. A*, vol. 115, no. 24, pp. 6603–6609, 2011, doi: 10.1021/jp110799m.

63. W. Olivares and D. A. McQuarrie, "On the theory of ionic solutions," *Biophys. J.*, vol. 15, no. 2, pp. 143–162, 1975, doi: 10.1016/S0006-3495(75)85798-5.

64. V. Manual, "VENUS96 A General Chemical Dynamics Computer Program by William L. Hase, Ronald J. Duchovic," pp. 1–51.
65. C. Schlier and A. Seiter, "High-order symplectic integration: An assessment," *Comput. Phys. Commun.*, vol. 130, no. 1, pp. 176–189, 2000, doi: 10.1016/S0010-4655(00)00011-4.
66. A. K. Paul, S. C. Kohale, and W. L. Hase, "Bath model for N_2 + C_6F_6 gas-phase collisions: Details of the intermolecular energy transfer dynamics," *J. Phys. Chem. C*, vol. 119, no. 26, pp. 14683–14691, 2015, doi: 10.1021/jp512931n.
67. T. Hirano, "MOPAC Manual (Seventh Edition)," *MOPAC Man.*, 1993.
68. M. Černohorský, M. Kutý, and J. Koča, "A multidimensional driver for quantum chemistry program mopac," *Comput. Chem.*, vol. 21, no. 1, pp. 35–44, 1997, doi: 10.1016/S0097-8485(96)00004-6.
69. L. Sun and W. L. Hase, *Born-Oppenheimer Direct Dynamics Classical Trajectory Simulations*, vol. 19. 2003. doi: 10.1002/0471466638.ch3.
70. K. Bolton, W. L. Hase, and G. H. Peslherbe, "Direct dynamics simulations of reactive systems," *Mod. Methods Multidimens. Dyn. Comput. Chem.*, no. 3, pp. 143–189, 1998, doi: 10.1142/9789812812162_0005.
71. A. K. Paul, N. A. West, J. D. Winner, R. D. W. Bowersox, S. W. North, and W. L. Hase, "Non-statistical intermolecular energy transfer from vibrationally excited benzene in a mixed nitrogen-benzene bath," *J. Chem. Phys.*, vol. 149, no. 13, pp. 134101, 2018, doi: 10.1063/1.5043139.
72. H. Hippler, J. Troe, and H. J. Wendelken, "Collisional deactivation of vibrationally highly excited polyatomic molecules. II. Direct observations for excited toluene," *J. Chem. Phys.*, vol. 78, no. 11, pp. 6709–6717, 1983, doi: 10.1063/1.444670.
73. D. C. Clary, R. G. Gilbert, V. Bernshtein, and I. Oref, "Mechanisms for supercollisions," *Faraday Discuss.*, vol. 102, pp. 423–433, 1995, doi: 10.1039/FD9950200423.
74. Z. R. Hren, C. R. Lazarock, T. A. Vincent, L. A. Rivera-Rivera, and A. F. Wagner, "Pressure effects on the relaxation of an excited ethane molecule in high-pressure bath gases," *J. Phys. Chem. A*, vol. 125, no. 39, pp. 8680–8690, 2021, doi: 10.1021/acs.jpca.1c05838.
75. R. Chitsazi and A. F. Wagner, "Pressure effects on the vibrational and rotational relaxation of vibrationally excited OH (v, J) in an argon bath," *J. Chem. Phys.*, vol. 150, no. 11, pp. 114303, 2019, doi: 10.1063/1.5063923.
76. L. A. Rivera-Rivera, A. F. Wagner, and J. W. Perry, "Normal mode analysis on the relaxation of an excited nitromethane molecule in argon bath," *J. Chem. Phys.*, vol. 151, no. 3, pp. 034303, 2019, doi: 10.1063/1.5099050.

77. H. Kim, H. N. Bhandari, S. Pratihar, and W. L. Hase, "Chemical dynamics simulation of energy transfer: Propylbenzene cation and N_2 collisions," *J. Phys. Chem. A*, vol. 123, no. 12, pp. 2301–2309, 2019, doi: 10.1021/acs.jpca.9b00111.

78. W. L. Jorgensen and D. L. Severance, "Aromatic aromatic interactions: Free-energy profiles for the benzene dimer in water, chloroform, and liquid benzene," *J. Am. Chem. Soc.*, vol. 112, no. 12, pp. 4768–4774, 1990.

79. W. Hresko, "Effect of π-bridges on electrochemical and photovoltaic properties of benzothiadiazole and benzodithiophene based conjugated polymers," *Remedial Spec. Educ.*, vol. 10, no. 3, p. 8, 1989, doi: 10.1177/074193258901000304.

80. V. Bernshtein and I. Oref, "Energy transfer between polyatomic molecules. 1. Gateway modes, energy transfer quantities and energy transfer probability density functions in benzene-benzene and Ar-benzene collisions," *J. Phys. Chem. B*, vol. 109, no. 17, pp. 8310–8319, 2005, doi: 10.1021/jp046693d.

81. N. A. West, J. D. Winner, R. D. W. Bowersox, and S. W. North, "Resolving the energy and temperature dependence of $C_6H_6^*$ collisional relaxation via time-dependent bath temperature measurements resolving the energy and temperature dependence of $C_6H_6^*$ collisional relaxation via time-dependent bath temperature measu," *J. Chem. Phys.* vol. 145, no. 1, pp. 014308, 2016, doi: 10.1063/1.4954896.

82. N. A. West and S. W. North, "Non-statistical intermolecular energy transfer from vibrationally excited benzene in a mixed nitrogen-benzene bath", *J. Chem. Phys.*, vol. 149, no. 13, pp. 134101, 2018, doi: 10.1063/1.5043139.

83. S. Hamon et al., "Low temperature measurements of the rate of association to benzene dimers in helium Low temperature measurements of the rate of association to benzene dimers in helium" *J. Chem. Phys.*, vol. 112, no. 10, pp. 4506–4516, 2000, doi: 10.1063/1.481080.

84. H. Kim, A. K. Paul, S. Pratihar, and W. L. Hase, "Chemical dynamics simulations of intermolecular energy transfer : Azulene + N_2 collisions" *J. Phys. Chem. A*, vol. 120, no. 27, pp. 5187–5196, 2016, doi: 10.1021/acs.jpca.6b00893.

85. H. Kim, B. Saha, S. Pratihar, M. Majumder, and W. L. Hase, "Chemical dynamics simulations of energy transfer for propylbenzene cation and he collisions," *J. Phys. Chem. A*, vol. 121, no. 40, pp. 7494–7502, 2017, doi: 10.1021/acs.jpca.7b07982.

86. D. L. Bunker, "Monte carlo calculations. IV. Further studies of unimolecular dissociation," *J. Chem. Phys.*, vol. 40, no. 7, pp. 1946–1957, 1964, doi: 10.1063/1.1725427.

87. R. A. Marcus, W. L. Hase, and K. N. Swamy, "RRKM and Non-RRKM behavior in chemical activation and related studies," *J. Phys. Chem.*, vol. 88, no. 26, pp. 6717–6720, 1984, doi: 10.1021/j150670a040.

88. Y. Yao, S. Lakshmanan, S. Pratihar, and W. L. Hase, "Direct dynamics simulations of the unimolecular decomposition of the randomly excited $^1CH_2O_2$ criegee intermediate: Comparison with $^3CH_2$ + $^3O_2$ reaction dynamics," *J. Phys. Chem. A*, vol. 124, no. 9, pp. 1821–1828, 2020, doi: 10.1021/acs.jpca.9b11513.

89. S. Pratihar, X. Ma, Z. Homayoon, G. L. Barnes, and W. L. Hase, "Direct chemical dynamics simulations," *J. Am. Chem. Soc.*, vol. 139, no. 10, pp. 3570–3590, 2017, doi: 10.1021/jacs.6b12017.

Chapter 3

Molecular Docking and Dynamics: Insights into Drug Discovery and Development

Mittal Thakkar,[a] Dhaval Patel,[b] Mukul Jain,[c] Nil Patil,[c] and Rupal Dhariwal[c]

[a]*Department of Chemical Engineering, Dharmsinh Desai University, Nadiad, Gujarat, India*

[b]*Gujarat Biotechnology University, Gandhinagar, Gujarat, India*

[c]*Research and Development Cell, Parul University, Vadodara, Gujarat, India*

mittalthakkar.ch@ddu.ac.in

In modern computational drug development, molecular docking and molecular dynamics (MD) simulations are cornerstone methodologies. The purpose of this chapter is to offer a complete understanding of their methodology, applications, and concepts. The chapter begins with a brief overview of the drug development pipeline and then moves on to discuss how computational tools are becoming more important in streamlining and improving the process. The discussion delves into molecular docking, examining its various types, scoring functions, and commonly utilized software

Computational Methods for Molecular Modelling and Docking

Edited by Arup Kumar Ghosh and Suban Kumar Sahoo

ISBN 978-981-5352-25-2 (Hardcover), 978-1-042-00204-7 (eBook)

www.jennystanford.com

DOI: 10.1201/9781042002047-3

tools. Additionally, MD simulations are elucidated through their theoretical underpinnings, force fields, and simulation workflows. This chapter highlights the combined effectiveness of docking and MD, demonstrating how their integration improves the accuracy of binding affinity predictions, fine-tunes ligand positioning, and aids in the optimization of lead compounds. System complexity, timescale limitations, and scoring inaccuracies present significant challenges that are being tackled. Emerging solutions include hybrid methods, enhanced sampling techniques, and AI-driven approaches. Drug repurposing, targeting particular biomolecular systems, and modelling protein-ligand interactions are some of the practical applications discussed in this chapter. This chapter highlights recent advancements and anticipates future trends, emphasizing the significant impact of molecular docking and dynamics on the evolution of drug-discovery methodologies.

3.1 Introduction

3.1.1 Overview of Drug Discovery and Development

From the early 19th century to the present, a variety of conditions have influenced drug development and discovery. Drug development in the past relied on conventional methods and treatments, such as fortuitous discoveries like penicillin [1]. The first phase of drug discovery began at this time. Indian cultures are renowned for using traditional medicine to cure a variety of illnesses in accordance with spiritual healing and religious beliefs [2]. Roots, herbs, and other important portions of plants that are recognized to heal illnesses are used in traditional systems. Some well-known choices include Curcuma longa, Bacopa monnieri, and Coleus forskohlii [3]. Due to the use of molecular modelling and combinatorial chemistry techniques, the second stage of drug discovery saw significant advancement as drug structures, their mechanisms, and associated pharmacology were recognized [4]. Antibiotics were developed during this period, and the understanding of DNA, RNA, recombinant technology, and molecular biology came to be understood. The development of biopharmaceutical products authorized by regulatory bodies worldwide marks the beginning of the third stage of drug research [4].

This stage saw the discovery of antibiotics and the beginning of our understanding of DNA, RNA, recombinant technologies, and molecular biology. The creation of biopharmaceutical products authorized by regulatory bodies worldwide marks the beginning of the third stage of drug research [5]. These days, medications must be created and assessed using a large library of analogues that are identical. This will enable the use of drug screening for research purposes. An optimization phase to increase the proposed drug's effectiveness by combining a mechanistic profile of safety and dependability with few or no side effects comes next [6]. Drugs nowadays must be created and assessed using a large library of analogues that are comparable. This will enable the use of a particular drug's screening in research. After this, an optimization stage is conducted to enhance the candidate drug's effectiveness by combining a mechanistic profile of safety and dependability with few or no side effects [7–9].

The whole healthcare system has gone through many highs and lows since the abrupt onset of the worldwide pandemic, including uncertainty and several issues with medicine research and discovery [10, 11]. Therefore, to stay up to date with contemporary advancements to support drug discovery, medicinal chemists and pharmaceutical corporations must continue their efforts [10–12].

3.1.2 Role of Computational Methods in Drug Discovery

A cover story headlined "Next Industrial Revolution: Designing Drugs by Computer at Merck" appeared in Fortune magazine on October 5, 1981 [13]. Several people believe that this event marked the beginning of a fervent interest in the potential applications of computer-aided drug design (CADD) during this time period [14]. Drug screening and design have benefited tremendously from a wide range of computational tools, which have considerably reduced the amount of time and cost required for drug development. These techniques have been made possible by the rapid advancement of computer hardware, software, and algorithms. The field of bioinformatics, in general, has the potential to aid in the discovery of significant genes amid huge volumes of genomic data [15, 16]. This might ultimately result in the identification of prospective target proteins for the purpose of drug design and screening.

Drug development using computers has been practiced for many years, but in recent years, there has been a seismic shift toward embracing computational tools in both academia and pharmaceutical companies. This transition is essentially defined by a flood of data on ligand characteristics and binding to therapeutic targets, as well as their 3D structures, plentiful computational capabilities, and the availability of billions of on-demand virtual libraries of drug-like small molecules [17]. In addition to virtual screening, de novo drug design techniques [18] provide an additional form of computer-aided drug design guidance by generating synthesizable small-molecule candidates with high binding affinity.

The use of computational techniques in drug design and screening has grown significantly. Multiscale biomolecular simulations can be used to clarify the processes of drug action and locate drug binding sites on target macromolecules. Drug development will be sped up by integrated computational methodologies, which will also aid the discovery of new therapeutics with efficient action mechanisms that may eventually be applied to a wide range of intricate biological systems [17]. Drug targets and candidates against degenerative diseases like diabetes, cancer, and bacterial infections have been efficiently identified through the adoption of novel computational technologies; this concept shows great promise for future breakthroughs in drug discovery, design, and development [19].

3.1.3 Importance of Molecular Docking and MD in Modern Drug Development

Molecular docking and MD stand out as some of the more recent applications in modern medicine. These facilitate research and development on medicines. These computational methods will facilitate the drug by predicting the interaction of small molecules with the target organism. The following sections elaborate on their significance.

3.1.3.1 Molecular docking in drug discovery

- **Predictive modeling:** Molecular docking simulates how small molecules bind to target proteins. It allows researchers to examine binding relations and structures [20].

- **High-throughput screening:** It enables virtual screening of large compound libraries. It identifies a holding stockpile from where drugs can be produced very quickly and relatively inexpensively [21].
- **Toward target discovery:** Connectivity may reveal novel targets for already marketed drugs. It can boost drug repurposing efforts [20].

3.1.3.2 MD study

- **Dynamic interaction analysis:** MD simulations elucidate protein conformational flexibility and ligand binding dynamics. This sheds light on drug efficacy [22].
- **Long-time stability assessment:** This helps to evaluate the stability of protein-ligand complexes over time. So, this indicates that the identified leader does stay stable under physiological conditions [23].

Although molecular docking and dynamic coupling enhance drug development at all stages, challenges still exist, such as the computational bottleneck and the need for experimental validation. This further emphasizes the need for such a combination of computational and experimental assessment methods [24].

3.2 Fundamentals of Molecular Docking

3.2.1 What Is Molecular Docking?

The computational approach known as docking was first proposed by Kuntz et al. [25]. It essentially aims to anticipate a complex consisting of two binding partners, typically. Biological macromolecules (such as proteins, DNA/RNA, or peptides) or tiny molecules (such as endogenous ligands or medicines) are the usual candidates for these binding partners. Molecular docking is very important for predicting the right pose of small molecules toward target proteins, as it forms part of the drug screening and design process. It is aimed at better understanding molecular and protein interactions, thereby aiding the development and production of new drugs. The origins of molecular docking date back to the 1980s, with ongoing improvements in algorithms and computational capabilities contributing to its precision and effectiveness [26].

Molecular docking has progressed from a basic tool to a sophisticated computational method, applicable in molecular modelling, structural biology, and drug development. Molecular docking has become a critical component of computational drug development processes. It employs experimental methods to identify leads and inform lead optimization efforts. Machine learning, advanced docking methods, and powerful computational tools have made docking models more accurate and faster. This lets researchers study additional chemistry and biology problems. Molecular docking continues to reveal pharmacological action mechanisms and protein-ligand interactions. Docking studies elucidate how to design drugs that target protein-protein interactions (PPIs), inhibit enzymes, modify receptors, and recognize ligands [27].

3.2.2 Types of Docking

There are three distinct kinds of docking studies, distinguished by the degree of flexibility of the molecules, receptor, and ligand that interact with one another: rigid, flexible-rigid, and flexible docking [28].

- **Rigid docking**: The ligand and receptor molecules are both seen as stiff entities in rigid-type docking. The interior geometry of each molecule remains unchanged since its form remains unchanged. The only degrees of freedom considered are translational and rotational, since the location may be changed. The outcome is less precise and dependable; hence, it is not commonly utilized in modern docking investigations, but this was an early docking approach that could be used to macromolecules like two protein molecules. Applying the lock-and-key principle is possible with this approach [29].
- **Flexible-rigid docking:** This docking approach is semi-flexible. The ligand or receptor is viewed here as a solid object. In most cases, when the ligand's conformation is altered, the receptor's shape remains constant. This approach is commonly utilized because it produces more precise and trustworthy results compared to the rigid docking method [30].

- **Flexible (soft) docking:** The goal of this docking method is to make it as flexible as possible, so the receptor and ligand are both seen as malleable entities. This technique is known as flexible docking or soft docking. An extensive rotation of the molecules is executed to determine the optimal shape and orientation for their interaction. By adjusting rotatable bonds and torsion angles, the form of the molecule may be changed. It may require substantial time and computational resources, but this technique predicts docked conformations with high precision, likely comparable to experimental data. The induced-fit approach may be applied in both semi-flexible and completely flexible docking methods. Docking gets more complex when the interacting molecule has several conformational degrees of freedom [29, 30].

3.2.3 Scoring Functions

Throughout the course of history, scoring functions for the purpose of binding affinity prediction and virtual screening have been divided into three distinct categories: force-field-based, empirical, and knowledge-based [31, 32]. However, recently Liu and Wang (2015) argued that this historical classification overlooks more recent developments in the field and thus proposed an updated classification scheme with four classes of scoring functions: force-field-based or physics-based, empirical or regression-based, knowledge-based or potential of mean-force-based, and descriptor-based or machine-learning-based [33]. This classification is useful to distinguish different methodologies and ideas appearing in the development of scoring functions.

In order to evaluate the direct interactions that occur between a ligand and a protein, force-field-based methods make use of molecular mechanics functions. On the other hand, solvent effects are typically evaluated using a generalized Born/surface area (GB/SA) type of approach [34], which is frequently based on the work that Wesson and Eisenberg [35] have done. In the form of distance-dependent atom-pair potentials, knowledge-based techniques rely on statistical information that is generated from the complex structures associated with ligands and receptors that are already in existence [36] Empirical scoring functions [37] are the third method, and they are based on the concept that all of the important elements that impact the binding are described in the form of equations, ideally simple equations. These equations include those that describe H-bonding, rotational/translational degrees of freedom, and polar/lipophilic effects. Furthermore, these equations are balanced using a regression-type technique, which is occasionally called regression-based scoring in the literature.

A significant obstacle in docking is the conformational flexibility of ligands and proteins. The majority of docking algorithms treat ligands as completely flexible, which causes them to produce a wide number of conformations. Force field and conformation generation approaches affect docking quality. All frequently used docking approaches can detect the right bioactive conformation of a ligand (i.e., replicate the X-ray pose) in numerous instances, indicating that the conformational part of the procedure is highly optimized. Consider the works of Li et al. [38], Warren et al. [39], and the CASP2 competition [40], all of which demonstrated the same thing. While these results are promising, they do not prove that pose generation always yields the optimal binding pose and ligand conformation. These findings make it quite clear that improving the docking quality requires concentrating on the scoring function.

The pose prediction studies utilized a subset of the Astex non-native set. Each ligand was docked into all protein models for each target. We eliminated the native protein model of each ligand to study non-native docking. Positions having a root mean square deviation (RMSD) less than 2.0 Å were deemed correct. Simulations used sequential docking findings for ensemble docking. We included all ensembles and compared the poorest, average, and best performers for all ensemble sizes. Typical single-protein cross-docking experiments use an ensemble size of one [41].

3.2.4 Overview of Popular Docking Software and Tools

Table 3.1 contains a summary of the most popular molecular docking applications used in the computerized drug design process [42].

Table 3.1 Overview of the most prominent molecular docking software utilized in the computational drug design process

No.	Programs	Application	Accessibility	Ref.
1	AutoDock	It is employed in molecular docking. It predicts the binding capacity of a tiny chemical and assigns a target protein to a 3D structure	https://autodock.scripps.edu/	[43]
2.	LPCCSU	Based on a comprehensive investigation of interatomic interactions and interface complementarity	https://oca.weizmann.ac.il/oca-bin/lpccsu	[44]
3.	PatchDock	The method performs rigid docking, with surface variability	https://bioinfo3d.cs.tau.ac.il/PatchDock/php	[45]
4.	Hex	For docking studies	http://hex.loria.fr/	[46]
5.	Glide Schrödinger	Comprehensive molecular modeling and CADD tool	https://www.schrodinger.com/	[47]
6.	Molecular operating environment	Comprehensive molecular modeling and CADD tool	https://www.chemcomp.com/	[48]

(*Continued*)

Table 3.1 (*Continued*)

No.	Programs	Application	Accessibility	Ref.
7.	DockingServer	A user-friendly web-based interface that manages all elements of molecular docking.	https://www.dockingserver.com/web	[49]
8.	SwissDock	A web service for predicting a protein's association with a small molecule ligand.	http://www.swissdock.ch/	[50]
9.	LeDock	A molecular docking program for docking ligands with protein targets	http://www.lephar.com/software.htm	[51]
10.	MedusaDock 2.0	Fast, flexible docking with a discrete rotamer library of ligands	https://dokhlab.med.psu.edu/cpi/#/MedusaDock	[52]
11.	Molegro Virtual Docker	Based on a novel heuristic search method that integrates differential evolution and a cavity prediction algorithm	http://molexus.io/molegro-virtual-docker/	[53]
12.	MOLS 2.0	Using mutually orthogonal Latin squares, induced-fit peptide–protein, and small molecule–protein docking	https://sourceforge.net/projects/mols2-0/	[54]
13.	ParaDockS	Metaheuristics for population-based molecular docking	http://www.paradocks.at/	[55]

3.3 MD Simulations

MD models explain the behavior of biomolecular systems on timeframes from femtoseconds to milliseconds. This method, which involves solving Newton's equations of motion for every atom in an MD system, allows us to see the atomic-level structure and functional dynamics of macromolecules, changes in structure, folding of molecules, binding of ligands, and interactions between proteins. They all fall into this category. Predicting ligand binding affinity and identifying binding site flexibility are common tasks in drug development. protein engineering and materials science Root mean square deviation (RMSD), root mean square fluctuation (RMSF), solvent accessible surface area (SASA), radial distribution function, free energy visualization, analysis. Principal components and other indicators have been calculated to provide a comprehensive MD view of biomolecular behavior. Because these models combine static structural models with the dynamic processes of biological functioning, MD simulations are important in computational biology.

3.3.1 Basic Principles of MD

In the fields of physics, chemistry, biology, and materials science, MD is an important computational modeling tool used to simulate interactions between atoms and molecules over time. Structural, dynamic, and thermodynamic properties [56]. The main idea is to apply a force from the potential energy function to the atoms to calculate their trajectories. This tool is very useful in studying phase transitions. spread Intermolecular interactions and protein folding, etc. The basic workflow of an MD simulation involves first creating an initial system configuration. This is often based on experimental data using techniques such as NMR (nuclear magnetic resonance) spectroscopy or X-ray crystallography [57]. To ensure charge neutrality, this initial structure next goes into a simulated environment that frequently contains ions and a solvent like water. A force field is a mathematical model made up of potential energy functions that take into consideration both non-bonded (van der Waals forces and electrostatic interactions) and bound (bond stretching, angle bending, and dihedral torsions) interactions. While MD simulation is dependent on several parameters, its most important factor is the reliability of the force fields used; with any significant discrepancies within the potential energy functions leading to a simulation of errors for the same reason, it is essential that time step strike a balance between computational efficiency and numerical stability- that the time step value is not too large (conducive to instability and erroneous, unphysical results) while also being not too small (which raises an unacceptably high computational burden for only marginal increases in accuracy). The majority of MD simulations are with PBCs; they are used to suppress edge effects and emulate an infinite system. With PBCs, particles that move out of the enclosure re-enter from one side and come out through the opposite side, so that a continuous and uniform environment is created [58]. Self-consistent temperature and pressure control are kept by employing algorithms such as the Berendsen thermostat and barostat, or the Nosé-Hoover method. Therefore, MD is a powerful tool for studying atomic-level molecular systems. By strictly adhering to the elementary principles of classical mechanics and employing the most recent and sophisticated computational techniques, this field of MD makes visible dynamic processes that would otherwise be invisible and challenging to observe experimentally.

3.3.2 Force Fields and Potential Energy Functions

MD, force fields, and potential energy functions offer a framework for characterizing the forces acting on particles in computational chemistry, enabling the prediction of atomic-level molecular structures, dynamics, thermodynamics, and interactions. A force field is a collection of mathematical equations and parameters that together calculate the potential energy of a molecular system by summing various energy components, including bond stretching, angle bending, torsional rotations, van der Waals forces, and electrostatic interactions [59]. A potential energy surface (PES), which illustrates how a system's energy varies with changes in atomic locations, is defined by the potential energy function. Stable molecule structures are represented as local minima on the PES. Newton's equations of motion can be integrated to trace the time development of the molecular system through the use of force fields and potential energy functions to simulate atom-to-atom interactions in MD simulations.

3.3.3 MD Simulation Setup and Workflow

Setting up a molecular system, allocating force field parameters, and conducting simulations under particular circumstances are all steps in the computer process known as MD simulation, which simulates the physical movements of atoms and molecules over time. The first step in MD simulation involves preparing the molecular system by selecting or constructing the model to be studied, which may include proteins, nucleic acids, membranes, or small molecules. It may be necessary to alter the original structure to fill in any gaps in the backbone, add missing atoms or residues, and eliminate crystallographic water molecules. Using established models like TIP3P, SPC, or TIP4P, the initial molecular structure must be constructed and then placed in a box that represents the solvent environment. Usually, this box is solvated with explicit water molecules. The potential energy of the system as a function of atomic locations is described by a force field, which is a collection of mathematical formulas and parameters [60]. OPLS-AA, AMBER, CHARMM, GROMOS, and other popular force fields are required since they have a direct impact on the caliber of the simulation results. Energy minimization is a crucial step that reduces unfavorable steric clashes and bad contacts by lowering the system's energy, allowing atoms to settle into a more favorable configuration without significantly altering the overall structure [61]. Typically, steepest descent or conjugate gradient algorithms are used for minimization. In order to achieve equilibrium, the system must be steadily heated to the appropriate pressure and temperature. This stage guarantees that the system achieves a stable thermodynamic state prior to the simulation's production phase. Following system equilibration, Newton's equations of motion are integrated using algorithms such as the Verlet integrator or Leapfrog integrator with a tiny time step during the manufacturing phase. The production run generates a trajectory file, which contains the atomic coordinates as a function of time. Herein, multiple properties such as RMSD, RMSF, SASA, Hydrogen Bonding Analysis, etc., could be calculated.

3.3.4 Commonly Used MD Software and Techniques

MD simulations are a potent computer technique for examining the atomic-level behavior of biomolecules. The accuracy, effectiveness, and usefulness of these simulations have improved over time through the development of numerous MD software programs and specialized methods. One of the most popular MD software programs is GROMACS (GROningen MAchine for Chemical Simulations), which has become noted for its excellent performance and scalability across a range of computer architectures, CPUs, and GPUs. It enables users to compute structural and energetic characteristics effectively, supports a variety of force fields, including AMBER, CHARMM, and OPLS, and employs highly optimized algorithms for force calculation and parallelization [62]. Another is the AMBER suite, which is particularly popular for simulations of proteins and nucleic acids. PMemd for high-performance simulations on CPUs and GPUs, cpptraj for sophisticated trajectory analysis, and Sander for energy minimization and MD are some of its essential parts. AMBER force fields, like ff14SB for proteins and OL3 for nucleic acids, are widely used in computational biology because of their ability to accurately reproduce experimental results [63]. A popular software program and family of force fields for MD simulations of biomolecular systems, such as proteins, lipids, carbohydrates, and small molecules, is called CHARMM (Chemistry at Harvard Macromolecular Mechanics). Known for its rigorous parameter development, CHARMM provides highly accurate results and supports advanced methods such as QM/MM simulations and enhanced sampling techniques. Its extensive and well-validated force fields, particularly CHARMM36 for proteins and lipids, are trusted for high-quality simulations in both academic and industrial research [64]. The high-performance MD engine Desmond, created by D. E. Shaw Research, is well known for its efficiency and potency in simulating biomolecular systems. It ensures precise and scalable simulations by using cutting-edge parallelization techniques and creative algorithms for long-range force computation. For deeper investigations, Desmond can be used with QM/MM methodologies and improved sampling strategies. It also supports a variety of force fields, including CHARMM [65]. Apart from the MD software, the following techniques commonly prevail in the field of molecular simulations and dynamics:

- **Energy minimization**: It is the process of finding a local minimum on the PES, an essential step before starting an MD simulation to remove bad contacts and steric clashes. Steepest Descent and Conjugate Gradient are the most commonly used algorithms.
- **Temperature & pressure coupling**: It is used to maintain the desired thermodynamic conditions during simulation by using Berendsen thermostat, velocity rescaling, and the Nosé-Hoover thermostat [66].
- **Interactions calculation**: Long-term interaction calculation of electrostatic bonds is done in MD simulations using Ewald Summation and Particle Mesh Ewald (PME).

3.4 Docking and MD Integration in Drug Discovery

While docking experiments are simple to implement and do not necessitate extensive computational resources, the conclusion that may be derived from them is a significant methodological challenge. For instance, the use of molecular docking in MD simulations would lend more credence to such findings. This facilitates the study of complicated events, such as protein folding and ligand binding, by allowing the generation and observation of dynamics of molecular structures at different temporal scales. Many scholarly articles have discussed MD and its medically supported uses, contributing to the technique's rising profile in the academic community. The model considers the impact of the receptor's induced fit around the ligand, accurately describes the energy landscapes investigated by MD simulations, and includes in silico predictions of ligand binding modes [67–75].

3.4.1 Advantages of Combining Docking and MD for Lead Optimization

Lead optimization using molecular docking and MD simulations is an efficient drug discovery method because it improves binding accuracy, stability, and selectivity. Docking is imprecise because it often assumes protein structures are rigid [76], yet it remains a strong drug-discovery method despite producing only initial binding poses and a high-throughput method for screening potential ligands. MD can detect induced-fit changes that affect pose predictions because the ligand and protein assume more pliable conformations over time [77].

While docking score algorithms perform well at providing rapid affinity estimates, their reliance on simplified energy models leads to errors. MD simulations provide a thorough evaluation of ligand binding, including solvent and entropy, using free-energy definitions from several models, such as MM-PBSA and MM-GBSA, to overcome this issue [78]. Similarly, MD is a way to find and remove molecules that do not bind stably in physiological contexts but have high docking scores [79]. Additionally, it depicts critical residues for binding stability and provides details on the dynamics and stability of important interactions, such as salt bridges and hydrogen bonds, which contribute to lead refining [77].

This flexibility is especially useful for dynamic binding pockets because MD can capture conformational changes that rigid docking cannot [80]. MD helps select compounds across homologous protein targets, prioritizing selective molecules and minimizing off-target effects [81]. MD simulations can also assess ligand stability relative to mutant forms for mutation-prone targets, including cancer-related proteins and antimicrobial resistance, thereby enabling the design of mutation-resistant inhibitors [82]. This combination strategy reduces experimental testing and optimizes computation by getting a fine-tuned subset from MD and rapidly screening enormous libraries with docking [83].

Finally, MD simulations at specific protein variant levels enable precision medicine lead optimization, making them possibly better for individual patients [84]. In conclusion, docking and MD improve hit identification, lead optimization, and candidate selection, maximizing drug discovery accuracy and efficiency.

3.4.2 Workflow: From Docking to MD Validation

The combination of molecular docking and MD modelling allows drug development lead optimization by combining high-throughput screening with rigorous refinement and validation. AutoDock Tools [85] and PyMOL remove water molecules, alter protonation states, add hydrogens, and apply force fields to create proteins from the Protein Data Bank (PDB). From databases such as PubChem [86] or ZINC15 [87], ligands are standardized and reduced using RDKit or Open Babel. Software like AutoDock Vina [88], Gold [89], or MOE can construct binding poses ranked by affinity scores after docking simulations. Finally, these high-scoring poses are selected for additional validation of the MD simulation using LigPlot+ and PyMOL, based on expected interactions [90].

System setup involves loading top-scoring docking poses onto MD platforms such as GROMACS [91], AMBER [92], or CHARMM [93]. The system is solvated using TIP3P water models and ion neutralization. An energy minimization technique is applied to the entire system to eliminate steric clashes. To further establish the conditions, NVT and NPT ensembles equilibrate temperature and pressure. Long production simulations ranging from 50–100 ns or longer capture the dynamic behavior and conformational stability of the ligand-protein complex [77]. Short-term runs ranging from 1 to 5 ns are used for initial MD equilibration, depending on the system's stability. Data are collected for several reasons; for example, RMSD is used to measure stability, and RMSF to measure residue flexibility. Finally, free energy calculations are typically performed using MM-PBSA or MM-GBSA methods. These approaches provide a more accurate estimate of binding affinity by considering solvent and entropy effects.

Visualizing MD trajectories using VMD [94] or PyMOL displays ligand dynamics and profiles of conformational changes and interactions, while hydrogen-bond and contact analysis quantify the stability of critical contacts during the simulation. In response to MD data, ligands undergo modifications to enhance selectivity, decrease flexibility, or boost particular interactions. According to E. Lionta et al. [83], these adjusted ligands can be docked once again and verified using additional MD simulations to show stability improvements prior to experimental validation. This combined method, which incorporates both docking and MD validation, improves and ranks ligands, decreases false positives, and increases the likelihood of discovering suitable drug candidates.

3.4.3 Case Studies: Successful Applications in Drug Discovery

3.4.3.1 SARS-CoV-2 main protease inhibitors

The COVID-19 pandemic necessitated a prompt examination of inhibitors aimed at the main protease (Mpro) of SARS-CoV-2, an essential enzyme for viral replication. MD simulations were validated by *in vitro* assays that inhibit Mpro activity. Nelfinavir has shown potential in MD simulations, exhibiting notable binding affinity and stable hydrophobic and hydrogen-bonding interactions. The binding free energy was validated via MM-PBSA analysis, and *in vitro* experiments confirmed that nelfinavir effectively inhibited viral replication. This demonstrates the speed at which computational methods can identify and validate potential new drugs for repurposing in pandemic situations [95, 96].

3.4.3.2 Identification of BACE-1 inhibitors for Alzheimer's disease

Inhibiting β-secretase (BACE-1), which promotes the formation of amyloid-β plaques that exacerbate Alzheimer's disease, has been a key aim in Alzheimer's disease treatment. We performed high-throughput docking tests on large chemical libraries in search of possible BACE-1 inhibitors. After analyzing docking scores and notable interactions with the enzyme's active site residues, we selected the top hits. To further corroborate these findings, MD simulations were employed to study the stability of binding and the conformational changes of BACE-1 during ligand binding. Stable interactions were observed in several small molecules, including H-bond engaging with catalytic residues Asp32 and Asp228, among others. Additional confirmation of their excellent binding affinity was done using MM-PBSA calculations, and one medication, PF-06751979, moved on to the preclinical phases. Finding selective and stable BACE-1 inhibitors for the treatment of Alzheimer's disease can be achieved by combining docking and MD, as demonstrated in this study [97, 98].

3.4.3.3 Development of selective inhibitors for CDK2 in cancer therapy

Cyclin-dependent kinase 2 (CDK2) is a significant therapeutic target owing to its essential function in cell-cycle regulation, with documented overactivity in various cancers. Compounds were evaluated for selective binding to CDK2 compared to other kinases using docking techniques; the most promising top-ranking candidates were further validated through MD simulations. The simulations assessed binding stability, conformational dynamics, and selectivity. By focusing on CDK2's ATP-binding region, SNS-032 proved to be an effective inhibitor. Free energy calculations and MD simulations showed that SNS-032 stabilized CDK2's active conformation via hydrophobic interactions and hydrogen bonds. Inhibition of human cancer cell lines was shown to be effective in preclinical trials. This is a perfect example of how docking and MD simulations can be used to find cancer treatments that target CDK2 specifically [99].

3.4.3.4 Design of novel inhibitors for influenza neuraminidase

To find new inhibitors of the influenza neuraminidase (NA), the virus's primary enzyme for replication and dissemination, the researchers used MD simulations for virtual screening. Then, docked hits with high binding scores were tested in MD simulations to verify binding properties like stability, orientation, and interaction profile within the active site. NA Subsequent confirmation of the high binding affinity findings from the MD simulation. enough stability of the reaction and the compound's binding. 4-GU-DANA was identified as a promising lead drug for the effective inhibition of NA in cell-based experiments that utilized MM-PBSA calculations. The lead optimization phase is eliminated in this study [100] by combining docking and MD to build and assess a novel influenza NA inhibitor.

3.4.3.5 Targeting HIV-1 protease with darunavir analogs

As far as HIV replication enzymes go, HIV-1 protease stands head and shoulders above the rest. This enzyme's inhibition is crucial to antiretroviral treatment. This study followed the success of darunavir as a protease inhibitor and used docking to create an analogue that was more resistant to HIV-1 protease mutations and had a higher affinity for the enzyme. To ensure the stability of the binding and the strength of the interactions, these analogues were subjected to MD simulations. The docking investigation produced binding affinity profiles that were better, but MD showed less flexibility and more stable binding in the presence of mutant protease forms. While MM-PBSA calculations verified higher binding affinities, numerous compounds showed potent *in vitro* activity against both wild-type and mutant HIV-1 proteases. Docking and MD, as proposed in this work [101, 102], can be used to create powerful inhibitors that can combat medication resistance in HIV treatment.

3.4.3.6 Discovery of novel inhibitors for the Ebola virus glycoprotein

Since the Ebola virus glycoprotein (GP) facilitates viral entrance into host cells, it has become a target for the creation of antiviral medications. To facilitate GP entrance into host cells, the researchers coupled it to the fusion loop using a high-throughput docking approach. The stability of the connection was confirmed by examining conformational changes in the GP using further MD simulations. MD models consistently reveal conformational and binding changes. For a large number of potential GP inhibitors, only a few chemicals likely to block Ebola cell entry were identified through free-energy calculations, confirming the tight link. Here, GP has shown the efficacy of docking-MD in *in vitro* testing, where it has made strides. Identified and confirmed viral protein inhibitors. along with contributing to the creation of possible Ebola therapeutics [103, 104].

3.5 Challenges and Limitations

3.5.1 Limitations of Docking Methods

Molecular docking is a very powerful computational approach for discovering new drugs, but it has shortcomings that can get in the way. Some of the possible problems that prevent optimal utilization are due to algorithmic limitations, protein flexibility, and realistic predictions of binding energies. Improving docking techniques and making better use of them in drug discovery requires a thorough understanding of these constraints. Molecular docking encounters constraints due to the inadequacy of scoring functions in precisely predicting binding energies, as specific intermolecular interaction factors, such as solvation effects and entropy changes, are not adequately estimated. Moreover, certain intermolecular interactions, such as halogen bonding and guanidine-arginine interactions, are hardly incorporated into scoring systems, despite their considerable impact on protein-ligand binding affinity [105–108]. There are two primary reasons why the problem of correctly addressing water molecules in the binding pocket during docking has not been solved. To begin with, X-ray crystal structures do not always accurately identify water molecules that bridge the receptor and the ligand because of the limited scattering from smaller atoms. Second, the strength and nature of the effects that ligands exert on water molecules cannot be predicted by any theoretically accurate method. In addition, the amount of water molecules that may be replaced by various ligands in the binding pocket and the disruption to the hydrogen-bonding network caused by ligand binding are both unknown [109].

3.5.2 Challenges in MD Simulations

MD simulations for the purpose of understanding molecular interactions. Structural dynamics and binding affinity are powerful tools in computational chemistry and biology, but MD simulations are hampered by factors such as system complexity, system size, and time scale. These limitations hinder the validity and practicality of MD. Specific problems include:

3.5.2.1 Timescale limitations

Time ranges from microseconds to seconds for various biologically relevant processes, including protein folding and significant structural changes, which have proven to exceed the simulation capabilities of MD experiments. The traditional constraints of MD simulation are evident: even with advanced computational resources, the simulation duration is ultimately limited by the need to compute forces at each short time step. Atomic simulations at femtosecond timescales (10^{-15} seconds) can accurately compute all significant interactions within a molecular network. A comparison of rapid oscillations of atoms and bonds facilitates MD simulations at sub-microsecond timescales. Many biological activities, such as ligand affinity binding and protein folding, occur over extended time frames, requiring microseconds or even seconds to complete. One has faced various challenges in achieving a deeper understanding of capturing long-term events.

3.5.2.2 System size constraints

The accuracy and efficiency of MD simulations are highly dependent on the system size because the processing requirements rise exponentially with the number of atoms. Explicit solvation simulations of large biomolecular complexes and membrane-protein systems can involve millions of atoms, resulting in computationally expensive solutions [110]. Coarse-grained MD and hybrid approaches, which cluster atoms or concentrate on particular areas to reduce degrees of freedom but permit research of considerably larger systems, are examples of low-cost and fast procedures [111]. While these artefacts arise from periodic boundary conditions (PBCs), small systems may also experience finite-size effects. This is particularly true when trying to describe bulk attributes or global dynamics [112]. These issues highlight the critical need for well-optimized, laboratory-validated MD research.

3.5.2.3 System complexity and realistic biological modeling

Because the computational cost of MD simulations scales exponentially with the number of atoms, system size critically affects their accuracy and efficiency. Explicit solvation simulations of giant biomolecular complexes or membrane-protein systems can involve millions of atoms, leading to computations that could require supercomputers [110]. Whereas not resolving atomic detail, techniques such as coarse-grained MD or hybrid approaches limit the degrees of freedom by clustering atoms or by focusing on particular regions, thereby allowing larger systems to be studied [111]. Moreover, finite-size effects, such as artifacts due to PBCs, can affect the features in smaller systems, especially in modeling bulk or global dynamics [112]. All these problems indicate how important optimization and validation should be for MD investigations.

3.5.3 Limitations of MD Simulations

In order to circumvent the limitations of MD simulations, particularly with respect to timescales, system size, and complexity, hybrid methods and enhanced sampling procedures have been developed. These strategies aim to deepen our understanding of biological systems by making computation more efficient and by documenting rare or long-timescale phenomena. A comprehensive analysis of these strategies is presented here:

3.5.3.1 Hybrid methods

To address specific issues with MD simulations, such as system size and computational costs. Hybrid methods use the capabilities of multiple computational methods. By combining specific modeling methods.

3.5.3.1.1 *Quantum mechanics/molecular mechanics (QM/MM)*

This approach combines molecular mechanics (MM) for the macromolecular context with quantum mechanics (QM) for local chemically reactive domains. This hybrid technique is very useful for studying enzymatic activities involving covalent bonds and pharmacological interactions with metal cofactors or reactive intermediates. Without the computational cost of applying QM to the entire system, it provides atomic-level accuracy for critical active sites. However, there are challenges, such as precisely calibrating the QM and MM zones to ensure appropriate coupling and to parameterize their transitions [111].

3.5.3.1.2 *Coarse-grained (CG) modeling*

In CG modelling, atomic collections are represented as single particles. This allows for the simplification of molecular processes, the reduction of computer complexity, and the facilitation of longer-term simulation capabilities for larger systems. Ribosomes, viral capsids, lipid bilayers, and protein-membrane interactions are some examples of the big biomolecular assemblies that are frequently investigated with the help of this technique. Biomolecular components are commonly investigated using it. Although CG modelling allows simulations with millions of particles throughout microsecond-to-millisecond timeframes, careful validation is still required to ensure critical interactions are correctly represented due to the loss of atomic resolution. Regardless of this drawback, CG modelling is still a very useful tool for studying biomolecular systems on a grand scale [113–115].

3.5.3.1.3 *Multiscale modeling*

By combining frameworks for continuum, coarse-grained, and atomistic models, multi-scale modelling finds a happy medium between computational efficiency and fine-grained resolution. We use coarser models for less important parts and apply atomistic detail to sections that require great precision to lessen processing demands. This approach is particularly useful for modelling cellular-scale activities, such as protein translocation across membranes, and large macromolecular systems, such as DNA polymerases. Complex algorithms and extensive validation are required to overcome challenges in integrating seamlessly and maintaining accuracy across scales, even while it captures large-scale behavior and preserves crucial atomic properties [115].

3.5.3.2 Enhanced sampling techniques

Enhanced sampling techniques, in a fundamental sense, would address the temporal limitations inherent in conventional MD simulations by augmenting the likelihood of certain rare events and expediting transitions between stable states.

3.5.3.2.1 *Meta-dynamics*

Meta-dynamics accelerates the modeling of molecular systems by providing a history-based bias in the free energy landscape, encouraging the discovery of new structures, and avoiding common situations. It is often used to investigate ligand binding/unbinding mechanisms, energy barriers, and protein folding pathways. This approach provides insight into the dynamics of molecular activity through efficient computation of low-energy terrain and capture of unusual events. Its efforts to obtain an efficient and accurate model rely on careful selection of clustered variables (CVs) [116].

3.5.3.2.2 *Replica exchange MD*

Replica Exchange Molecular Dynamics (REMD) uses numerous simulations (replicas) at different temperatures to increase conformational space sampling. It is useful for protein folding route exploration and ligand-binding event sampling. REMD efficiently overcomes energy barriers, allowing conformational landscape exploration. However, several replicas make it computationally expensive [117].

3.5.3.2.3 *Accelerated molecular dynamics (aMD)*

aMD alters the PES and accelerates conformational transitions by reducing energy barriers. Large-scale biomolecule domain motions and protein structural alterations are often studied with it. This method enhances sampling efficiency without sacrificing physical precision; however, parameter tweaking is needed to avoid artefacts [118].

3.5.3.2.4 *Markov state models (MSMs)*

MSMs evaluate MD trajectories by partitioning conformational space into discrete states and representing transitions among these states as a Markov process. They are especially valuable for reconstructing protein-folding pathways and for predicting long-timescale behaviors based on short MD simulations. MSMs effectively capture long-timescale dynamics through the application of statistical models; however, they necessitate substantial trajectory data and precisely defined states to ensure reliable predictions [119].

3.6 Recent Advances and Trends

The field of computational drug discovery has grown rapidly in recent years, particularly in molecular docking and molecular dynamics (MD) simulations. Rate-limiting techniques, such as accurate selection of artificial intelligence (AI) and machine learning (ML), enhanced sampling techniques, and the enormous use of cloud and high-performance computing in large-scale simulations.

3.6.1 AI and ML in Docking and MD

AI and ML are integrated in different fields, making a significant impact on drug discovery, among others. Primarily in molecular docking alignment and other MD simulations, AI/ML has been applied for secure and swift prediction of protein-ligand binding. Most of these technologies are capable of analyzing large datasets and picking out patterns that normal analytical methods cannot detect [120]. In particular, the findings of prior investigations indicate the use of AI and ML approaches to enhance docking algorithms. Most docking techniques used in the past fail to account for the possibility of protein conformational changes. Nonetheless, AI-based strategies allow for flexible docking because ligands can accommodate conformational changes in target proteins. For example, a recent development in ML is the ability to refine binding affinities by learning from prior docking results, thereby improving the scoring functions used in docking [121]. Molecular docking, when integrated with MD simulations, offers additional strength to the existing drug discovery models. Several extensions of AI in MD simulations can enhance exploration of conformational space, so that the parameters that are used can be optimized. This integration enables the extension of the interaction of ligands and proteins over time so that the researcher can gain better insights into the stability as well as the dynamics of ligand binding [122]. Moreover, AI can help select specific conformations from large MD trajectories, as selecting potential drug candidates is time-consuming [123]. Despite these advancements, there are challenges that exist, such as the need for high-quality training data for ML models and the computational cost incurred in running complex simulations.

These technologies are more and more employed to predict protein-ligand interactions with higher precision and productivity. For example, last year, researchers developed methods to predict docking scores using attention-based long short-term memory (LSTM) networks, enabling virtual screening of millions of compounds in a short time [124]. Besides, with the integration of deep learning, the modeling of complex molecular structures has been enhanced, and hence provides better prediction in drug discovery. The next generation innovations persistently supporting AI/ML techniques will help the growth and improvement of molecular docking and MD simulations.

3.6.2 Enhanced Sampling Techniques: Meta-Dynamics, Umbrella Sampling

The importance of extended sampling methods is a critical factor in eliminating energy humps in MD and expanding conformational space. Two important types of enhanced damaging procedures are metadynamics and umbrella sampling. An interesting technique is metadynamics, which is the addition of a bias potential for the free-energy landscape during the simulation. This approach helps the system to come out of the local minima by expanding the search space in phase space [125]. Thus, to the extent that metadynamics refills and restarts the bias potential based on previously visited configurations and reorients the system to other regions of configuration space, thereby suppressing convergence in the system, metadynamics also effectively speeds up convergence toward equilibrium states. This has led to several advancements by integrating adaptive biasing methods that depend on the metadynamics simulation feedback data. This flexibility improves accuracy by reducing the number of calculations needed to get free energy profiles [126]. Moreover, relating the method to AI methods, which can enhance the parameter of bias, may improve or develop this technique further. Another of the enhanced sampling methods discussed here is umbrella sampling, where the system uses a set of windows with an overlap along the reaction coordinate. The outcomes from all these windows are then averaged through histogram comparisons matched with weights to reconstitute the complete free energy landscape. Some of the new advances in umbrella sampling consist of better window positioning techniques employing a predictive algorithm based on previous simulations regarding reaction coordinates. This approach reduces the level of window overlap as well as increases the convergence rate [127]. Moreover, when combined with other methods, umbrella sampling also leads to a better understanding of the complex structure of various systems. Metadynamics and umbrella sampling have been successfully applied in drug discovery, and possibly in understanding ligand binding mechanisms and conformational changes in target proteins. These techniques allow the researchers to obtain important information on binding constants and rates essential for improving lead compounds. In the context of increasing computational capacity, improved sampling techniques will become more and more valuable in the investigation of biomolecular interactions [128]. These techniques have become fundamental approaches to comprehend biomolecular systems and interaction energetics and have given an understanding of binding and conformational alterations that are essential for protein-ligand interact and diffusion.

3.6.3 Utilizing Cloud and High-Performance Computing for Large-Scale Simulations

High-throughput computing has become a unique technology in scientific research since cloud computing offers a scalable resource for simulation that is parallel with its scale. In drug discovery and development, the cloud platform helps researchers perform numerous molecular docking analyses and MD Simulations without physical limitations. With the help of cloud infrastructure, one can get better access to high-performance computing (HPC) architecture that encompasses parallel computing operations. This capability is especially advantageous for performing several docking experiments or long-haul MD simulations concurrently, thus enhancing the time efficiency of the process [129]. More recent developments in HPC technologies have gone on to add more improvements to the simulation inputs. New architectures with embedded GPUs enable fast computations that are required in the case of big data sets characteristic of drug discovery processes. These systems can carry out millions of calculations in a fraction of a second to allow the researchers to search through big chemical space. Further, algorithm-specific software optimizations, which are particular to HPC, have boosted the performance of algorithms. For instance, parallel implementations of widely used MD software allow an optimal usage of available resources without a decrease in accuracy [125]. Implementing AI-associated tools in HPC workflows can also benefit from automating many parameters in the simulation setup and analysis process. As cloud computing continues to evolve, its applications in drug discovery will increase. Future development in computing hierarchy, like quantum computing, may, over time, support classical high-performance computing techniques by opening new ways of simulating molecular behavior at deeper quantitative levels. The continued incorporation of AI/ML into these platforms will also enhance their capabilities for simulating large-scale analytics and facilitate scaling. In light of future improvements in computational capacity, the possibility of achieving new advances in deciphering intricate biological phenomena increases.

Recent advancements in AI/ML applications for molecular docking and MD simulations mark a giant leap in computational drug discovery. Enhanced sampling techniques, such as metadynamics and umbrella sampling, provide the power to probe complex biomolecular interactions more extensively than ever before. Moreover, using cloud computing and HPC resources allows researchers to perform large-scale simulations efficiently. As these trends evolve, they will accelerate the speed of drug discovery by providing more profound insights into ligand-protein interactions (LPIs) and overcoming the limitations that exist with traditional methodologies. The future will likely be characterized by increasingly sophisticated computational tools that harness the power of AI in combination with advanced sampling techniques within robust cloud infrastructures.

3.7 Applications in Drug Development

Molecular docking may illustrate the viability of a biological process, since it is conducted before the experimental phase of any inquiry. There are several domains in which molecular docking has brought about a transformative change in results. The activation or drug's binding qualities to nucleic acid may be predicted, in particular, by the interaction between small molecules (the ligand) and the protein target (which might be an enzyme). The molecular structure of the medicine is correlated with its cytotoxicity, as demonstrated in [130].

3.7.1 Case Studies: Targeting Enzymes, GPCRs, and Other Biomolecular Targets

Molecular docking was used as a consistent approach for virtual compound screening in the research. This method enabled the screening of many quinolone derivatives for their ability to inhibit monoamine oxidases (MAO A & B) and cholinesterase (AChE and BChE, respectively). A number of novel inhibitors with promising potential as lead molecules were discovered via the molecular docking investigations. These substances may one day be refined into new enzyme inhibitors, opening up exciting new avenues for the treatment of Alzheimer's disease [131].

This research shows that molecular docking is a powerful tool for finding new anticancer drugs that target CDKs. Cancer progression is associated with CDKs, which play an essential role in cell cycle control. Scientists can anticipate the binding efficacy of prospective inhibitors using molecular docking, an efficient in silico method for comprehending interactions between CDKs and inhibitors. Improving inhibitor design requires a thorough understanding of inhibitor-target interactions. More potent and selective CDK inhibitors have been enabled by recent advances in molecular docking methods, which have substantially aided drug discovery. The development of new anticancer drugs can be influenced by the findings of future studies [132].

The successful use of molecular docking in drug development is discussed in this study, together with its application in treating diseases including HIV, tuberculosis, influenza, malaria, and malignancies. It emphasizes several drug design techniques, including structure-based, ligand-based, fragment-based, reverse drug design, and drug repurposing. These techniques provide lead molecules, elucidate interactions, help identify potential therapeutic candidates, and highlight new uses for current medications. Furthermore, the study suggests that incorporating molecular docking into drug development will significantly reduce the time and costs associated with launching new medications to market [133].

Human cytochrome P450 (CYP) enzyme drug-binding pockets were found utilizing experimental and computational methods. The Gaussian program suite optimized all drug structures using B3LYP/6-31 theory. Docking demonstrated that most medications bind to CYP enzymes in the same pocket. Due to changes in intermolecular interactions and distances from the heme prosthetic group, which includes the transition metal iron as a chelating ion, drug binding affinities vary. Predicting drug behavior in biological systems and designing more effective drugs requires understanding intermolecular interactions. Understanding these binding locations is crucial for medication design [134].

A study evaluating homology models of 19 human G-protein-coupled receptors (GPCRs) in virtual ligand screening, in comparison to X-ray crystal structures, revealed that 10 of the 19 GPCRs exhibited models that performed better or comparably to the X-ray structures. Homology models constructed with templates exhibiting more sequence identity (84%) enhanced virtual screening, equal to 83% of X-ray structures (83%). The research investigated consensus enrichment, which amalgamated optimal docking scores from various homology models, illustrating that consensus scoring can improve ligand recognition. The research indicated that regenerating sidechain conformations in homology models could enhance their ligand recognition capability, particularly when several ligands were docked to the model [135].

The research reveals significant findings from a dynamic docking analysis of the β2-adrenergic receptor and its antagonist, alprenolol. The researchers employed multicanonical molecular dynamics (McMD) for dynamic docking, facilitating improved conformational sampling without prior knowledge of the reaction coordinate. A 48-microsecond manufacturing run was executed, documenting 427 binding events, offering an extensive perspective on the ligand's interactions with the receptor. The research identified multiple stable and metastable configurations, with the most stable structure corresponding to the global free-energy minimum. Supplementary canonical molecular dynamics simulations validated the stability of intermediates, corroborating the precision of the predictions. The results indicate that the methodology can yield significant insights for rational drug design aimed at GPCRs, including the optimization of compounds for the second-ranking configuration, resulting in the creation of high-affinity allosteric inhibitors [136].

Auto-docking and cross-docking simulations were evaluated for ligand pose prediction. Cross-docking can anticipate ligand poses, with a subset performing similarly to self-docking. Top-scored poses had median RMSD values of 1.2 Å for self-docking and 2.0 Å for StateMatch/FunctionMatch cross-docking. The Kolmogorov-Smirnov test found no significant difference between self-docking and cross-docking. The study also showed how receptor activation and ligand type affect docking. Docking agonists into active-state structures and inverse agonists or antagonists into inactive-state structures should improve accuracy. Receptor conformational sampling may improve docking, especially for mismatched ligand–receptor activation states [137].

Research into GPCR-related malignancies has uncovered 265 potential therapeutic targets linked to bioactive chemicals extracted from Gnetum gnemon (GG). To discover important hub targets, a PPI network was created to map shared targets between GG-derived drugs and GPCR-related malignancies. Some of the promising hub targets identified by the analysis for the development of therapeutics employing GG molecules are MAPK3, SRC, EGFR, STAT3, ESR1, MTOR, CCND1, and PPARG. Results from pathway enrichment analyses using Gene Ontology and the Kyoto Encyclopedia of Genes and Genomes (KEGG) showed that the selected targets have therapeutic potential by predicting their action mechanisms. The molecular docking validation process verified that several drugs had a strong binding affinity to important targets, suggesting that they could be useful for blocking these targets. Further evidence for the therapeutic potential of chemicals produced from GG was provided by survival analysis, which had clinical prognostic value [138].

3.7.2 Drug Repurposing Through Docking and MD

In 2004, Ashburn and Thor first proposed the drug repurposing strategy—also called "drug repositioning" or "drug re-profiling"—to identify diseases that currently have no therapy options. To put it more technically, the method for determining the new uses for medications has already been greenlit. Examples of such situations include rare diseases or a rapidly spreading pandemic for which a cure is not yet available [139–141]. To find potential targets, molecule docking is a common method. It requires only the drug's and target's structural inputs to discover probable targets for a certain medicine or pharmaceutical [142].

Kumar et al. (2017) conducted a study using this approach to identify leading antipsychotic medications—bromperidol, pimozide, anisoperidone, melperone, benperidol, and anisopirol—as potential candidates for drug repurposing against various targets related to Alzheimer's disease. The results demonstrated that benperidol was the most effective medication for engaging with various target proteins linked to Alzheimer's disease [143].

Most of the globe is still worried about the COVID-19 pandemic, but the drug repurposing strategy has helped find a treatment that can fight the virus. To overcome time constraints, several researchers employed in silico methods, which involve molecular docking of existing medications and molecules studied *in vivo* [143–145]. According to docking scores, the most promising compounds were identified as zafirlukast and simeprevir, which were reported to have leukotriene receptor antagonist and protease inhibitor properties, respectively [146].

In 2023, Baby K. et al. made a significant accomplishment in the fight against non-small cell lung carcinoma (NSCLC). NSCLC is the most common kind of lung cancer and has a high morbidity rate because of the dysregulation of the amino acid Akt. Among FDA-approved medications, a computational study was conducted to identify allosteric Akt-1 inhibitors. The pharmaceuticals that demonstrated the highest binding affinity to the Akt-1 allosteric site were dasatinib, valganciclovir, novobiocin, and indacaterol [147].

A study has identified four potential drug candidates that inhibit the interaction between the SARS-CoV-2 S-glycoprotein and the ACE2 receptor. The selection of these compounds is based on their robust and stable interactions at the viral-host interface, essential for inhibiting viral entry into host cells. The research employed computational analysis and MD simulations to examine the binding interactions of these compounds with the S-glycoprotein-ACE2 interface. The findings demonstrate that these compounds can effectively bind to the target, indicating their potential as therapeutic agents for COVID-19. The research identified specific viral binding residues and host residues essential for viral-host protein interactions, indicating that inhibiting these interactions through the identified compounds may effectively prevent viral entry. Lymecycline and hexoprenaline were identified as potential candidates for initial clinical studies [148].

3.7.3 Insights into Protein-Protein and Protein-Ligand Interactions

The impact of PPIs on LPIs is examined in this research utilizing a novel approach known as multi-chain dynamics perturbation analysis (mcDPA). This approach accurately predicts, with 50% or higher accuracy rate, the locations of small-molecule binding sites in protein-protein complexes. When PPIs are vital, the McDPA method could prove to be a useful tool for creating drug-target interactions, thereby enhancing the efficiency of drug development processes [149].

"From Multiple Protein Docking to Protein-Protein Docking at Interactome Level" examines the practical applications of protein docking in enhancing our understanding of molecular interactions and facilitating drug discovery. This illustrates the significance of protein-protein docking in identifying protein interactions, which facilitates novel insights into the functioning of biological systems. The prediction of complex protein structures represents a significant advancement in research methodologies. This development facilitates the creation of models that can be utilized to examine protein interactions and activities more effectively. Protein docking is a critical step in drug development, facilitating the prediction of drug-like properties of compounds and contributing to the development of safer and more effective medications. It also lets them measure inhibitory efficacy, which is great for studying cancer and neurological diseases. By highlighting a number of docking methods suitable for PPIs, this publication provides a structured methodology for researchers to systematically study and assess protein interactions. This research has important implications for the fields of molecular biology and pharmacology, showing how basic biological information may be applied in new ways to discover and develop drugs and therapies [150].

3.8 Future Directions

3.8.1 Emerging Trends in Docking and MD

Molecular modelling is essential to medicinal chemistry, (enantio) separation science, pharmaceutical and biomedical research, and various other important fields, as it relies on a sound, contemporary scientific inquiry methodology. Contemporary approaches in medicinal chemistry construct novel drugs by identifying and characterizing the dynamics of proteins through the integration of cutting-edge techniques from molecular modelling, organic chemistry, chemical and structural biology, and medicinal chemistry. In addition, researchers were able to construct trustworthy virtual models of biomacromolecules by making use of the Protein Data Bank's crystallographic structures, which are available for a large number of proteins and associated complexes [151].

General-purpose computing on GPUs has improved molecular docking efficiency, making complex simulations easier to manage. The rapid advancement of protein engineering and computer-aided drug creation requires this advancement. A novel bee swarm method offers an alternative to genetic algorithms and shows how computational tools can improve the accuracy and speed of forecasting biological macromolecule-ligand interactions [152].

Drug development for new infectious illnesses, such as HIV/AIDS, H1N1, and Coronavirus, relies heavily on molecular docking, as discussed in this article. Ritonavir, Saquinavir, and Amprenavir are HIV/AIDS drugs; Stilbenoids_19 and Zanamivir are H1N1 drugs; and Darunavir, Lopinavir, and Baricitinib are Coronavirus drugs. It finds medicinal compounds that demonstrate potential that interact favorably with target proteins. The results of the study show that molecular docking can cut down on the expense and time needed for conventional drug discovery methods [153].

3.8.2 Key Challenges and Opportunities for Integrating Experimental Methods with Theoretical Models

The integration of experimental methodologies and theoretical models presents numerous opportunities and challenges in the drug development process. The efficiency of drug development can be improved by integrating various methodologies; nonetheless, there are challenges that need to be addressed for successful implementation.

3.8.2.1 *Key challenges*

- **Complexity of biological systems**: Compounding concerns, moreover, are assumptions around the specialized biological interactions that are difficult to model and will potentially cause discrepancies between theory and practice in regard to some kinds of practices [154].

- **Data integration**: Data integration is difficult because data reside in different domains-experimental and computational-making it hard for them to be assimilated in a meaningful way, requiring newer robust frameworks for data sharing and interpretation [155].
- **High costs and time:** People often argue that because it is pricey and takes too long to get a drug to the market (estimated at 10–12 years and costs in the hundreds of millions), it may deter the integration of new methodologies [156].

3.8.2.2 *Opportunities*

- **Improved predictive power:** Computational methods such as MD simulations. It can provide insights into drug-target interactions. This will help improve the lead-to-lead process [156].
- **Translational research:** Integrating pharmacokinetics and pharmacodynamics with computational modeling can reduce late-stage failures in clinical trials by identifying potential payoffs early [156].
- **Drug design innovation:** Revolutions in molecular modelling have opened the door to new drug designs that can treat difficult-to-treat diseases like cancer and neurological disorders [157].

Meeting the inherent constraints of realizing a medicine's full potential requires a balance between creativity and practical restrictions. Future progress in this area will depend on successfully merging experimental and theoretical methods into drug discovery.

3.8.3 Future Perspectives on Computational Drug Discovery

The future of computational drug discovery promises major changes driven by advances in AI, ML, and emerging technologies such as quantum computing. This will improve the drug development process/usability, as well as make them more efficient and effective with personalized therapies. Key areas shaping this future landscape are as follows:

- **Advancements in AI and ML:** Text-based AI and ML are reshaping drug discovery and development practices by enhancing predictive models of drug-target interactions, synthetic route optimization, and pharmacokinetics [158, 159]. AI also enables more rapid peptide drug development through case studies [160]. This streamlines the drug development pipeline as ML methods fill the gaps among candidate selection, toxicity prediction, and multi-omics data integration [159].
- **Integration of emerging technologies**: Vlasiou (2024) predicts that quantum computing will significantly improve molecular simulations, giving drug development unparalleled computational powers. Because of the global nature of the cloud, researchers may interact from anywhere in the world and have access to computational tools that help democratize the area of drug development and assist innovation [158, 160].
- **Ethical and regulatory considerations**: As computational methods evolve, ethical considerations, data privacy, and regulatory challenges become increasingly important [158, 160]. The need for robust ethical frameworks and transparent reporting is essential to ensure responsible AI and ML applications in drug discovery [159].

While the advancements in computational drug discovery are promising, challenges such as algorithm optimization, bias in AI, and ensuring equitable access to these technologies remain critical issues that must be addressed to fully realize their potential.

3.9 Conclusions

Molecular docking and MD simulations provide a potent combination of computational efficiency and prediction accuracy, making them vital tools in current drug discovery and development. These methods help reconcile theoretical predictions with experimental evidence by providing detailed insight into conformational flexibility, binding affinities, and protein-ligand interactions.

Integrating docking with MD simulations provides a potent approach, allowing for more precise and reliable research of lead compounds. By combining these methods, we can improve the pharmacokinetic and pharmacodynamic properties of newly discovered drug candidates. The increasing use of AI, ML, and improved sampling methods greatly enhances the efficacy of these computational approaches, thereby promoting advances in lead validation, optimization, and identification.

Despite the potential, computing cost, temporal constraints, and biological system complexity remain. Hybrid techniques, cloud computing, and algorithm development must evolve to meet these problems. Docking and MD's predictive capacity and applicability to drug development will improve with greater integration of experimental approaches.

The future of computational drug discovery is bright, thanks in large part to molecular docking and MD simulations. Innovations in drug design, repurposing, and development are on the horizon, owing to advances in biomolecular systems, state-of-the-art technologies, and interdisciplinary collaboration. These tools are poised to revolutionize the drug discovery industry by making it easier to find targets and create treatments that save lives.

References

1. Jones, A.W. (2011) 'Early drug discovery and the rise of pharmaceutical chemistry,' *Drug Testing and Analysis*, 3(6), pp. 337–344. https://doi.org/10.1002/dta.301.
2. Takebe, T., Imai, R. and Ono, S. (2018) 'The current status of drug discovery and development as originated in United States Academia: The influence of industrial and academic collaboration on drug discovery and development,' *Clinical and Translational Science*, 11(6), pp. 597–606. https://doi.org/10.1111/cts.12577.
3. Balganesh, T. et al. (2014) 'Drug discovery research in India: Current state and future prospects,' *ACS Medicinal Chemistry Letters*, 5(7), pp. 724–726. https://doi.org/10.1021/ml500183c.
4. Pina, A.S., Hussain, A. and Roque, A.C.A. (2009) 'An historical overview of drug discovery,' *Methods in Molecular Biology*, pp. 3–12. https://doi.org/10.1007/978-1-60761-244-5_1.
5. Colombo, M. and Peretto, I. (2008) 'Chemistry strategies in early drug discovery: An overview of recent trends,' *Drug Discovery Today*, 13(15–16), pp. 677–684. https://doi.org/10.1016/j.drudis.2008.03.007.

6. MacInnes, R., Lumley, C.E. and Walker, S.R. (1994) 'New chemical entity output of the international pharmaceutical industry from 1970 to 1992,' *Clinical Pharmacology & Therapeutics*, 56(3), pp. 339–349. https://doi.org/10.1038/clpt.1994.145.
7. Lombardino, J.G. and Lowe, J.A. (2004) 'The role of the medicinal chemist in drug discovery — Then and now,' *Nature Reviews Drug Discovery*, 3(10), pp. 853–862. https://doi.org/10.1038/nrd1523.
8. Hoffer, L. et al. (2018) 'Chemistry-driven hit-to-lead optimization guided by structure-based approaches,' *Molecular Informatics*, 37(9–10). https://doi.org/10.1002/minf.201800059.
9. De La Torre, B.G. and Albericio, F. (2022) 'The pharmaceutical industry in 2021: An analysis of FDA drug approvals from the perspective of molecules,' *Molecules*, 27(3), p. 1075. https://doi.org/10.3390/molecules27031075.
10. Kaitin, K.I. (2010) 'Deconstructing the drug development Process: The new face of innovation,' *Clinical Pharmacology & Therapeutics*, 87(3), pp. 356–361. https://doi.org/10.1038/clpt.2009.293.
11. Cha, Y. et al. (2017) 'Drug repurposing from the perspective of pharmaceutical companies,' *British Journal of Pharmacology*, 175(2), pp. 168–180. https://doi.org/10.1111/bph.13798.
12. Leeson, P.D. and Springthorpe, B. (2007) 'The influence of drug-like concepts on decision-making in medicinal chemistry,' *Nature Reviews Drug Discovery*, 6(11), pp. 881–890. https://doi.org/10.1038/nrd2445.
13. Van Drie, J.H. (2007) 'Computer-aided drug design: The next 20 years,' *Journal of Computer-Aided Molecular Design*, 21(10–11), pp. 591–601. https://doi.org/10.1007/s10822-007-9142-y.
14. Lin, Xiaoqian, Li, X. and Lin, Xubo (2020) 'A review on applications of computational methods in drug screening and design,' *Molecules*, 25(6), p. 1375. https://doi.org/10.3390/molecules25061375.
15. Yamanishi, Y. et al. (2008) 'Prediction of drug–target interaction networks from the integration of chemical and genomic spaces,' *Bioinformatics*, 24(13), pp. i232–i240. https://doi.org/10.1093/bioinformatics/btn162.
16. Bakheet, T.M. and Doig, A.J. (2009) 'Properties and identification of human protein drug targets,' *Bioinformatics*, 25(4), pp. 451–457. https://doi.org/10.1093/bioinformatics/btp002.
17. Sadybekov, A.V. and Katritch, V. (2023) 'Computational approaches streamlining drug discovery,' *Nature*, 616(7958), pp. 673–685. https://doi.org/10.1038/s41586-023-05905-z.

18. Schneider, G. and Clark, D.E. (2019) 'Automated de novo drug design: Are we nearly there yet?' *Angewandte Chemie International Edition*, 58(32), pp. 10792–10803. https://doi.org/10.1002/anie.201814681.
19. Kaka, M.O. et al. (2022) 'Computational applications in the drug discovery and development processes,' in *BENTHAM SCIENCE PUBLISHERS eBooks*, pp. 365–401. https://doi.org/10.2174/9789815050622122010018.
20. Shamim, S. et al. (2024b) 'Molecular docking: An insight from drug discovery to drug repurposing approach,' in *IntechOpen eBooks*. https://doi.org/10.5772/intechopen.1005526.
21. Ahmed, M., Sharma, A. and Bora, R. (2024b) 'Unveiling novel drug leads: Utilizing molecular docking studies for efficient drug discovery,' *Research Journal of Chemistry and Environment*, 28(9), pp. 92–100. https://doi.org/10.25303/289rjce920100.
22. Cvijetić, I., Petrović, D. and Zloh, M. (2024b) 'Dynamics of biomolecular ligand recognition,' in *BENTHAM SCIENCE PUBLISHERS eBooks*, pp. 103–139. https://doi.org/10.2174/9789815165463124010008.
23. Shamsi, A. et al. (2024) 'Structure-based drug-development study against fibroblast growth factor receptor 2: Molecular docking and molecular dynamics simulation approaches,' *Scientific Reports*, 14(1). https://doi.org/10.1038/s41598-024-69850-1.
24. Sahu, D. et al. (2024) 'A review on molecular docking as an interpretative tool for molecular targets in disease management,' *Assay and Drug Development Technologies*, 22(1), pp. 40–50. https://doi.org/10.1089/adt.2023.060.
25. Kuntz, I.D. et al. (1982) 'A geometric approach to macromolecule-ligand interactions,' *Journal of Molecular Biology*, 161(2), pp. 269–288. https://doi.org/10.1016/0022-2836(82)90153-x.
26. Muhammed, M.T. and Aki-Yalcin, E. (2022) 'Molecular docking: Principles, advances, and its applications in drug discovery,' *Letters in Drug Design & Discovery*, 21(3), pp. 480–495. https://doi.org/10.2174/1570180819666220922103109.
27. Sahu, M.K. et al. (2024b) 'Exploring recent updates on molecular docking: Types, method, application, limitation & future prospects,' *International Journal of Pharmaceutical Research and Allied Sciences*, 13(2), pp. 24–40. https://doi.org/10.51847/une9jqjucl.
28. Fan, J., Fu, A. and Zhang, L. (2019) 'Progress in molecular docking,' *Quantitative Biology*, 7(2), pp. 83–89. https://doi.org/10.1007/s40484-019-0172-y.

29. Tripathi, A. and Bankaitis, V.A. (2017) 'Molecular docking: From lock and key to combination lock,' *Journal of Molecular Medicine and Clinical Applications*, 2(1). https://doi.org/10.16966/2575-0305.106.
30. Mohanty, M. and Mohanty, P.S. (2023) 'Molecular docking in organic, inorganic, and hybrid systems: A tutorial review,' *Monatshefte Für Chemie - Chemical Monthly*, 154(7), pp. 683–707. https://doi.org/10.1007/s00706-023-03076-1.
31. Muegge, I. and Rarey, M. (2001) 'Small molecule docking and scoring,' *Reviews in Computational Chemistry*, pp. 1–60. https://doi.org/10.1002/0471224413.ch1.
32. Böhm, H.-J. and Stahl, M. (2003) 'The use of scoring functions in drug discovery applications,' in *John Wiley & Sons, Inc. eBooks*, pp. 41–87. https://doi.org/10.1002/0471433519.ch2.
33. Liu, J. and Wang, R. (2015) 'Classification of current scoring functions,' *Journal of Chemical Information and Modeling*, 55(3), pp. 475–482. https://doi.org/10.1021/ci500731a.
34. Morris, G.M. et al. (1998) 'Automated docking using a Lamarckian genetic algorithm and an empirical binding free energy function,' *Journal of Computational Chemistry*, 19(14), pp. 1639–1662. https://doi.org/10.1002/(sici)1096-987x(19981115)19:14.
35. Wesson, L. and Eisenberg, D. (1992) 'Atomic solvation parameters applied to molecular dynamics of proteins in solution,' *Protein Science*, 1(2), pp. 227–235. https://doi.org/10.1002/pro.5560010204.
36. Gohlke, H., Hendlich, M. and Klebe, G. (2000) 'Knowledge-based scoring function to predict protein-ligand interactions,' *Journal of Molecular Biology*, 295(2), pp. 337–356. https://doi.org/10.1006/jmbi.1999.3371.
37. Friesner, R.A. et al. (2004) 'Glide: A new approach for rapid, accurate docking and scoring. 1. Method and assessment of docking accuracy,' *Journal of Medicinal Chemistry*, 47(7), pp. 1739–1749. https://doi.org/10.1021/jm0306430.
38. Li, Y. et al. (2014) 'Comparative assessment of scoring functions on an updated benchmark: 2. Evaluation Methods and general results,' *Journal of Chemical Information and Modeling*, 54(6), pp. 1717–1736. https://doi.org/10.1021/ci500081m.
39. Warren, G.L. et al. (2005) 'A critical assessment of docking programs and scoring functions,' *Journal of Medicinal Chemistry*, 49(20), pp. 5912–5931. https://doi.org/10.1021/jm050362n.

106. Yang, Y. et al. (2015) 'Like-charge guanidinium pairing between ligand and receptor: An unusual interaction for drug discovery and design?' *The Journal of Physical Chemistry B*, 119(36), pp. 11988–11997. https://doi.org/10.1021/acs.jpcb.5b04130.

107. Verkhivker, G.M. et al. (2000) 'Deciphering common failures in molecular docking of ligand-protein complexes,' *Journal of Computer-Aided Molecular Design*, 14(8), pp. 731–751. https://doi.org/10.1023/a:1008158231558.

108. Spyrakis, F. and Cavasotto, C.N. (2015) 'Open challenges in structure-based virtual screening: Receptor modeling, target flexibility consideration and active site water molecules description,' *Archives of Biochemistry and Biophysics*, 583, pp. 105–119. https://doi.org/10.1016/j.abb.2015.08.002.

109. Shoichet, B.K., Leach, A.R. and Kuntz, I.D. (1999) 'Ligand solvation in molecular docking,' *Proteins Structure Function and Bioinformatics*, 34(1), pp. 4–16. https://doi.org/10.1002/(sici)1097-0134(19990101)34:1.

110. Dror, R.O. et al. (2011) 'Pathway and mechanism of drug binding to G-protein-coupled receptors,' *Proceedings of the National Academy of Sciences*, 108(32), pp. 13118–13123. https://doi.org/10.1073/pnas.1104614108.

111. Warshel, A. and Levitt, M. (1976) 'Theoretical studies of enzymic reactions: Dielectric, electrostatic and steric stabilization of the carbonium ion in the reaction of lysozyme,' *Journal of Molecular Biology*, 103(2), pp. 227–249. https://doi.org/10.1016/0022-2836(76)90311-9.

112. Allen, M.P. and Tildesley, D.J. (2017) 'Computer simulation of liquids,' *Oxford University Press eBooks*. https://doi.org/10.1093/oso/9780198803195.001.0001.

113. Lindorff-Larsen, K. et al. (2011) 'How fast-folding proteins fold,' *Science*, 334(6055), pp. 517–520. https://doi.org/10.1126/science.1208351.

114. Klauda, J.B. et al. (2010) 'Update of the CHARMM all-atom additive force field for lipids: Validation on six lipid types,' *The Journal of Physical Chemistry B*, 114(23), pp. 7830–7843. https://doi.org/10.1021/jp101759q.

115. Baaden, M. and Marrink, S.J. (2013) 'Coarse-grain modelling of protein–protein interactions,' *Current Opinion in Structural Biology*, 23(6), pp. 878–886. https://doi.org/10.1016/j.sbi.2013.09.004.

116. Laio, A. and Parrinello, M. (2002) 'Escaping free-energy minima,' *Proceedings of the National Academy of Sciences*, 99(20), pp. 12562–12566. https://doi.org/10.1073/pnas.202427399.
117. Sugita, Y. and Okamoto, Y. (1999) 'Replica-exchange molecular dynamics method for protein folding,' *Chemical Physics Letters*, 314(1–2), pp. 141–151. https://doi.org/10.1016/s0009-2614(99)01123-9.
118. Hamelberg, D., Mongan, J. and McCammon, J.A. (2004) 'Accelerated molecular dynamics: A promising and efficient simulation method for biomolecules,' *The Journal of Chemical Physics*, 120(24), pp. 11919–11929. https://doi.org/10.1063/1.1755656.
119. Bowman, G.R. et al. (2009) 'Progress and challenges in the automated construction of Markov state models for full protein systems,' *The Journal of Chemical Physics*, 131(12). https://doi.org/10.1063/1.3216567.
120. Gholap, A.D. et al. (2024) 'Advances in artificial intelligence for drug delivery and development: A comprehensive review,' *Computers in Biology and Medicine*, 178, p. 108702. https://doi.org/10.1016/j.compbiomed.2024.108702.
121. Zhang, X. et al. (2024) 'Advancing ligand docking through deep learning: Challenges and prospects in virtual screening,' *Accounts of Chemical Research*, 57(10), pp. 1500–1509. https://doi.org/10.1021/acs.accounts.4c00093.
122. Du, X. et al. (2016) 'Insights into protein–ligand interactions: Mechanisms, models, and methods,' *International Journal of Molecular Sciences*, 17(2), p. 144. https://doi.org/10.3390/ijms17020144.
123. Bai, Q. et al. (2021) 'Application advances of deep learning methods for de novo drug design and molecular dynamics simulation,' *Wiley Interdisciplinary Reviews Computational Molecular Science,* 12(3). https://doi.org/10.1002/wcms.1581.
124. Catacutan, D.B. et al. (2024) 'Machine learning in preclinical drug discovery,' *Nature Chemical Biology*, 20(8), pp. 960–973. https://doi.org/10.1038/s41589-024-01679-1.
125. Lukauskis, D. (2024). Calculating absolute ligand-binding free energies and ranking ligand-binding poses using metadynamics enabled molecular dynamics simulations, *Doctoral Dissertation*, UCL (University College London).
126. Ries, B. et al. (2024) 'Kartograf: A geometrically accurate atom mapper for hybrid-topology relative free energy calculations,' *Journal of Chemical Theory and Computation*, 20(5), pp. 1862–1877. https://doi.org/10.1021/acs.jctc.3c01206.

127. Macuglia, D. (2024) 'Free-energy calculations in condensed matter: From early challenges to the advent of umbrella sampling,' *Archive for History of Exact Sciences*, 78(5), pp. 479–522. https://doi.org/10.1007/s00407-024-00327-3.

128. Grassmann, G. et al. (2024) 'Computational approaches to predict protein–protein interactions in crowded cellular environments,' *Chemical Reviews*, 124(7), pp. 3932–3977. https://doi.org/10.1021/acs.chemrev.3c00550.

129. Ellingson, S.R. and Baudry, J. (2012) 'High-throughput virtual molecular docking with AutoDockCloud,' *Concurrency and Computation Practice and Experience*, 26(4), pp. 907–916. https://doi.org/10.1002/cpe.2926.

130. Dar, A.M. and Mir, S. (2017) 'Molecular docking: Approaches, types, applications and basic challenges,' *Journal of Analytical & Bioanalytical Techniques,* 08(02). https://doi.org/10.4172/2155-9872.1000356.

131. Ejaz, S. et al. (2023) 'Computer-aided approach for the identification of lead molecules as the inhibitors of cholinesterase's and monoamine oxidases: Novel target for the treatment of Alzheimer's disease,' *Journal of the Serbian Chemical Society*, 89(2), pp. 177–194. https://doi.org/10.2298/jsc230307050e.

132. Solanki, P. et al. (2023) 'A comprehensive analysis of the role of molecular docking in the development of anticancer agents against the cell cycle CDK enzyme,' *Biocell*, 47(4), pp. 707–729. https://doi.org/10.32604/biocell.2023.026615.

133. Gupta, D. and Khan, A.U. (2021) 'Advances in docking-based drug design for microbial and cancer drug targets,' in *Elsevier eBooks*, pp. 407–424. https://doi.org/10.1016/b978-0-12-822312-3.00020-5.

134. Yadav, R. et al. (2020) 'Molecular docking studies of enzyme binding drugs on family of cytochrome P450,' *Advanced Science Engineering and Medicine*, 12(1), pp. 83–87. https://doi.org/10.1166/asem.2020.2520.

135. Lim, V.J.Y. et al. (2018) 'A benchmarking study on virtual ligand screening against homology models of human GPCRs,' *Proteins Structure Function and Bioinformatics*, 86(9), pp. 978–989. https://doi.org/10.1002/prot.25533.

136. Bekker, G.-J. et al. (2021) 'Accurate binding configuration prediction of a G-protein-coupled receptor to its antagonist using multicanonical molecular dynamics-based dynamic docking,' *Journal of Chemical Information and Modeling*, 61(10), pp. 5161–5171. https://doi.org/10.1021/acs.jcim.1c00712.

137. Thomas, B.N., Parrill, A.L. and Baker, D.L. (2021) 'Self-docking and cross-docking simulations of G protein-coupled receptor-ligand complexes: Impact of ligand type and receptor activation state,' *Journal of Molecular Graphics and Modelling*, 112, p. 108119. https://doi.org/10.1016/j.jmgm.2021.108119.

138. Chatatikun, M. et al. (2024) 'Mechanistic exploration of bioactive constituents in Gnetum gnemon for GPCR-related cancer treatment through network pharmacology and molecular docking,' *Scientific Reports*, 14(1). https://doi.org/10.1038/s41598-024-75240-4.

139. Kumar, S. and Roy, V. (2023) 'Repurposing drugs: An empowering approach to drug discovery and development,' *Drug Research*, 73(09), pp. 481–490. https://doi.org/10.1055/a-2095-0826.

140. Park, K. (2019) 'A review of computational drug repurposing,' *Translational and Clinical Pharmacology*, 27(2), p. 59. https://doi.org/10.12793/tcp.2019.27.2.59.

141. Oprea, T.I. et al. (2011b) 'Drug repurposing from an academic perspective,' *Drug Discovery Today Therapeutic Strategies*, 8(3–4), pp. 61–69. https://doi.org/10.1016/j.ddstr.2011.10.002.

142. Kumar, S., Chowdhury, S. and Kumar, S. (2017) 'In silico repurposing of antipsychotic drugs for Alzheimer's disease,' *BMC Neuroscience*, 18(1). https://doi.org/10.1186/s12868-017-0394-8.

143. Shah, B., Modi, P. and Sagar, S.R. (2020) 'In silico studies on therapeutic agents for COVID-19: Drug repurposing approach,' *Life Sciences*, 252, p. 117652. https://doi.org/10.1016/j.lfs.2020.117652.

144. Cavasotto, C.N. and Di Filippo, J.I. (2020) 'In silico drug repurposing for COVID-19: Targeting SARS-CoV-2 proteins through docking and consensus ranking,' *Molecular Informatics*, 40(1). https://doi.org/10.1002/minf.202000115.

145. Li, Q. and Kang, C. (2020) 'Progress in developing inhibitors of SARS-COV-2 3C-like protease,' *Microorganisms*, 8(8), p. 1250. https://doi.org/10.3390/microorganisms8081250.

146. Lazniewski, M. et al. (2022) 'Drug repurposing for identification of potential spike inhibitors for SARS-CoV-2 using molecular docking and molecular dynamics simulations,' *Methods*, 203, pp. 498–510. https://doi.org/10.1016/j.ymeth.2022.02.004.

147. Baby, K. et al. (2023) 'Computational drug repurposing of Akt-1 allosteric inhibitors for non-small cell lung cancer,' *Scientific Reports*, 13(1). https://doi.org/10.1038/s41598-023-35122-7.

148. Hussein, D. et al. (2023) 'In silico drug repurposing using molecular docking and dynamics to target the protein interaction between the SARS-CoV-2 S-glycoprotein and the ACE2 receptor,' *F1000Research*, 12, p. 1452. https://doi.org/10.12688/f1000research.131508.1.

149. Cieślak, D., Kabelka, I. and Bartuzi, D. (2024) 'Molecular dynamics simulations in protein–protein docking,' *Methods in Molecular Biology*, pp. 91–106. https://doi.org/10.1007/978-1-0716-3985-6_6.

150. Gabrani, R. et al. (2024) 'From multiple protein docking to protein–protein docking at interactome level,' *Methods in Molecular Biology*, pp. 69–89. https://doi.org/10.1007/978-1-0716-3985-6_5.

151. Peluso, P. and Chankvetadze, B. (2023) 'Recent developments in molecular modeling tools and applications related to pharmaceutical and biomedical research,' *Journal of Pharmaceutical and Biomedical Analysis*, 238, p. 115836. https://doi.org/10.1016/j.jpba.2023.115836.

152. Al-Hussaniy, H.A. (2023) 'The development of molecular docking and molecular dynamics and their application in the field of chemistry and computer simulation,' *Journal of Medical Pharmaceutical and Allied Sciences*, 12(1), pp. 5552–5562. https://doi.org/10.55522/jmpas.v12i1.4137.

153. Gouthami, K. et al. (2022) 'WITHDRAWN: Molecular docking used as an advanced tool to determine novel compounds on emerging infectious diseases: A systematic review,' *Progress in Biophysics and Molecular Biology*. https://doi.org/10.1016/j.pbiomolbio.2022.10.001.

154. Vicini, P. (2010) 'Multiscale modeling in drug discovery and development: Future opportunities and present challenges,' *Clinical Pharmacology & Therapeutics*, 88(1), pp. 126–129. https://doi.org/10.1038/clpt.2010.87.

155. Vangala, S. et al. (2012) 'Translational drug discovery research: Integration of medicinal chemistry, computational modeling, pharmacology, ADME, and toxicology,' *Encyclopedia of Drug Metabolism and Interactions*, pp. 1–54. https://doi.org/10.1002/9780470921920.edm038.

156. Shukla, R. and Tripathi, T. (2021) 'Molecular dynamics simulation in drug discovery: Opportunities and challenges,' *in Springer eBooks*, pp. 295–316. https://doi.org/10.1007/978-981-15-8936-2_12.

157. Iyengar, S.M. (2023) 'Applications of molecular modeling techniques to drug discovery and structural genomics: Development of molecular probes for neurological disorders and SARS-CoV-2.' https://doi.org/10.17760/d20560791.

158. Vlasiou, M.C. (2024) 'The future of computer-aided drug discovery methods: Advancements and opportunities,' in *BENTHAM SCIENCE PUBLISHERS eBooks*, pp. 126–130. https://doi.org/10.2174/9789815305036124010009.
159. Prabhu, M.R. et al. (2024) 'Exploring machine learning applications and future prospects in drug discovery,' *10th International Conference on Communication and Signal Processing* (ICCSP), Melmaruvathur, India. https://doi.org/10.1109/iccsp60870.2024.10543411.
160. Niazi, S.K. and Mariam, Z. (2023) 'Computer-aided drug design and drug discovery: A prospective analysis,' *Pharmaceuticals* [Preprint]. https://doi.org/10.20944/preprints202311.0852.v1.

Chapter 4

Molecular Modelling and Docking Approaches to Study Biomolecular Interactions

Sana Quraishi, Amarjyoti Mondal,
Surendra Nath Barman, and Atanu Singha Roy
Department of Chemical and Biological Sciences, National Institute of Technology Meghalaya, Saitsohpen, Sohra, Meghalaya, India
asroy86@nitm.ac.in

Molecular docking, a prominent computational technique in bioinformatics, analyses the interactions between two or more compounds to predict the resulting complex's structure with high precision. This chapter offers a comprehensive review of molecular docking techniques, detailing their development and applications in drug discovery. We discussed the foundational theory, including sampling algorithms, scoring functions, and the software tools used in molecular docking. This discussion encompassed detailed applications in studying biomolecular interactions, specifically,

Computational Methods for Molecular Modelling and Docking
Edited by Arup Kumar Ghosh and Suban Kumar Sahoo

ISBN 978-981-5352-25-2 (Hardcover), 978-1-042-00204-7 (eBook)
www.jennystanford.com

DOI: 10.1201/9781042002047-4

protein–ligand, deoxyribonucleic acid (DNA)–ligand, and single-walled carbon nanotube (SWCNT)-protein interactions, alongside an in-depth description of molecular dynamics (MD) simulations utilizing GROMACS software, enriched with practical examples. These methods simulate the association between small molecules and biomolecular targets, providing key insights into molecular recognition, binding affinities, and structural conformations. Ongoing advancements in software and computational power have revolutionized molecular modeling and docking. This progress has significantly enhanced biomolecular interaction studies and accelerated the development of new drugs.

4.1 Introduction

In recent years, computational methods have significantly improved our understanding of the structure and function of intricate biological networks. Advancements in high-performance computing have made in silico virtual experimentation an invaluable tool, bridging the gap between laboratory experiments and theoretical research. Computational modeling has been essential in efficiently identifying lead compounds for subsequent experimental testing *in vivo* and *in vitro*. In recent years, *in silico* approaches have been highly demanding due to the fact that they save time and are economically efficient [1, 2]. For example, conventional microscopy lacks sufficient resolution to capture real-time molecular interactions, whereas computational methods can predict optimal binding and likely trajectories. The evaluation of the efficacy of the synthesized and purified proteins can span over several years and incur costs in the millions.

In contrast, the use of in silico methods significantly reduces the costs associated with this process. It possesses a wide array of purposes and applications. Following its introduction in the mid-1970s, this tool has been employed in various fields, such as drug discovery and development, structural biology, medicinal chemistry, and bioinformatics [3–8]. They are utilized for conducting structure-activity studies, identifying initial hits, refining lead compounds, proposing binding hypotheses to assist mutation studies, helping fit substrates and inhibitors into X-ray crystallographic electron density maps, exploring chemical processes, and constructing combinatorial libraries. This technique is utilized in molecular biology to uncover novel chemical probes and in drug development to pinpoint promising candidates that can be further developed into lead compounds and, eventually, candidate drugs [9, 10]. Techniques like molecular docking and modeling are invaluable for predicting molecular interactions (how macromolecules interact with ligands) and understanding the 3D-structural organization of biological macromolecules. Molecular docking explicitly evaluates the favored orientation of one molecule against another, resulting in a stable complex formation [11]. This information is used to estimate the binding affinity between two molecules through a scoring function that identifies the optimal docking pose for each molecule and ranks them accordingly. Thus, molecular docking refers to the study of how two or more structures interact and align. Scheme 4.1 illustrates the formation of a stable receptor-ligand complex, which represents the molecular recognition event where the ligand binds to the receptor. This binding can stabilize or alter the receptor's structure, initiating downstream effects, such as activation or inhibition of signaling pathways, regulation of gene expression, modulation of cellular functions, or changes in enzymatic activity. These downstream effects collectively represent the biological response resulting from the receptor-ligand interaction, which is critical to understanding the functional implications of molecular docking studies. By analyzing these interactions through molecular docking, one can predict how specific ligands may influence physiological processes or therapeutic outcomes. Molecular docking is based on the principle of molecular complementarity, where two structures interact as pieces of a puzzle fitting together. This shape complementarity is the primary principle for estimating the binding and fitting of two molecular structures [12]. Moreover, chemical and physicochemical complementarities significantly influence interaction and binding between two molecules. Docking can be utilized to investigate a wide range of binding interactions, including protein–ligand, protein–protein, protein–nucleic acid, lipid–protein, lipid–lipid, and enzyme–substrate, playing crucial roles throughout all phases of molecular biology and biochemistry, in addition to structural interactions. Figure 4.1 depicts a flowchart of the molecular docking workflow.

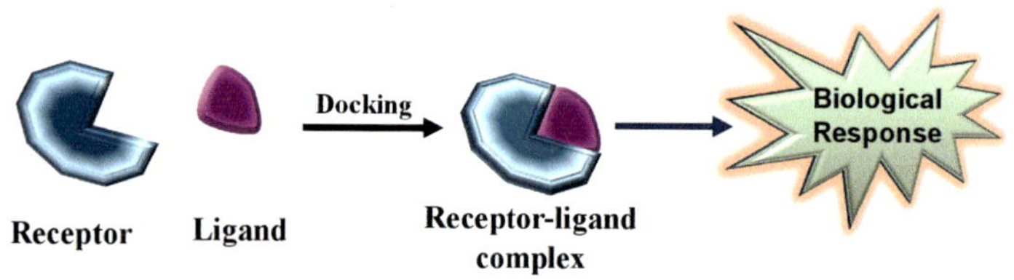

Scheme 4.1 A general illustration of a receptor docking with a ligand.

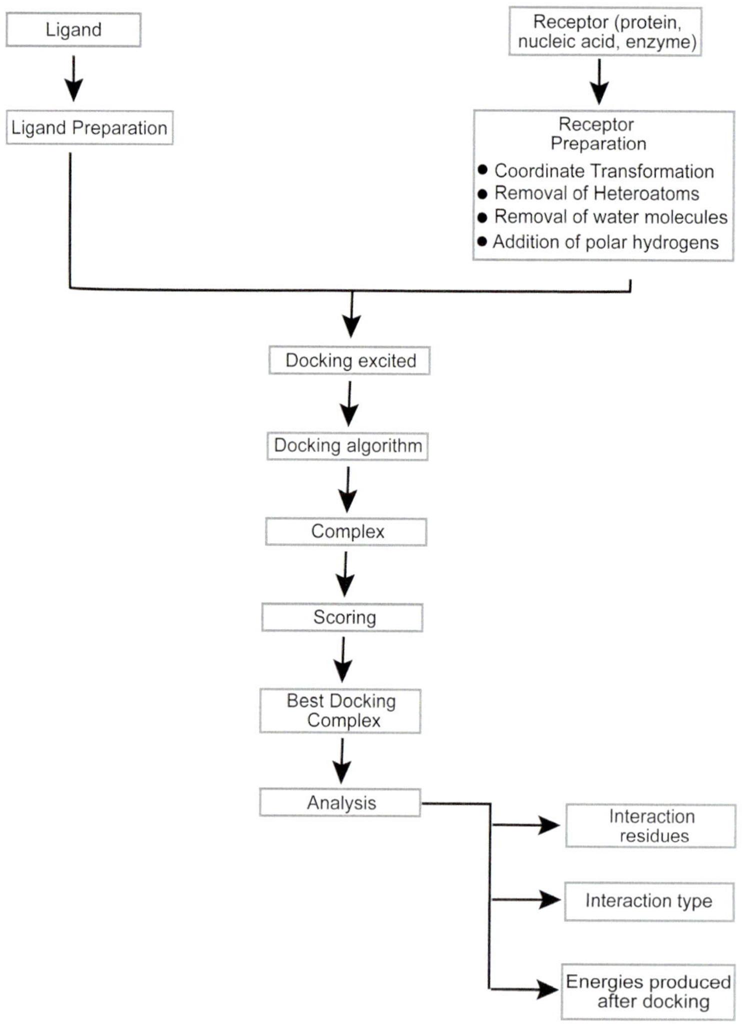

Figure 4.1 Flowchart representing the overview of the molecular docking workflow process.

Molecular docking provides an initial understanding of how ligands bind to targets, receptors, or macromolecules. MD simulations expand on this by assessing the stability of the complexes and capturing the dynamic, time-dependent behavior of macromolecules at the scale of individual atoms. MD simulations predict the movement of atoms in a molecular system over time, relying on physics-based intermolecular interactions to provide detailed insights into the motion and positioning of each atom in a biomolecular system [13]. The first ever MD simulation of a protein, specifically the bovine pancreatic trypsin inhibitor (BPTI), was performed in 1977 by *McCammon and Karplus* [14]. This groundbreaking work marked the introduction of MD simulations into biomolecular research [14]. MD simulations permit the examination of flexibility, structural changes, and interaction stability, taking into account environmental factors like temperature, solvent conditions, and ionic concentration. It also enables the calculation of key thermodynamic properties like binding free energies, which are essential for assessing ligand affinity and specificity.

4.2 Fundamentals of Molecular Docking

4.2.1 Theory

Docking is a computational method that evaluates how effectively a designed compound fits into a specific binding cavity and interacts with adjacent residues. Most docking software is based on molecular mechanics (MM), which employs classical physics to describe polyatomic systems. To minimize discrepancies between theoretical predictions and experimental data, experimental parameters are incorporated. Since experimental techniques have limitations, quantum mechanics, through semiempirical and ab initio methods, is used to derive useful parameters from mathematical equations [15]. In this context, the approach involves a system of equations with parameters that define molecular force fields, such as the Lennard-Jones potential, torsional properties, potential energy, electrostatic forces, and bond geometry. Prominent examples of force fields include AMBER, CHARMM, GROMOS, OPLS-AA, and UFF [16]. Extending these approaches, molecular processes like ligand binding to targets are modeled. The docking process seeks to imitate the natural interaction between a ligand and its receptor by following the lowest-energy pathway. Performing docking studies with precision requires high-resolution structural data from NMR, X-ray crystallography, Cryo-EM, or homology modeling with a recognized or predicted binding site in the biomolecule [16]. Docking methods integrate steric, hydrophobic, and electrostatic interactions to enhance ligand fit within a macromolecular binding pocket while simultaneously estimating binding free energy.

4.2.2 Types of Docking

Molecular docking techniques are classified into three categories according to the molecular flexibility during calculations: rigid docking, semi-flexible docking, and flexible docking (Scheme 4.2). Table 4.1 illustrates the features of different types of docking.

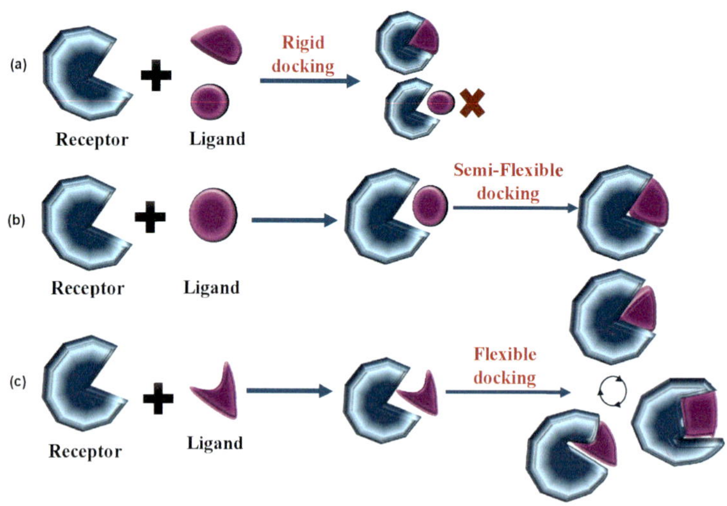

Scheme 4.2 Types of molecular docking: (a) rigid docking in which both receptor and ligand are rigid, (b) semi-flexible docking in which the receptor is rigid and the ligand is flexible, and (c) flexible docking in which both receptor and ligand are flexible.

Table 4.1 Key features of the different types of docking

Feature	Rigid docking	Semi-flexible docking	Flexible docking
Flexibility of Receptor	Rigid	Rigid	Flexible
Flexibility of ligand	Rigid	Flexible	Flexible
Tools/Software	Dock (initial versions): FLOG, FTDOCK	AutoDock, FlexX, GOLD	AutoDock Vina, GLIDE
Efficiency and accuracy	Docking approach is suitable for docking between macromolecules	commonly used with higher accuracy than rigid docking	Significantly increases docking accuracy but requires high computational power

4.2.2.1 Rigid docking

Rigid docking is grounded in Emil Fischer's 1894 'Lock and Key' hypothesis [17]. This method assumes both the receptor and ligand are inflexible, considering only their translational and rotational movements and ignoring any conformational changes. As a result, interactions are more difficult to observe, and determining the optimal ligand confirmation is not straightforward [18]. At times, ligands may not fit adequately into the receptor pocket, resulting in weak interactions that do not produce satisfactory outcomes. Internal flexibility is crucial for effective docking interactions. These methods primarily involve optimizing both the binding interface and the ligand, identifying key interaction points, and assessing the alignment of corresponding features [18]. This approach is primarily utilized in protein–protein docking, where extensive conformational degrees of freedom present sampling challenges. Rigid docking is adequate for observing interactions [18]. Early versions of docking programs, including DOCK [19–22], FLOG [23], and certain tools for protein–protein docking, such as FTDOCK [24], used this method, treating both the ligand and receptor as rigid entities throughout the docking process.

4.2.2.2 Semi-flexible docking

This docking strategy operates on the principle that the ligand molecule can exhibit flexibility, whereas the target macromolecule remains fixed [25]. Ligand flexibility is facilitated by rotatable bonds, which allow the ligand to assume multiple conformations during docking. These methods are based on the hypothesis that the rigid target framework can effectively recognize the docking ligand [15]. Semi-flexible docking is often employed to explore molecular interactions because it requires more computational resources than rigid docking but fewer than flexible docking. However, this approach's underlying assumption does not always hold true [26]. Nonetheless, most docking programs, including AutoDock [27] and FlexX [28], have adopted this approach.

4.2.2.3 Flexible docking

This docking analysis relies on the induced fit model proposed by *Daniel Koshland* in 1958 [18]. Flexible docking considers the flexibility of both the receptor and the ligand during the binding process, allowing for a more accurate representation of their interactions. Flexibility can be incorporated either implicitly through sidechain and/or backbone adjustments or directly through conducting several docking runs from various conformations, a process referred to as cross or ensemble docking. While several potential conformations of the ligand are predicted, from which binding energies are calculated for each conformation in the receptor pockets, this method is highly regarded for its accuracy, but can be quite time-consuming [29]. AutoDock Vina, GLIDE [30] uses this approach.

4.2.3 Docking Approaches

4.2.3.1 Monte Carlo approach

Monte Carlo (MC) methods represent a widely used class of stochastic optimization methods. These algorithms gradually modify a ligand by rotating bonds or moving and rotating the entire ligand within an active site, producing randomized conformations. Several parameters can be modified simultaneously to achieve the desired conformation [31]. The resulting conformation undergoes an energy evaluation via MM, and Boltzmann's probability constant determines its acceptance for further iterations. The decision is based on energy change relative to a parameter often interpreted as temperature in simulated annealing. The MC method's reliance on this acceptance criterion makes it distinct from other techniques. If a conformation meets the criteria, it is stored and then subsequently modified to generate additional conformations. The repetitive process continues until a predetermined number of conformations have been obtained. One major advantage of the MC approach is that it allows for significant changes, enabling the molecule to overcome energy barriers on the potential energy landscape [32]. MC-based approaches are utilized in programs such as AutoDock [7], ICM [33], QXP [34], FDS [35], GlamDock [36], MCDOCK [37], PRODOCK [38], ROSETTA LIGAND [39], and RiboDock [40].

4.2.3.2 Fragment-based method

Fragment-based approaches in molecular docking involve using small fragments of ligands to identify potential binding interactions. In 1986, DesJarlais and his team proposed a strategy for fragmenting the ligand into smaller sections [41]. In this method, ligands are divided into smaller fragments or protons, which are docked separately within the receptor binding pocket. The results are then used to reconstruct various potential docking conformations, ultimately optimizing binding affinity and specificity. By breaking the ligand into smaller fragments, the algorithm can more accurately accommodate ligand flexibility. Rigid fragments are docked first, acting as "anchors," and are then connected by the more flexible segments of the ligand, characterized by their rotatable bonds. This enables the ligand to be progressively "constructed" within the receptor's binding site. It has been integrated into many docking programs such as DOCK 4.0 [20], FLEXX [27], FLOG [23], LUDI [42], and eHiTs [43].

4.2.3.3 Ligand fit approach

The ligand fitting method delivers a quick and reliable approach for docking small molecule ligands into the receptor binding pocket by evaluating shape matching between the ligand and the receptor's binding region [44]. The process involves two key steps: first, detecting cavities to locate and define the active docking site; and second, fitting the ligands within the selected docking site.

4.2.3.4 Blind docking

In blind docking, a ligand is docked across the receptor's full surface without knowing the location of the binding pocket beforehand. In this approach, the entire receptor is considered a possible interaction site for the ligand, resulting in a significantly larger search space and an extended runtime [45]. The level of complexity in blind docking increases exponentially in relation to the number of potential binding sites, which restricts its practical use. Nonetheless, blind docking approaches that can predict ligand conformations without prior insight into the binding site are necessary for fully automated *in silico* drug design. This method was initially introduced to identify potential interaction sites and configurations of peptide ligands by analyzing the whole surface of protein targets. This has been shown to be successful with AutoDock [46, 47].

4.2.4 Scoring Function

Scoring functions typically estimate a complex's binding energy quickly by employing many assumptions and simplifications, aiming to approximate the actual binding energy as closely as possible within a minimal time [48–50]. The primary goal of the scoring function is to quickly differentiate the most accurate, low-energy poses from incorrect ones and to identify active compounds from inactive ones [51]. Assessing and ranking predicted ligand conformations is essential in virtual screening. Primarily, this function estimates and ranks conformational binding facilitated by search algorithms. The score derived from the ranking reflects the ligand's binding affinity, with higher scores indicating better binders [51–53]. Thus, designing reliable scoring functions and ranking systems is crucial. Popular scoring functions balance the accuracy in estimating binding energy and computational efficiency. Over the years, many scoring functions have been developed, falling into three main categories: force field-based, empirical, and knowledge-based.

As previously mentioned, while different scoring functions employ distinct specific formulas, a typical empirical scoring function can be generally expressed as:

$$\text{Score} = \Sigma \text{ van der Waals} + \text{Electrostatic} + \text{Hydrogen Bonding} - \text{Desolvanation Penalty} - \text{Entropy Loss} \quad (4.1)$$

4.2.4.1 Force-field (physics)-based scoring function

Physics-based scoring functions utilize energy components from MM models. These parameters are derived from empirical data or quantum mechanical computations to evaluate receptor-ligand interactions. They account for physical interactions at the atomic scale, such as van der Waals forces, bond angles, bond lengths, and torsions. Different scoring functions employ diverse potentials to characterize these interactions. For instance, when accounting for electrostatic screening caused by the solvent, the dielectric constant can vary with distance, especially in protein–ligand binding sites [54–56]. Prominent examples of physics-based (force-field) scoring functions include DOCK [57–60] and AutoDock (versions 2, 3, and 4) [7, 59, 61, 62], which employ hybrid scoring functions like GoldScore [63, 64] and GalaxyDock [65]. Furthermore, the precision of binding energy estimations from force-field-based docking results can be enhanced through advanced methods such as linear interaction energy and free-energy perturbation (FEP) techniques.

Empirical terms addressing the loss of torsional degrees of freedom upon complexation can be included. Usually, simple approximations based on the number of rotatable bonds are used, although more advanced methods have been proposed. Both empirical and knowledge-based scoring functions apply these adjustments, and these two functions are discussed as follows:

4.2.4.1.1 *Empirical scoring function*

The empirical approach presumes that the total binding free energy (ΔG_{bind}) of a receptor-ligand complex can be broken down into multiple components, each corresponding to a different type of molecular interaction. Each of these components represents physical interaction, such as electrostatic interactions, hydrogen bonds, and van der Waals forces. This function combines these components linearly, with coefficients fitted from experimental data (e.g., crystallography, calorimetry, and binding assays). Empirical scoring functions, which depend on experimental data, are typically easier to implement than physics-based methods (such as field-based functions). However, their applicability may be limited to the specific conditions or systems used in their training datasets [66]. Empirical scoring functions typically evaluate a set of physical interactions that contribute to protein–ligand binding. Common components include:

1. van der Waals (Steric) Interactions

These are non-bonded, weak interactions between atoms arising from induced dipoles. In empirical scoring functions, the van der Waals term often models attractive interactions between complementary surfaces of the receptor and ligand, as well as repulsive forces when atoms are in close proximity. Attractive van der Waals interactions occur at optimal distances. Repulsive van der Waals interactions occur when atoms are too close, and their electron clouds overlap.

$$E_{VdW} = \sum \frac{A}{r^6} - \frac{B}{r^{12}} \tag{4.2}$$

In Eq. (4.2), r represents the distance between atoms, while A and B are constants.

2. Electrostatic Interactions

These interactions arise from the attraction or repulsion between charged atoms in the ligand and the protein, with charged amino acid residues (such as lysine and aspartate) and charged ligand groups playing a significant role.

Coulomb's law is commonly used to demonstrate electrostatic interactions:

$$E_{\text{elec}} = \sum \frac{q_i q_j}{4\pi\varepsilon_0 r_{ij}} \tag{4.3}$$

In Eq. (4.3), i and j represent two atoms, while q_i and q_j denote their respective charges, and ε_0 is the dielectric constant. r_{ij} is the distance between two atoms (i and j).

3. Hydrogen Bonding

Hydrogen bonds are significant contributors to the binding of polar ligands to the receptor. In this scoring function, hydrogen bonding is displayed considering the angle and distance between the hydrogen bond donor and acceptor atoms.

$$E_{\text{H - Bond}} = \Sigma(\text{Energy contribution based on donar acceptor distance and angle}) \quad (4.4)$$

Hydrogen bonds are strongest when the donor-acceptor distance is optimal (usually between 2.5 and 3.5 Å), and the bond angle is close to 180°.

4. Hydrophobic (Desolvation) Interactions

This interaction arises when nonpolar regions of the ligand and receptor converge, decreasing their exposure to water. Hydrophobic interactions are entropically favorable because they reduce the structured water layer around nonpolar surfaces.

Empirical scoring functions often model the hydrophobic effect by assigning energy penalties to exposed hydrophobic surface areas or by rewarding hydrophobic contacts.

$$E_{\text{hydrophobic}} = \Sigma\text{Hydrophobic surface area buried upon binding} \quad (4.5)$$

When a ligand binds to a receptor, it loses some of its freedom of movement. This entropy loss is generally unfavorable to binding. Some empirical scoring functions include a term that penalizes the ligand's loss of entropy upon binding.

5. Rotational and Translational Entropy (Entropy Loss)

When ligands bind to receptors (e.g., proteins, enzymes, DNA, or RNA), the interaction reduces the ligands' degrees of freedom. This entropy loss is generally unfavorable to binding. Some empirical scoring functions include a term that accounts for the entropy penalty associated with ligand binding. This is often done by limiting the ligand's rotatable bonds:

$$E_{\text{entropy}} = C \times (\text{Number of rotable bond}) \quad (4.6)$$

where C is a constant derived from empirical data.

6. Desolvation Penalty

Desolvation refers to the removal of H_2O molecules from both the receptors and ligand binding sites during complex formation. An energy cost is often linked to disrupting favorable bonds between H_2O and polar atoms, particularly at the ligand and receptor active sites. Empirical scoring functions incorporate a penalty for the desolvation of polar or charged groups that are solvent-exposed in the unbound state and become buried upon binding.

$$E_{\text{desolv}} = \Sigma\text{Penalty based on polar atom burial} \tag{4.7}$$

If we consider all the above-discussed terms for the empirical scoring function [67], the overall formula for the empirical scoring function will be:

$$\Delta G_{\text{bind}} = \alpha_1 E_{vdW} + \alpha_2 E_{\text{elec}}$$
$$+\alpha_3 E_{H-\text{bond}} + \alpha_4 E_{\text{hydrophobic}} + \alpha_5 E_{\text{entropy}} + \alpha_6 E_{\text{desolv}} + \dots \tag{4.8}$$

In Eq. (4.8), α_1, α_2, and α_3 are empirically fitted coefficients for each term, and the individual energy terms correspond to van der Waals interactions (E_{vdW}), electrostatic interactions (E_{elec}), hydrogen bonding ($E_{\text{H-bond}}$), hydrophobic effects ($E_{\text{hydrophobic}}$), entropy loss (E_{entropy}), and desolvation penalties (E_{desolv}) are examples of Empirical Scoring Functions. Common examples of empirical scoring functions include ChemScore (a widely used scoring function in structure-based drug design) and its components, which include hydrogen-bonding, van der Waals, metal-coordination, and desolvation terms.

4.2.4.1.2 *Knowledge-based scoring function*

This scoring function uses data from experimentally determined atomic structures. They analyze crystal structures of receptor-ligand complexes to see how often different atoms come into contact within the interaction range. The idea is that stronger interactions are more common. The total binding score is calculated by evaluating attractive and repulsive interactions between every ligand and receptor atom within a specified distance.

The pairwise potential $w_{ij}(r)$, which is dependent on the distance between the receptor and the ligand and is given by Eq. (4.9):

$$w_{ij}(r) = -K_b T \ln \frac{\rho_{ij(r)}}{\rho^*_{ij}(r)} S \tag{4.9}$$

The term S represents the pairwise statistical potentials between the receptor and the ligand, expressed as:

$$S = \sum_{i \in \text{ligand},\ j \in \text{receptor}} \sum w_{ij} \tag{4.10}$$

In Eq. (4.10), the Boltzmann constant is represented by K_B. The system's absolute temperature is T. $\rho(r)$ denotes the density of receptor-ligand atoms at a distance r, while $\rho^*(r)$ is the pair density in the absence of interatomic interactions between the receptor and ligand. One advantage of knowledge-based scoring functions is that they implicitly consider solvation and entropic contributions [68]. However, certain knowledge-based scoring functions explicitly account for solvation and entropy effects [68].

Common examples of knowledge-based scoring functions include (i) potential of mean force (PMF) discovered by Muegge and co-workers, one of the earliest deriving atom-atom potentials by analyzing interaction frequencies between atom pairs at various distances in known macromolecules-ligand complexes [69, 70], (ii) DrugScore (This scoring function derives statistical potentials from many macromolecules-ligand complexes, covering hydrogen bonding, van der Waals, and electrostatic interactions. It is often used in drug discovery to score potential candidates.) [71–73], (iii) statistically optimized atomic potential (SOAP) (uses statistical potentials from known complexes to predict binding affinities, designed to be transferable across different macromolecules and ligands), and (iv) ITScore (Incorporates statistical potentials for various interaction types, performing well in virtual screening. It uses distance-dependent potentials based on observed contact frequencies in the Protein Data Bank (PDB) [74, 75].

4.2.5 Docking Software

Kuntz and his team pioneered the creation of the first docking program in the mid-1980s [19]. Since then, there has been a continuous push to advance docking computations. For the last 30 years, various docking programs have emerged. Table 4.2 outlines the essential features, including supported platforms, licensing details, algorithms, and scoring systems, of current docking tools and web servers.

Table 4.2 Properties of currently available molecular docking tools

Different tools	Name of designer/ company	Licensing terms	Approach for docking	Scoring function	Ref.
AutoDock	D. S. Goodsell and A. J. Olson. The Scripps Research Institute	Free for academic Use	Lamarckian genetic algorithm, simulated annealing genetic algorithm	Semi-empirical	[76–78]
Dock	I. Kuntz University of California, San Francisco	Free for academic use	Fragment-based method	Chem Score, GB/SA solvation scoring, and other	[19]
AutoDock Vina	Dr. Oleg Trott. The Scripps Research Institute	Free for academic Use	Monte Carlo Sampling	Empirical/ Knowledge-Based	[79, 80]
FlexX	T. Lengauer and M. Rarey Bio SolveIT	Commercial-Free evaluation	Fragment-based	Empirical	[81, 82]
Glide	Schrödinger Inc.	Commercial	Monte Carlo sampling	Empirical	[83, 84]
GOLD	Cambridge Crystallographic Data Centre	Commercial-free evaluation (2 months)	Genetic algorithm	Physics-based (Gold Score), Empirical (ChemScore, ChemPLP): and Knowledge-based (ASP)	[85, 86]
Ligand Fit	Accelrys Inc.	Commercial	Monte Carlo Sampling	Lig Score, PLP, PMF	[44, 87]
Z-Dock	Zhang lab. University of Massachusetts Medical School	Free for academic use	Fast Fourier Transform (FFT) algorithm	Physics-based	[88]

Different tools	Name of designer/ company	Licensing terms	Approach for docking	Scoring function	Ref.
Swiss Dock	Aurélien Grosdidier, Vincent Zoete, and Olivier Michielin, Swiss Institute of Bioinformatics	Free for academic use	CHARMM force field	Physics-based	[89, 90]
Achilles	Catholic University of Murcia, South East Spain	Free for academic use	Monte Carlo sampling	Physics-based	[91]
Arguslab	Mark Thompson, Medio System Inc., Seattle, USA	Free for academic use	Genetic algorithm	Empirical	[92, 93]

4.3 Practical Application of Molecular Docking

4.3.1 Preparation of the Macromolecules and Ligands

Prior to initiating docking studies, it is crucial to identify the most suitable receptor and ligand structures [94]. Numerous databases exist that offer experimentally determined 3D structures of target receptors, such as the RCSB Protein Data Bank (RCSB PDB), Uniprot, and the Therapeutic Target Database (TTD). Following the download of macromolecules (proteins or DNA) in PDB format from the website, it is essential to prepare them for docking studies to address the absence of hydrogen atoms. In recent times, numerous tools are available for the preliminary preparation of macromolecules, including Open Babel [95], Chimera, and AutoDock tools [61] (Figure 4.2). Different sources for downloading crystallographic coordinates of ligands include DrugBank [96], PubChem [97], ZINC [98], ChEMBL [99], and ChemSpider [100]. To achieve the ligand's minimized energy, the downloaded structure was further optimized using Gaussian software (Gaussian 09W). In computational chemistry, Gaussian is a popular tool for doing many kinds of quantum mechanical calculations on molecules. It offers tools for vibrational frequency analysis, energy calculations, molecular geometry optimization, and much more. Using Gaussian software and Density Functional Theory (DFT), the downloaded structure was optimized using the 3–21G basis set and the B3LYP functional. This configuration was selected to balance accuracy and computational cost while achieving a reliable, optimized geometry for the ligand. Using the 3–21G basis set and the B3LYP functional enables effective optimization; however, there are accurate limitations compared to larger basis sets. The choice depends on how the author optimizes the ligand structure. If the lower basis set does not provide the proper orientation or optimized geometry, the author can use a higher basis set like 6–311G. The optimized ligand structure elucidates its stable geometry and can serve as a foundation for subsequent investigations of its interactions.

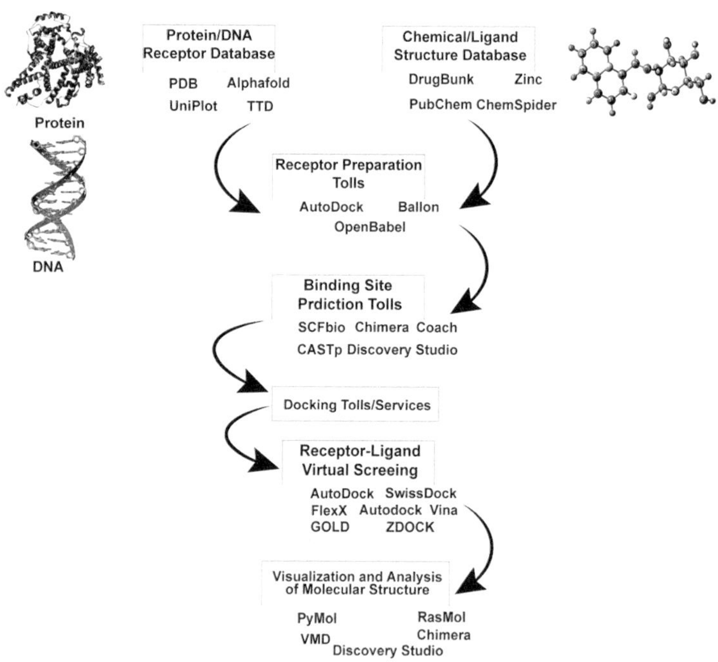

Figure 4.2 Flowchart depicting the steps for the molecular docking process and different software that can be used during the whole docking process.

Guiding docking studies with preliminary information on the binding site is crucial. Without this, blind docking occurs due to the lack of binding site constraints, making it more challenging to identify the correct binding poses in a large ligand search space.

When a specific binding site is known for macromolecules, blind docking is unnecessary. Instead, the specific site can be chosen before starting the docking process, allowing the ligand to bind directly to the predetermined site. Nowadays, different types of algorithms are available to predict the specific binding site of the macromolecules, such as SurfNet [101], GRID [102], SCFbio [103], COACH [104], DeepSite [105], CASTp [106], and PUResNet [107] (Figure 4.2). Docking capabilities vary among different software, so it is essential to choose the right and appropriate algorithm before starting the docking process.

4.3.2 Protein–Ligand Docking

In living organisms, proteins belong to a key group of molecules involved in various biological functions and are essential for sustaining life. Their functions primarily depend on interactions with various groups of bioactive molecules within our body. The association of therapeutic compounds and carrier proteins has sparked considerable attention in medicinal chemistry worldwide. The interaction of proteins and ligands is governed by various forces, including ionic and non-covalent interactions such as H-bonding, van der Waals forces, and hydrophobic interactions. In this section, we examine the molecular interactions between selected proteins: Human Serum Albumin (HSA), Hen Egg White Lysozyme (HEWL), and Bovine Hemoglobin (BHb) with their corresponding ligands. Analyzing the results of protein–ligand docking to determine the ligand's position within the protein's interaction site and to assess the binding energy and forces involved in the interaction.

HSA is the predominant protein within blood plasma, functioning as the main transporter of various substances. The first crystallographic structure of HSA, described by Carter et al. in 1989 [108], has a heart-shaped appearance and dimensions of approximately 80 × 80 × 30 Å. Figure 4.3a depicts the crystal structure of HSA. It is composed of a single polypeptide chain containing 585 amino acids, structured into three homologous domains: domain I (residue 1–195), II (residue 196–383), and III (residue 384–585). Each of these domains is subdivided into subdomains A and B, consisting of six and four α-helices, respectively. Sudlow's site I and site II, which are hydrophobic pockets found in subdomains IIA and

IIIA (Figure 4.3a) [109], are responsible for binding small aromatic and heterocyclic molecules.

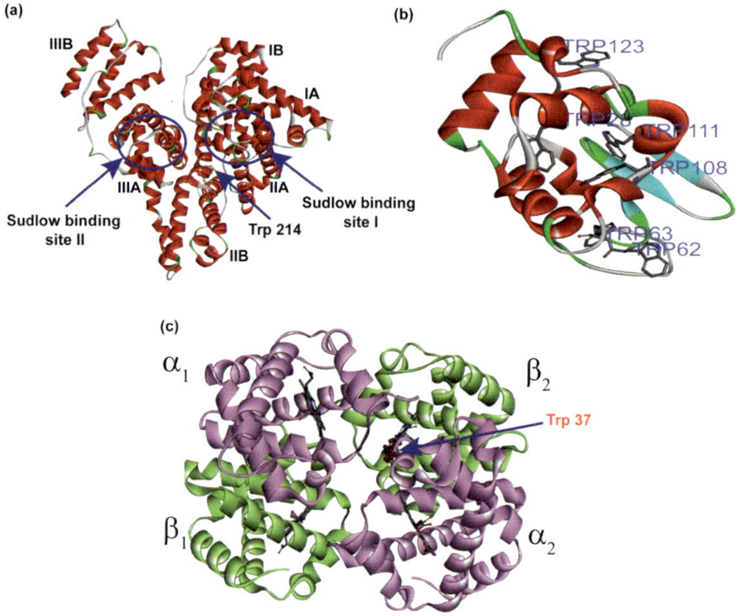

Figure 4.3 Crystal structure of **(a)** HSA (PDB ID: 1AO6) showing Sudlow binding site I and II, **(b)** HEWL (PDB ID: 6LYZ) illustrates all six Trp residues: Trp28, Trp62, Trp63, Trp108, Trp111, and Trp123, and **(c)** BHb (PDB ID: 1G09) showing the dominant fluorophore Trp37.

Lysozyme, a small and single-chain protein, possesses a molecular weight of ~14.3 kDa. The primary sequence of lysozyme was discovered in 1963, followed by David C. Phillips and coworkers determining its 3D structure at 2 Å resolution in 1965 [110]. This protein has 129 amino acid residues and contains four disulfide bonds. As depicted in Figure 4.3b, its active site is a deep groove that splits the protein into two distinct domains, linked by an α-helix (residues 89–99). One of these domains is primarily composed of β-sheet structures (residues 40–85), while the other domain is formed by the N-terminal and C-terminal segments (spanning residues 1–39 and 101–129), which are more helical [111]. Lysozyme's cleft serves as a binding site for small molecules. In its crystal structure, residues Trp62, Trp63, and Trp108 are located near the substrate-binding region and contribute significantly to ligand or substrate binding [110]. HEWL is commonly employed as a model protein for investigating protein–ligand interactions owing to its ~60% sequence similarity with human lysozyme [111].

Hemoglobin (Hb) is a protein that contains iron, predominantly present in red blood cells, serving as a vital respiratory protein. Its crystal structure was first determined by Perutz in the 1960s (Perutz et al., 1960). Hemoglobin has a tetrameric structure with a molecular weight of 64.5 kDa, consisting of two identical $\alpha\beta$ dimers. Each α-chain consists of 141 amino acid residues arranged into seven α-helices, while each β-chain consists of 146 amino acids and contains 8 α-helices. These helices are folded into compact globules that form heterodimers and subsequently assemble into tetramers, as shown in Figure 4.3c. These dimeric conformations are commonly seen as $\alpha\beta$ subunits, which include α-Trp14, β-Trp15, and β-Trp37 serving as the main fluorophore in hemoglobin [113]. Each polypeptide chain contains a heme prosthetic group near the exterior of the protein surface, enabling reversible interactions with oxygen and small molecules.

Phytochemicals are bioactive compounds renowned for their various therapeutic properties, including anti-inflammatory, anti-oxidant, and anti-cancer effects. Their structural diversity and biological activities have sparked increasing interest in drug discovery and development. Their ability to interact with multiple molecular targets positions them as promising candidates for computational studies like molecular docking, which enable the assessment of their binding affinities and interactions with specific proteins. This provides valuable insights into their potential as lead compounds for developing new therapeutic agents. In this section, we focus on flavonoids and coumarins, which were selected for docking analysis to assess their interactions. Flavonoids are natural products that belong to polyphenolic secondary metabolites found in vegetables, fruits, flowers, honey, nuts, and certain beverages. The core structure of flavonoids is made up of three cyclic rings: A, B, and C. They are classified based on the position of the C ring relative to the oxygen atom, as shown in Figure 4.4a. They are associated with a broad range of health-enhancing effects. They are extensively used in pharmaceutical, medical, nutraceutical, and cosmetic products because of their ability to regulate key cellular enzyme functions

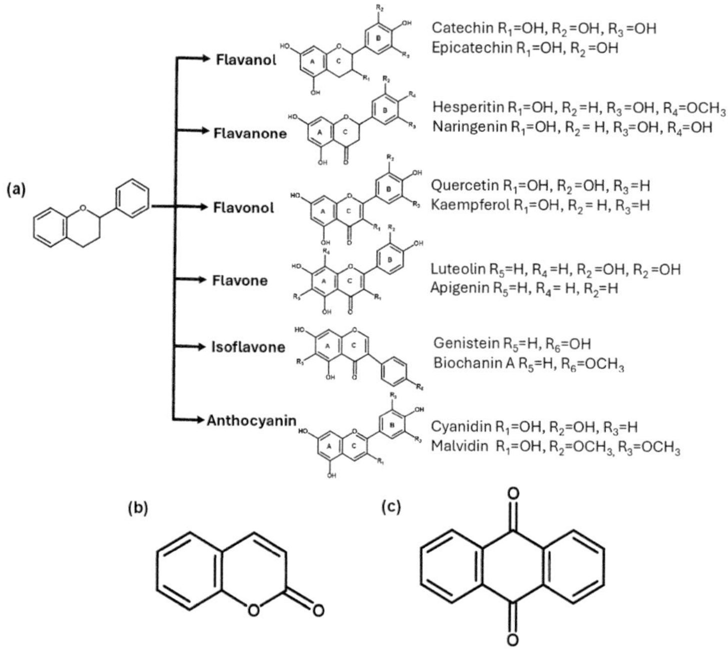

Figure 4.4 **(a)** The basic chemical structure of flavonoids includes different classes of flavonoids, the chemical structure of **(b)** coumarin, and **(c)** anthraquinone.

and their anti-mutagenic, anti-carcinogenic, anti-oxidative, and anti-inflammatory properties [114–118]. Coumarins are polyphenolic compounds that occur naturally and are classified as derivatives of 1,2-benzopyrones. Their structure features a benzene ring that is fused with a pyrone ring (Figure 4.4b). This fused ring system serves as the core skeleton of coumarins, which can undergo various substitutions at different positions on both rings. Depending on their structural variations, coumarin derivatives exhibit a remarkable range of pharmacological and therapeutic properties, including anti-inflammatory, anti-cancer, anti-coagulant, anti-microbial, anti-neurodegenerative, and anti-oxidant properties [119–122]. Anthraquinones (9,10-dioxoanthracenes) are widely occurring indigenous phenolic compounds that appear in the leaves and roots of senna, rhubarb, and aloe latex. Their structure comprises two aromatic rings linked by two carbonyl groups, forming a planar aromatic structure (Figure 4.4c). They are classified as secondary metabolites and have been significant in medicine for more than 4000 years. Many anthraquinone derivatives like mitoxantrone, daunomycin, and doxorubicin are known for their robust chemotherapeutic effects, primarily owing to their capacity to interact with DNA at the molecular level [123, 124]. Numerous reports suggest they exhibit antiproliferative effects with various cancer cell lines [125–127].

4.3.2.1 Docking studies of HSA with flavonoids

Docking analysis involves two main steps. First, it predicts the conformation and orientation of ligands within the active site of the protein, known as docked poses. Second, it evaluates the binding affinity. Figure 4.5 depicts the lowest energy docked poses for HSA and HEWL with few flavonoids and BHb with coumarin derivatives generated through molecular docking conducted with AutoDock 4.2 and AutoDock Vina. Meanwhile, Table 4.3 highlights the binding affinities, key interacting residues, and $\Delta ASA_{\text{total}}$ (macromolecule) for these proteins interacting with the respective molecules.

Molecular docking studies of kaempferol (KMP) with HSA demonstrated that KMP enters the binding pocket of subdomain IIA (Figure 4.5a). The binding energy was measured at −31.003 kJ mol^{-1}, with KMP positioned ~6 Å distant from the Trp214 fluorophore of HSA. It forms hydrogen bonds (within ~3.5 Å) with Tyr150, Ser192, Arg222, and Arg257. The A ring of KMP is situated deep within the hydrophobic region, while the C ring occupies an intermediate position, interacting efficiently with Lys195, Leu238, and Ala291 through hydrophobic interactions. Additionally, π–cation interactions are observed with residues, including Lys199, Lys195, and His242 [128].

Docking analysis of HSA and luteolin showed that luteolin can bind near subdomain II A and III A, with a determined binding energy of −36.82 kJ mol^{-1} and a binding constant of 25.69×10^{5} L mol^{-1}. In subdomain II A, luteolin forms hydrogen bonds with Ser192, Arg257, and Arg222, while in subdomain III A, Asp451 establishes H-bonds with the 5-OH group of luteolin. Additionally, Lys436 and Cys448 further stabilize the binding through van der Waals forces [129].

6-Hydroxyflavone (6-HF) was found to interact within subdomains II A and III A of HSA. The binding energy for the

HSA-6-HF complex was –36.819 kJ mol^{-1} in subdomain II A and –29.706 kJ mol^{-1} in subdomain IIIA, with a binding constant of 25.69 × 10^4 L mol^{-1}. In Subdomain II A, Arg257 and Tyr150 surrounded the 6-HF, forming hydrogen bonds. At the same time, the majority of interactions occurred with hydrophobic residues, implying that hydrophobic forces were the dominant forces behind the interaction [130].

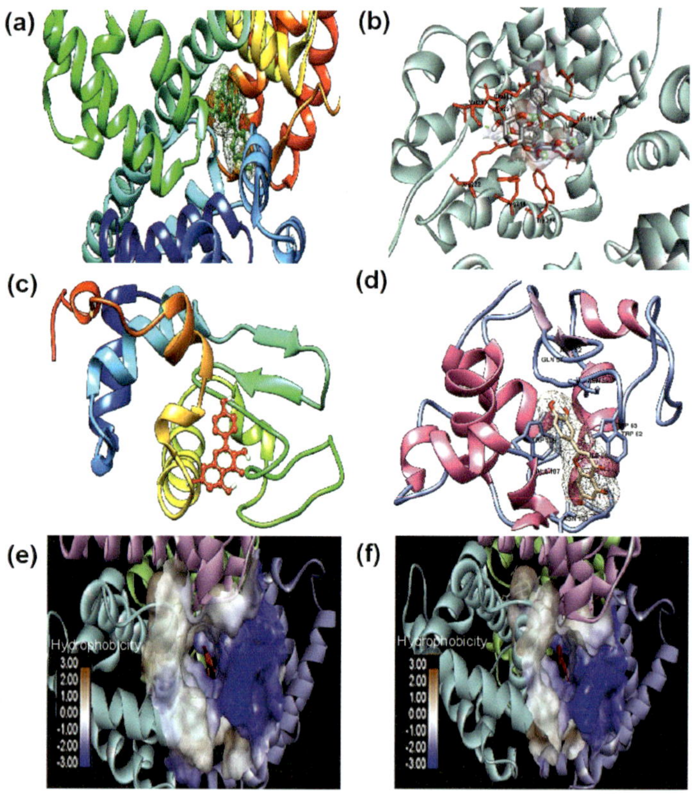

Figure 4.5 (a) Minimum energy docked pose of HSA–KMP complex (adapted with modification from [128]), (b) Surface representation of different types of interactions in subdomain IIA for the binding of HSA with naringenin (adapted with modification from [131]), (c) Represents the docked pose of HEWL with KMP (adapted with modification from [128]), (d) Interaction of luteolin in the binding site of HEWL (adapted with modification from [132]), (e) and (f) Visualization of the hydrophobic surface of 7-hydroxycoumarin (7-HC) and 4-Methyl-7-hydroxycoumarin (4-Me-7-HC), respectively in the binding pocket of BHb (adapted with modification from [133]).

Naringin demonstrated a greater affinity for subdomain II A of HSA with a calculated binding energy of −26.861 kJ mol^{-1}. Due to its bulkier size, naringin is associated with Sudlow's binding site I. Within the binding site, naringin formed hydrogen bonds with Ala291, Lys195, Arg222, Val293, and Glu292. Hydrophobic interaction was considered the primary driving force for naringin's binding to HSA [131].

Molecular docking studies of naringenin (NRG) with HSA revealed that naringenin binds to both subdomain IIA and IIIA regions of HSA. The binding energies for the HSA-NRG interactions were −29.329 kJ mol^{-1} in subdomain IIA and −24.811 kJ mol^{-1} in subdomain IIIA. The benzoyl moiety interacts with polar residues at the opening of the binding cavity. Meanwhile, the cinnamoyl moiety of naringenin penetrates deeply into the hydrophobic pocket of subdomain IIA of HSA. Naringenin forms a total of four hydrogen bonds with Arg222, Arg257, and Gln196 in subdomain IIA and three hydrogen bonds with Tyr411, Arg485, and Ser489 in subdomain IIIA (Figure 4.5b) [131].

4.3.2.2 Docking studies of Lysozyme (HEWL) with flavonoids and Coumarins

The HEWL matrix revealed that the active site of KMP is positioned within the cleft or orthosteric site adjacent to the catalytic region (Figure 4.5c). It was observed that the B and C rings of the KMP molecule are significantly embedded in the hydrophobic domain, whereas the solvent-exposed region near ring A facilitates electrostatic interaction with Asp101 and Asn103. Hydrogen bonds with residues Gln57, Trp108, Ile98, and Asn103, alongside hydrophobic interactions with Trp62, Trp63 (through π–π interactions), and Ala107 (through π–alkyl interactions), stabilize the HEWL-KMP complex [128].

The binding site of luteolin with HEWL, as shown by docking studies, is flanked by residues Asp52, Gln57, Trp63, Trp62, Ile98, Asn103, Trp108, Asn59, and Ala107 (Figure 4.5d). Hydrophobic forces appear to have a significant influence on the complexation process because of the presence of hydrophobic residues such as Ala,

Trp, and Ile close to luteolin's interaction site. Gln57 and Ile98 also form hydrogen bonds with luteolin [132]. HEWL serves as a well-established model for studying amyloid fibril formation [134, 135], with a critical 9-amino acid segment from residues 54–62, referred to as the K-peptide. This segment is positioned at the junction between the protein's α-helix and β-domain, playing a crucial role in amyloid fibrillation [136]. Among these residues, Trp62 is particularly significant for fibril assembly. Docking studies reveal that luteolin binds to a site flanked by residues such as Trp62, Asp52, Ile98, Gln57, Asn103, Ala107, Asn59, Trp108, and Trp63 residues associated with either the K-peptide or the amyloid-resistant (32–108) region [137], respectively. Thus, the binding of luteolin near the Trp62 residue, via hydrophobic interactions and other key aggregation residues, affects HEWL fibrillation.

Docking analysis of 6-HF demonstrated that it interacts with the binding site of HEWL, with distances of 3.50 Å, 4.00 Å, and 4.90 Å from Trp63, Trp62, and Trp108, respectively. These residues are crucial co-operators in the complexation process. Furthermore, 6-HF forms hydrogen bonds with Gln57 at 2.52 Å and engages in hydrophobic interactions with residues such as Trp108, Ala107, Trp63, and Trp62. Therefore, both hydrophobic interactions and hydrogen bonding are crucial for the binding of 6-HF to HEWL [128].

The docking evaluation of HEWL interacting with some of the coumarin derivatives performed using AutoDock 4.0 suggests that 7-HC and 4-Me-7-HC bind near a site surrounded by Ala107, Ile98, Asp52, Leu56, Trp62, Gln57, Asn46, Ile58, Asn59, Trp63, and Trp108. The hydrophobic surface of the binding pocket influences interaction. For esculin, the catalytic pocket involved residues like Leu56, Ser50, Ala107, Trp108, Trp63, Gln57, Asn103, Asp52, Asp101, Asn46, Ile58, Trp62, Asn59, and Ile98, while esculetin interacted with Ser50, Asn59, Leu56, Trp108, Gln57, Asp52, Ala107, Ile58, Asn46, Trp63, and Ile98. The results imply that these coumarins interact in a hydrophobic region, particularly with 7-HC interacting with Ala107 (π–alkyl) and Ile98 (π–alkyl), 4-Me-7-HC interacting with Ala107 (π–alkyl), Ile98 (π–alkyl), and Trp63 (π–alkyl), and esculetin interacting with Trp62, Trp63, and Ala107, and Ile98. Additionally, these coumarin derivatives are partially exposed to the solvent, facilitating hydrogen-bond formation. Specifically, Trp108, Asp52, and Asn59 residues form hydrogen bonds with 7-HC, 4-Me-7-HC, esculin, and esculetin, contributing to complex stability. Other types of interactions also contributed to the formation of the complex. Van der Waals interactions occur between 7-HC and residues Leu56, Trp63, Gln57, Leu56, Asp52, Gln57, and Asn46; between 4-Me-7-HC and Leu56, Gln57, Asn46, and Asp52; between esculin and Asn46, Ile98, Ser50, Asn103, Gln57, Asp101, and Trp108 residues; and between esculetin and Asn46, Leu56, Gln57, Ile58, and Trp63. Thus, it is essential to consider the presence of different interaction types in the HEWL-7-HC, HEWL-4-Me-7-HC, HEWL-esculin, and HEWL-esculetin complexes [138, 139].

4.3.2.3 BHb with coumarins

Docking studies of BHb with coumarin derivatives 7-HC, 4-Me-7-HC, esculin, and esculetin performed using AutoDock Vina demonstrated that these derivatives were positioned within the binding pocket formed by α_1, α_2, and β_2 subunits of BHb. The theoretical binding affinity values (ΔG) for each complex were calculated to be −28.45, −26.77, −32.635, and −28.451 kJ mol^{-1} for BHb-7-HC complex, BHb-4-Me-7-HC complex, BHb-esculin complex, and BHb-esculetin complex. Hydrogen bonds were formed between 7-HC and α_1-Thr137, α_2-Lys127, α_1-Tyr140, and α_1-Arg41, while 4-Me-7-HC formed hydrogen bonds with α_1-Thr137 and α_1-Arg141. Esculin established hydrogen bonds with β_2-Val34, α_1-Ser138, and α_2-Lys99, and esculetin interacted with α_1-Tyr140, α_1-Arg141, and α_2-Lys127. A significant portion of these derivatives was located in a hydrophobic region inside the BHb binding cavity (Figure 4.5e,f), suggesting hydrophobic interactions enhanced the stability of the complexes. In addition, the dominant fluorophore β_2-Trp37 was found to interact with the coumarin derivatives (7-HC, 4-Me-7-HC, esculin, and esculetin) at distances of 3.82 Å, 4.20 Å, 3.642 Å, and 3.111 Å, respectively [133, 140].

Table 4.3 Binding energies [ΔG (kJ mol^{-1})] between HSA/HEWL/BHb with few flavonoids and coumarin derivatives

HSA					
Flavonoids	**Binding site**	**Binding affinity (kJ mol^{-1})**	**Interacting amino acid residue**	**ΔASA_{total} Macromolecule (Å^2)**	**Ref.**
Luteolin	Subdomains II and III A	–36.82	Arg222, Ser192, Arg257, Asp451	189.7 (Subdomain IIA). 170.6 (Subdomain IIIA)	[129]
6-HF	Subdomains II and III A	–36.82	Arg257, Tyr150	159.6 (Subdomain IIA) 28,439.5 (Subdomain IIIA)	[130]
KMP	Subdomains II and III A	–31.00	Tyr150, Ser192, Arg222, Arg257, Trp214, Lys195, Leu238, and Ala291	153.8	[128]
Naringin	Subdomain II A	–26.86	Lys195, Arg222, Ala291, Glu292, and Val293	-	[131]
Naringenin	Subdomains II and III A	–29.32	Arg222, Arg257, Gln196, Tyr411, Arg485, and Ser489	-	[131]
HEWL					
Flavonoids	Luteolin	–31.05	Asp52, Asn59, Trp62, Trp63, Trp108, Ile98, Gln57, Ala107, and Asn103	49.9 Å2	[132]

HSA					
Flavonoids	**Binding site**	**Binding affinity (kJ mol^{-1})**	**Interacting amino acid residue**	**ΔASA_{total} Macromolecule (Å^2)**	**Ref.**
Flavonoids	6-HF	−25.24	Asp52, Trp62, Asn59, Trp63, Asp101, Gln57, Ile98, Asn103, Ala107, Gly102, and Trp108	39.2 Å	[141]
	KMP	−27.33	Gln57, Trp62, Ile98, Trp63, Trp108, Asp101, Asn103, Asn59, and Ala107	39.7 Å^2	[128]
Coumarins	7-HC	−22.17	Ala107, Leu56, Gln57, Asn46, Asn59, Trp108, Trp62, Trp63, Asp52, Ile58, and Ile98	145.86 Å^2	[138]
	4-methyl-7-HC	−23.01	Ala107, Leu56, Asp52, Ile58, Asn59, Trp62, Asn46, Gln57, Trp63, Trp108, and Ile98	161.08 Å^2	[138]
	Esculin	−30.12	Leu56, Ser50, Ala107, Trp108, Trp63, Gln57, Asn103, Asp52, Asp101, Asn46, Ile58, Trp62, Asn59, and Ile98	276.27 Å^2	[139]
	Esculetin	−22.59	Ser50, Leu56, Gln57, Asn59, Ala107, Ile58, Asn46, Asp52, Trp63, Trp108, and Ile98	276.27 Å^2	[139]

HSA					
Flavonoids	**Binding site**	**Binding affinity ($kJ\ mol^{-1}$)**	**Interacting amino acid residue**	**ΔASA_{total} Macromolecule ($Å^2$)**	**Ref.**
BHb					
Coumarins	7-HC	-28.45	α_1-Arg41, α_1-Thr137, α_1-Tyr140, α_2-Lys127	190.68 $Å^2$	[133]
	4-methyl-7-HC	-26.77	α_1-Arg141, α_1-Thr137	196.37 $Å^2$	[133]
	Esculin	-32.635	α_1-Ser138, β_2-Val34, α_2-Lys99	351.61 $Å^2$	[140]
	Esculetin	-28.451	α_1-Tyr140, α_1-Arg141, α_2-Lys127	203.72 $Å^2$	[140]

4.3.2.4 Change in accessible surface area

The changes in the accessible surface area (ΔASA) of protein residues during their binding with small ligand molecules provide important information about the packing efficiency of the amino acid residues and the ligands' binding affinity. The ΔASA values are obtained from the following equation:

$$\Delta ASA = ASA_{\text{free}} - ASA_{\text{complex}} \tag{4.11}$$

where ASA_{free} and ASA_{complex} refer to the ASA of the native protein and the protein bound to the ligand, respectively. The accessible surface area of the proteins in their native and complex forms has been calculated using a Linux-based program, NACCESS.

The total accessible surface area (ASA_{total}) of native HSA was 28600.6 $Å^2$. Following interaction with luteolin and 6-HF, the $\Delta ASA_{\text{total}}$ values were reduced to 189.7 $Å^2$ and 158 $Å^2$ for the Subdomain IIA complex, 170.2 Å2 and 161.1 $Å^2$ for the Subdomain IIIA complex, respectively [128, 130]. For KMP, the $\Delta ASA_{\text{total}}$ reduced to 156.8 $Å^2$ [128]. Native HEWL exhibited a $\Delta ASA_{\text{total}}$ of 6676.00 $Å^2$, and upon binding with luteolin, 6-HF, and KMP, the $\Delta ASA_{\text{total}}$ decreased by 49.9 $Å^2$, 39.2 $Å^2$, and 39.7 $Å^2$ [128, 132, 141], and when bound to 7-HC, 4-Me-7-HC, esculin, and esculetin, the $\text{ASA}_{\text{total}}$ reduced by 145.86 $Å^2$, 161.08 $Å^2$, 276.27 $Å^2$, and 276.27 $Å^2$, respectively [138, 139].

The total ASA of 23470.61 Å^2 was calculated for native BHb, and the reduction in ΔASA_{total} by 190.68 Å^2 (for BHb-7-HC), 196.37 Å^2 (for BHb-4-Me-7-HC), 351.61 Å^2 (for BHb-esculin), and 203.72 Å^2 (for BHb-esculetin) [133, 140]. Thus, the reduction in ΔASA_{total} indicated that each protein forms specific biochemical interactions with different flavonoids and coumarin derivatives.

4.3.2.5 Docking of a SWCNT with a protein

Docking analysis of HSA with oxidized SWCNTs using AutoDock 4.0 identified two potential binding sites: binding site 1 and binding site 2 on HSA. The amino acid residues across both binding sites are Arg160, Glu167, Ser435, and Tyr452. Electrostatic interaction between oxidized groups on SWCNTs and the positively charged residue Arg160 on HSA appears to be essential for stabilizing this binding. Furthermore, the Tyr452 residue facilitates π–π stacking interaction with SWCNTs, further stabilizing the binding [142]. The minimum energy docking analysis of HO-SWCNTs with Hb shows that it interacts with 20 amino acids from α_1 and β_2 subunits (Figure 4.6). The specific residues involved in binding are α_1His45,

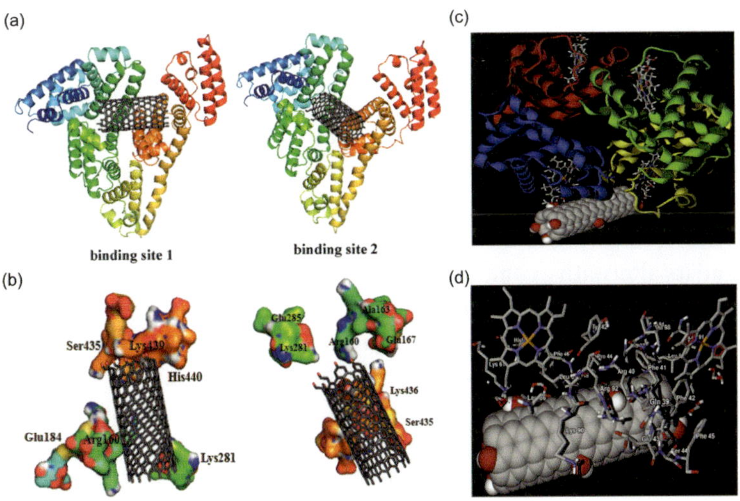

Figure 4.6 (a) Molecular docking illustrating the two potential binding sites of SWCNTs (Site 1 and Site 2) on HSA, (b) Amino acid residues in close proximity to SWCNTs contributing to the stabilization of both binding sites (Reproduced from [142]), (c) lowest energy docked pose of HO-SWCNTs with Hb, and (d) Visualization of the amino acid residues in Hb interacting with HO-SWCNTs (Reproduced from [143]).

α_1Leu86, α_1Arg92, α_1Lys61, α_1His58, α_1Phe46, α_1Leu91, α_1Arg40, α_1Tyr42, α_1Lys90, β_2 Arg40, β_2 Leu96, β_2 Gln39, β_2 His97, β_2 Glu43, β_2 Phe41, β_2 Pro44, β_2 Phe42, β_2 Phe45. The calculated binding energy was 73.30 kJ mol^{-1}. Hydrophobic and π–π interactions were observed, along with four hydrogen bonds formed by the OH group of SWCNT and the residues β_2 Arg40, β_2 Ser44, α_1Lys61, and α_1Lys90 at distances of 2.9377 Å, 3.0263 Å, 3.5082 Å, and 3.5914 Å, respectively, stabilizing the complex at the preferred binding site [143].

4.3.2.6 DNA–ligand interaction

DNA is a biological macromolecule composed of phosphate groups, deoxyribose sugars, and a mixture of purine bases, including adenine and guanine, as well as pyrimidine bases like thymine and cytosine (Figure 4.7) [144]. The purines and pyrimidines pair via hydrogen bonds, stabilizing the DNA's double helix through A–T and G–C base pairing. James Watson, Francis Crick, and Maurice Wilkins received the Nobel Prize in 1962 for discovering the double-helix structure [145]. DNA can exist in three double-helical forms: A, B, and Z-form, each characterized by differences in its major and minor grooves. The A-form consists of a deep, narrow major groove and a broad, shallow minor groove, while the Z-form helical shows a flat major groove and a narrow, deep minor groove. The B-form, commonly found in physiological conditions, exhibits a broad and deep major groove along with a narrow, deep minor groove. DNA plays an essential role in numerous cellular functions, including reproduction and protein synthesis. Owing to their essential function in cellular division, DNA and its alternative conformations have become prime targets in drug discovery, particularly for combating cancer and infectious diseases. DNA modulators function by interfering with the DNA duplication process, affecting cell growth, transcription, and ultimately resulting in suppression of gene expression. Metal-containing therapeutic agents are a prevalent category of drugs employed in the treatment of various types of cancer. Following the serendipitous discovery of cisplatin in 1965 [146] and its remarkable anti-cancer efficacy [147], there has been growing research interest in creating new metallo–organic complexes for use as chemotherapeutic agents. Platinum-based compounds are among the most commonly used agents in clinical settings. Nevertheless, clinical limitations are evident, including its restricted applicability, development of resistance, and significant side effects like nephrotoxicity and neurotoxicity [148, 149]. Consequently, researchers have been searching for alternative metal-based drugs that could offer a cost-effective metal complex, reduced side effects, and enhanced therapeutic efficacy. Ruthenium compounds stand out for their unique ability to mimic iron in interactions with biological molecules. This property reduces overall toxicity and shows a higher affinity for cancerous tissues over healthy ones, making them promising candidates for anti-cancer therapies [150–152].

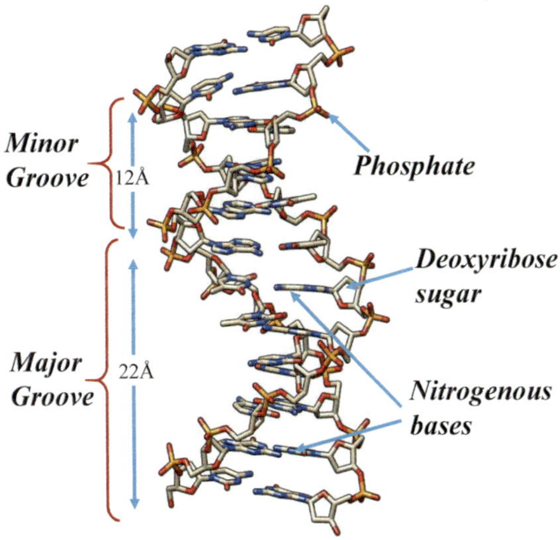

Figure 4.7 Crystal structure of the B-form of DNA (PDB ID: 1BNA) (adapted with modification from [159).

In addition, the development of copper-based metallodrugs is particularly significant due to their biocompatibility. Copper complexes have established themselves as the primary choices in the quest for biologically important metal complexes [153–155]. Recently, the association of metal complexes with DNA has received substantial attention, as it suggests that these complexes may exhibit potential biological activity, which is influenced by their binding mode and affinity for DNA [156–158].

Generally, small molecules interact with DNA in two different ways: (i) by intercalating between base pairs or (ii) through groove recognition. Minor groove binders possess a flat crescent shape or can adopt this configuration, enabling them to fit snugly into the groove with minimal steric hindrance [160]. Regions rich in adenine-thymine (AT) are narrower and slightly more profound than those rich in guanine-cytosine (GC), which allows ligands to fit more effectively. As a result, various minor groove binders have a preference for AT-rich sequences [161–163]. Interaction is believed to be influenced by three primary factors, with van der Waals interactions being essential and supported by hydrogen bonding and electrostatic interactions [164]. The recognition of the minor groove by small molecules is akin to a lock-and-key mechanism, as there is minimal to no observable distortion of DNA post-binding [165]. The insertion and stacking of a planar molecule among DNA base pairs, with the molecule kept perpendicular to the helical axis, is referred to as intercalation. Most classical intercalators, such as proflavine and ethidium bromide, lack sequence selectivity, as their binding to DNA fundamentally relies on π-stacking and reinforcing electrostatic interactions [165, 166]. Additionally, it is widely accepted that intercalation causes modification to the structure of DNA, creating an intercalation gap in successive base pairs. Thus, intercalative binding can be viewed as an induced-fit mechanism [167]. Comprehending the preferred interaction sites of biologically active compounds in nucleic acids is essential for determining their potential as therapeutic agents [168].

4.3.2.7 Docking of DNA with small molecules

This section analyzes docking studies conducted of B-form DNA and several biologically active phytochemicals using AutoDock 4.2. The docking results of DNA with esculetin (a coumarin derivative), aloin, and purpurin (anthraquinones) indicate that each of these molecules binds to the minor groove of DNA (Figure 4.8). It was found that binding energy values for the complexes of DNA with esculetin, aloin, and purpurin were −26.19 kJ mol^{-1}, −29.790 kJ mol^{-1}, and −34.601 kJ mol^{-1} (Table 4.4), respectively. In the esculetin-DNA complex, hydrogen bonds are formed between DG B:16 and DA B:18, while DT A:8, DG A:10, DC A:9, DA B:17, and DT B:19 interact through van der Waals forces. In the aloin-DNA complex, three hydrogen bonds form between aloin and DA B:18, DG A:10, and DC A:12, while DC A:9 and DA B:17 interact through van der Waals forces. DG A:12 in a

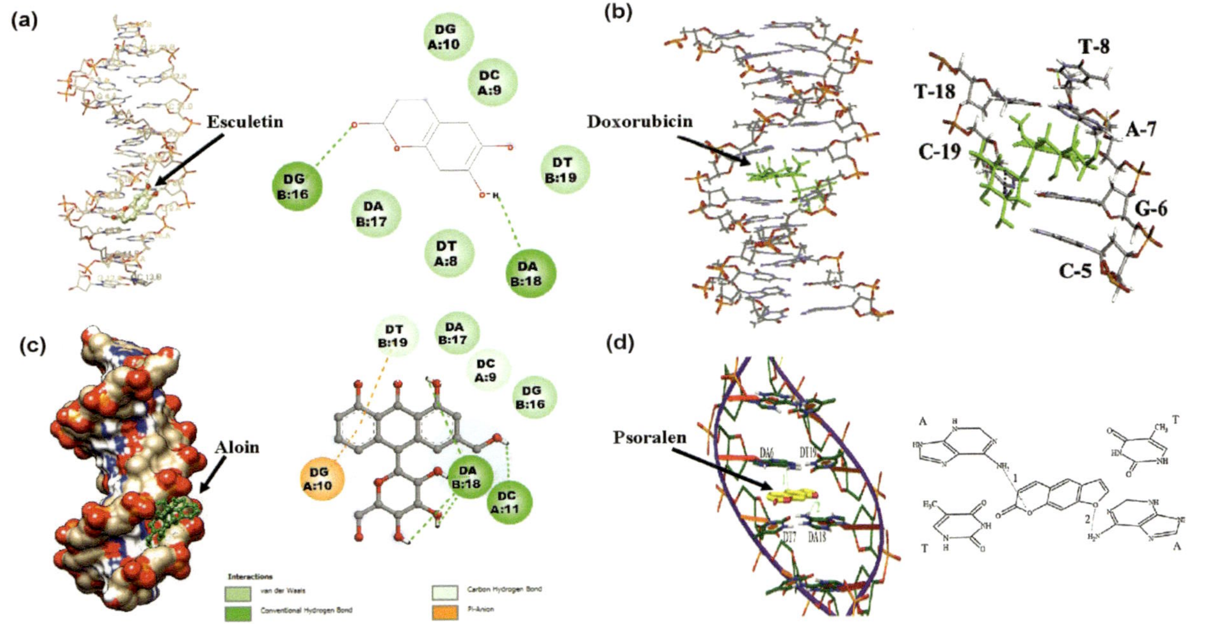

Figure 4.8 (a) Minimum energy docked pose of DNA-esculetin complex along with its 2D plot (adapted with modification from [159]), (b) DNA represented in a stick model, showing DOX intercalated within the DNA, along with the binding sites and corresponding bases (Reproduced from [170]), (c) Structural visualization of the DNA-aloin complex, accompanied by its 2D plot (adapted with modification from [169]), and (d) Probable docking configuration of psoralen binding to DNA, illustrating potential hydrogen bonds (dashed line) between psoralen and the nucleotide bases (Reproduced from [171]).

π–anion interaction and a carbon-hydrogen bond involves DC A:9. In the purpurin-DNA complex, hydrogen bonds are observed between DG B:10, DA B:18, and purpurin, while van der Waals interactions involve DT A:8 and DC A:9, and a carbon-hydrogen bond connects DT B:19 Thus, the complex formation is influenced by a range of binding interactions have been recognized, playing role in the formation of the complex, affecting its notable binding strength [159, 169].

Table 4.4 Value of ΔG, mode of binding, deoxyribonucleotides involved in interaction

Complex	Mode of binding	Deoxyribonucleotides involved in interaction	$\Delta G_{binding}$ (kJ mol^{-1})	Ref.
DNA-esculetin	Groove binding	DG B:16, DG A:10, DT B19, DA B:18, DT A:8, DC A:9, and DA B:17	−26.19	[159]
DNA-aloin	Groove binding	DG A:10, DC A:9, DC A:12, DA B:17, DA B:18, and DG A:12	−29.79	[169]
DNA-purpurin	Groove binding	DA B:18, DT B:19 DT A:8, DG B:10, and DC A:9	−34.60	[169]
DNA-psoralen	Intercalation binding	DA:6, DT:19, DT:7, and DA:18	-	[171]
DNA-doxorubicin	Intercalation binding	DA:7, DC:5, DC:19, DG:6, DT:8, and DT:18	−20.87	[170]
DNA-Ru(II) polypyridyl complexes	Partial intercalation binding in the major groove	DA:5, DT:19, DA:18	−37.18, −31.95, -9.70	[172]
DNA-Cu(II) complex	Intercalation binding	DC:5, DA:19, DA:18	–	[173]
DNA-Pt (II) complex	Minor groove binding	–	−45.31	[174]

Docking studies of doxorubicin (an anthracycline) revealed that this bioactive molecule intercalates into the DNA duplex, surrounded by A-7, T-8, C-19, G-6, C-5, and T-18 (Figure 4.8c). It was found that a hydrogen bond exists between the NH_2 group of doxorubicin and C-19. The free binding energy for the doxorubicin-DNA complex was calculated to be –20.878 kJ mol^{-1} (Table 4.4). Psoralen, with its entire planar structure, intercalates between two adjacent base pairs (DA6–DT19 and DT7–DA18) (Figure 4.8c). Two hydrogen bonds are established between psoralen and the adenine bases, suggesting that hydrogen bonding plays a crucial role in the intercalative binding of the bioactive molecule to DNA [170, 171].

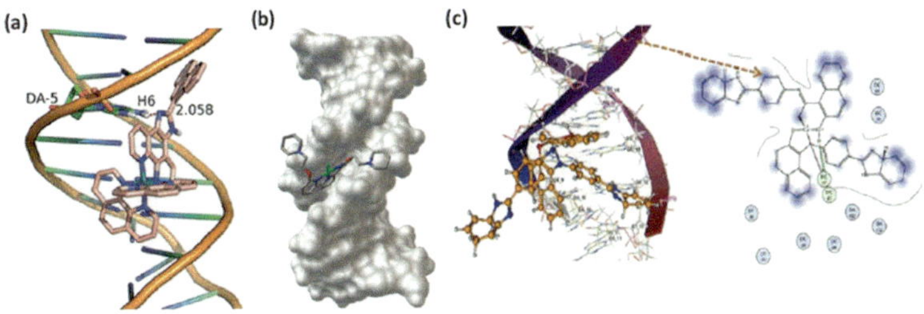

Figure 4.9 (a) Visualization of the docked Ru(II)-DNA polypyridyl complex, displaying hydrogen bonds indicated by red dashed lines bases (Reproduced from [172]), (b) DNA-binding docked pose of the Cu(II) complex bases (Reproduced from [173]), and (c) 3D docked complex of DNA Pt(II) terpyridine complex accompanied by its 2D plot bases (Reproduced from [174]).

Docking investigations of three Ru(II) polypyridyl complexes ([Ru(phen)2(PTCP)](ClO4)2, [Ru(dmp)2(PTCP)](ClO4)2, and [Ru(ttbpy)2(pttcp)](ClO4)2, with DNA, performed using AutoDock, reveal that each of the three complexes partially intercalates within the major groove of DNA. The corresponding relative binding energies for the docked complexes are –37.18, –31.95, and –9.70 kJ mol^{-1}, respectively [172]. The (E)-1-((4-(1H-benzo[d]imidazole-2-yl)phenylimino)methyl)naphthalene-2-ol Cu (II) complex docking results with DNA show that the complex inserts into stacked base pairs, suggesting intercalation binding. Two hydrogen bonds were identified in the complex, involving the phenolic-H and the nitrogen atom of the imidazole ring in the binding configuration. The N (imidazole) of all three metal complex forms hydrogen bonding with DA:5, DT:19, and DA:18 with distances of 2.058 Å, 2.523 Å, and 1.991 Å, respectively [173]. The computational docking results for the platinum(II) terpyridine [Pt(terpyridine)Cl](PF6) complex with DNA suggest that its planar structure fits into the minor groove in a parallel orientation relative to the DNA backbone. The complex's two alkane chains are positioned outside of the groove facing the DNA backbone. The estimated free energy of this binding interaction is -45.31 kcal mol^{-1}. A hydrogen bond of distance 2.10 Å was formed between the positively charged piperidines and the negatively charged phosphate groups of the backbone [174].

4.3.3 MD Simulation

With the help of advanced computer technology, the scientific approach to understanding biological chemistry has evolved, enabling a deeper exploration of cellular matrices' functional roles. This integration empowers researchers to model and simulate complex biomolecular interactions. The primary focus of contemporary biological research is to unravel atomic-level interactions, including protein–ligand, DNA-ligand, and protein–protein interactions.

Nowadays, computational methods for understanding biological systems are divided into two main categories: the MM approach and the MD approach. The MM approach analyzes systems at a static level, while the MD approach emphasizes the dynamic behavior of biomolecules over time. This section will address the dynamic behavior of biomolecules, including proteins and DNA, and all related discussions will be based on MD. Simulation studies.

MD refers to a computational approach employed to simulate the dynamic processes and time-dependent interactions of molecular systems. This approach treats every molecule within the simulation box as flexible, thereby enhancing conformational sampling significantly. MD is a powerful tool for addressing various complex issues spanning from theoretical to biological aspects. The analysis of structural changes in biological systems through MD simulation can be relevant to fields such as structure determination, changes in conformation, enzyme inhibition, protein folding, stability, molecular interactions, and ion/membrane transport mechanisms.

In MD, atomic movements are derived by integrating Newton's equations of motion, utilizing MM force fields to characterize interatomic forces and their corresponding potential energy. The forces present on a system at any given time can be calculated using the equation of motion ($F = ma$), and can also be represented as the negative gradient of the potential energy ($F = -\Delta V(r)$) within force field models. Thus, in a simulation context, the derivative of potential energy is linked to positional changes over time, as shown in the following equation:

$$-\frac{dV}{dr} = m\frac{d^2r(t)}{dt^2} = a(t) \tag{4.12}$$

In this context, V represents the system's potential energy, F represents the force applied to any specific atom at time t, $r(t)$ denotes the position vector of any atom at time t, and m indicates the mass of the atom. MD simulations also reveal cryptic wallet and allosteric-binding sites within receptors. GROMACS software is commonly employed for these simulations, using force fields like AMBER, CHARMM, OPLS, and GROMOS to facilitate them. The resulting trajectories allow docking across multiple receptor conformations. Analyses such as root mean square deviation (RMSD), root mean square fluctuation (RMSF), radius of gyration (Rg), hydrogen-bond formation, and others are performed to assess the structural stability and variability of the receptor over the simulation period [175].

RMSD tracks the usual atomic displacement from a reference shape at the end of the simulation, assessing the stability and time-based simple actions of the receptor-ligand complex. During the simulation, RMSF analysis was used to monitor changes in fluctuations at the level of individual residues. It shows residue-by-residue variations over time, which are essential because any substantial changes in the flexibility of critical functional residues will affect receptor function. Rg gauges the compactness of the receptor or complex, with decreased Rg values signifying a more compact form. Structural changes over the years display protein stability. MD simulations validate molecular docking results with the aid of analyzing ligand binding under physiological conditions, revealing metastable states, water positions, and conformational shifts, further to the opportunity for binding.

The RMSD plot reveals that native HSA and HEWL (Figure 4.10a,b) experience minor fluctuations until 70 ns and 50 ns, respectively, after which they converge and remain stable throughout the rest of the simulation. Conversely, in the presence of KMP, HSA demonstrates similar behavior up to 70 ns, while HEWL displays fluctuations until

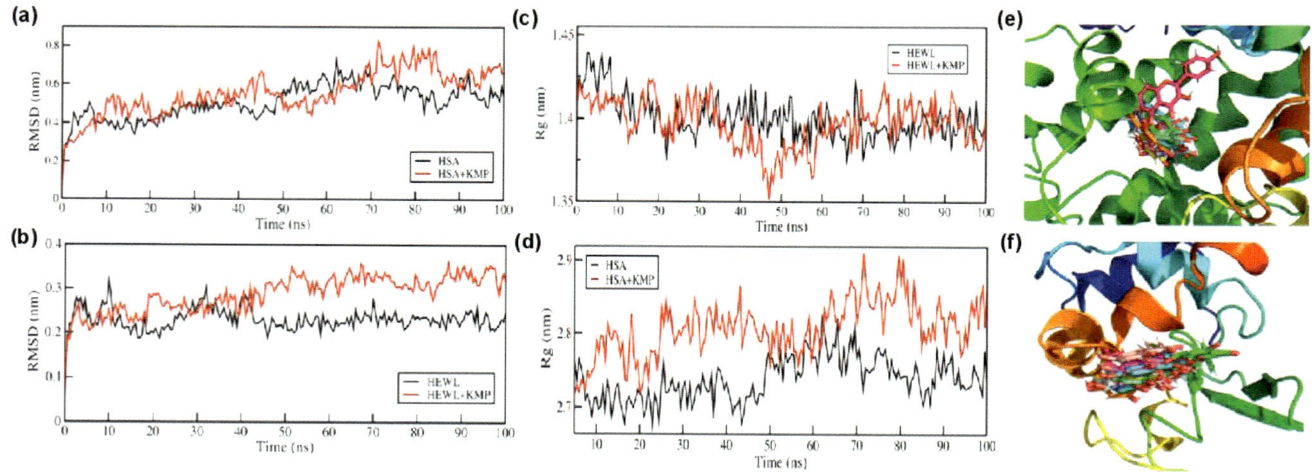

Figure 4.10 RMSD plot of (a) HSA, and (b) HEWL in the absence and presence of KMP. Rg plot of (c) HSA, and (d) HEWL with and without KMP (Reproduced from [128]). Superimposed snapshots of the heavy atoms of KMP within (e) HSA, and (f) HEWL as derived from simulated configurations of the protein-KMP complex observed at 10 ns intervals (Reproduced from [128]).

40 ns before achieving stability for the remainder of the simulation period. The marginally increased RMSD of the HSA-KMP and HEWL-KMP complexes, as compared to the native (Figure 4.10e,f), implies that KMP is striving for stability. The Rg values for HSA are somewhat higher than those for the unbound HSA, whereas HEWL maintains a consistent Rg throughout the simulation in both the absence and presence of KMP (Figure 4.10c,d). Thus, it can be concluded that the presence of the ligand does not significantly alter the compactness of the protein. The superimposed image of several configurations recorded from the simulated trajectories at uniform time intervals is represented in Figure 4.10e,f to visualize the structural variation of HSA and HEWL complexes with KMP. It was also observed that a variety of interactions, such as hydrogen bonding, van der Waals, π–π, and π-alkyl, play distinct roles in ligand stabilization within the binding cleft of proteins at intervals of 0 ns, 50 ns, and 100 ns. Observation at 100 ns indicates that for HSA, residues Ser287, Arg257, His242, and Arg222 stabilize KMP hydrogen bonding, while Arg218, Leu260, and Ala291 engage in π–alkyl interactions. In HEWL, Asn59 and Asp101 form hydrogen bonds with KMP; Trp62 and Trp108 participate in π–π interactions; and Ala107 in π–alkyl interaction. Moreover, the consistency in the total number of non-bonded contacts across both complexes indicates the ligand remains within the binding cavity through the simulation [128].

The RMSD analysis depicted in Figure 4.11a for BHb bound to 7-HC and 4-Me-7-HC reveals a peak for BHb at approximately 0.33 nm at 20 ns, followed by stability until around 70 ns, after which it begins to destabilize again. Conversely, the complexes of BHb with 4-Me-7 HC and 7-HC show RMSD values of 0.42 nm and 0.32 nm at 72 ns and 91 ns, respectively. The elevated RMSD value observed in the BHb-4-Me-7-HC complex suggests substantial structural changes in the protein caused by the binding of the ligand, whereas the lower RMSD value for the BHb-7-HC indicates greater stability. The RMSF pattern for BHb was largely similar across the two ligand complexes, with many regions overlapping with unbound BHb, suggesting that ligand binding did not significantly alter the protein's flexibility (Figure 4.11b). The Rg of BHb remained consistent with the complexes of 4-Me-7-HC and 7-HC throughout the entire simulation (Figure 4.11c). Notably, the Rg values for both BHb-4-Me-7-HC and BHb-7-HC complexes were lower than that of the unbounded BHb, indicating that the complexes are more compact than the unbounded form [133].

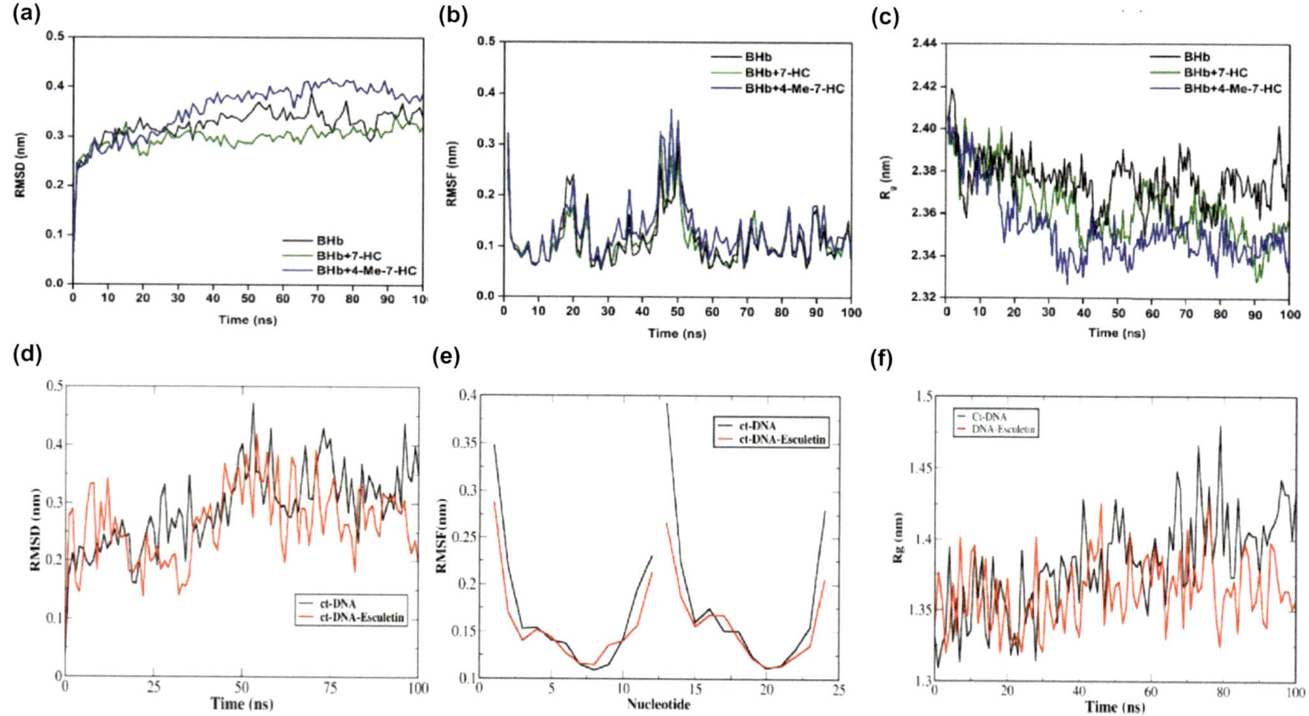

Figure 4.11 MD simulation plots of BHb: (a) RMSD, (b) RMSF, and (c) Rg, in the absence and presence of 7-HC and 4-Me-7-HC over the simulation time (Reproduced from [133]). MD simulation plots of DNA (d) RMSD, (e) RMSF, and (f) Rg in the absence and presence of esculetin (Reproduced from [159]).

The MD simulation studies of DNA with esculetin demonstrate that the RMSD of DNA in the absence of esculetin (Figure 4.11d). is nearly 0.42 nm while the DNA-esculetin complex stabilizes around 0.3 nm throughout the simulation, suggesting that esculetin binding enhances the stability of DNA. Furthermore, the RMSF plot in Figure 4.11e demonstrates reduced fluctuations in the DNA-esculetin complex compared to only DNA, supporting the stabilizing effect of esculetin. Additionally, Rg analysis (Figure 4.11f) reveals that only the DNA exhibits lower compactness, whereas the DNA-esculetin complex remains consistently compact during simulation, further indicating structural stabilization [159].

The MD simulation of a SWCNT was initiated using the configuration of the SWCNT-HSA complex that exhibited the lowest free binding energy. The RMSD values for both free HSA and HSA–SWCNT complex (Figure 4.12a) were calculated over the 10–30 ns trajectory, with values of 0.37 and 0.42 nm, respectively. It was observed that both systems stabilized and varied around their average mean RMSD values after approximately 10 ns, suggesting a stable conformation was achieved. Notably, the higher RMSD of the HSA-SWCNTs complex relative to free HSA suggests conformational changes occur in HSA upon binding to SWCNTs. The RMSF calculated from the final 20 ns of the trajectory (Figure 4.12b) revealed a similar atomic variation pattern for free HSA and the HSA–SWCNTs complex, suggesting that the binding residues remain relatively rigid throughout the simulation. The Rg profile for both systems (Figure 4.12c) reached equilibrium around 7.5 ns. The average Rg values calculated from 10–30 ns were 2.667 nm for native HSA and 2.660 nm for HSA-SWCNTs complex, showing a slight reduction upon complexation. This decrease in Rg suggests that SWCNTs binding alters the local environment of HSA, contributing to structural changes that stabilize the HSA-SWCNTs complex by 7.5 ns due to rearrangement in its structure [176].

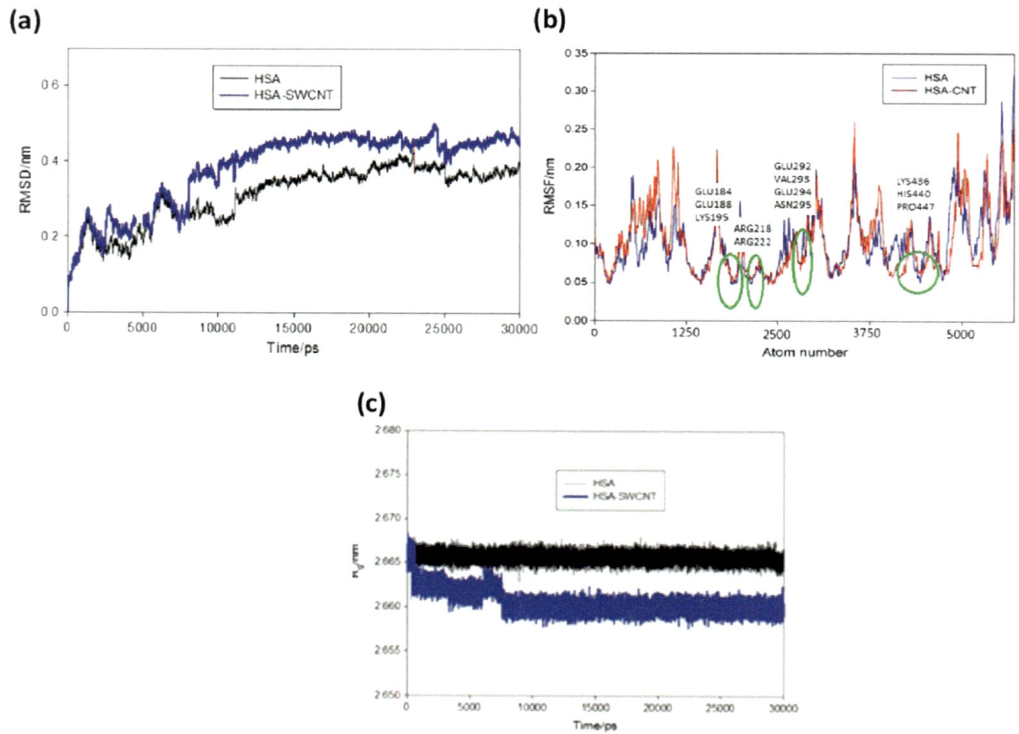

Figure 4.12 (a) RMSD values for free HSA and HSA–SWCNT complex, (b) The RMSF values versus atom number of HSA, and (c) Rg plot for HSA and HSA-SWCNT (Reproduced from [176]).

4.4 Conclusion

This chapter highlights that molecular docking has evolved into a vital and expanding field, providing a fundamental understanding of molecular recognition between biomolecules and ligands. Over the past several years, *in silico* approaches, when combined with biophysical data and experimental studies, have proven indispensable in drug discovery. Docking studies play an essential role in decision-making, rapidly testing hypotheses, and offering complex challenges that are either difficult or expensive to explore experimentally. With advancements in sampling algorithms, scoring functions, and the increasing availability of docking software, this approach has emerged as a cornerstone in pharmaceutical research, facilitating the design of novel therapeutic drugs. Docking enables the identification of potential binding sites within biomolecules, particularly in protein–ligand and DNA–ligand systems. These applications are vital for gaining insight, at a molecular level, into how small molecules, such as drugs, interact with biological targets, identifying specific residues involved, predicting binding affinities, and identifying the forces involved that drive these interactions. In protein–ligand interactions, docking offers insights into the molecular mechanisms by which a ligand modulates protein activity, such as inhibition, activation, and allosteric regulation. Likewise, in DNA-ligand interactions, docking reveals mechanisms of drug action that influence gene expression or DNA repair. By predicting these interactions and assessing the stability of ligand biomolecule complexes, molecular docking has emerged as an indispensable tool in the rational design of targeted therapeutic compounds.

References

1. Morris, G. M., and Lim-Wilby, M. (2008). *Molecular Docking*. 365–382.
2. Sethi, A., Joshi, K., Sasikala, K., and Alvala, M. (2020). Molecular Docking in Modern Drug Discovery: Principles and Recent Applications. In *Drug Discovery and Development – New Advances*, IntechOpen.
3. Pinzi, L., and Rastelli, G. (2019). Molecular Docking: Shifting Paradigms in Drug Discovery. *International Journal of Molecular Sciences*. **20 (18):** 4331.

4. Kitchen, D. B., Decornez, H., Furr, J. R., and Bajorath, J. (2004). Docking and Scoring in Virtual Screening for Drug Discovery: Methods and Applications. *Nature Reviews Drug Discovery.* **3 (11)**: 935–949.

5. DesJarlais, R. L., Sheridan, R. P., Dixon, J. S., Kuntz, I. D., and Venkataraghavan, R. (1986). Docking Flexible Ligands to Macromolecular Receptors by Molecular Shape. *Journal of Medicinal Chemistry.* **29 (11)**: 2149–2153.

6. Levinthal, C., Wodak, S. J., Kahn, P., and Dadivanian, A. K. (1975). Hemoglobin Interaction in Sickle Cell Fibers. I: Theoretical Approaches to the Molecular Contacts. *Proceedings of the National Academy of Sciences.* **72 (4)**: 1330–1334.

7. Goodsell, D. S., and Olson, A. J. (1990). Automated Docking of Substrates to Proteins by Simulated Annealing. *Proteins: Structure, Function, and Bioinformatics.* **8 (3)**: 195–202.

8. Wodak, S. J., and Janin, J. (1978). Computer Analysis of Protein-Protein Interaction. *Journal of Molecular Biology.* **124 (2)**: 323–342.

9. Maruca, A., Ambrosio, F. A., Lupia, A., Romeo, I., Rocca, R., Moraca, F., Talarico, C., Bagetta, D., Catalano, R., Costa, G., Artese, A., and Alcaro, S. (2019). Computer-Based Techniques for Lead Identification and Optimization I: Basics. *Physical Sciences Reviews.* **4 (6)**.

10. Kontoyianni, M. (2017). *Docking and Virtual Screening in Drug Discovery*. 255–266.

11. Mukesh, B., and Rakesh, K. (2011). Molecular Docking: A Review. *International Journal of Research in Ayurveda & Pharmacy.* **2 (6)**: 1746–1751.

12. Lamb, M. L., and Jorgensen, W. L. (1997). Computational Approaches to Molecular Recognition. *Current Opinion in Chemical Biology.* **1 (4)**: 449–457.

13. Drahansky, M., Paridah, M., Moradbak, A., Mohamed, A., Owolabi, F. A. T., Asniza, M., and Abdul Khalid, S. H. (2017). We Are IntechOpen, the World's Leading Publisher of Open Access Books Built by Scientists, for Scientists TOP 1%. *Intech.*

14. McCammon, J. A., Gelin, B. R., and Karplus, M. (1977). Dynamics of Folded Proteins. *Nature.* **267 (5612)**: 585–590.

15. Meng, X.-Y., Zhang, H.-X., Mezei, M., and Cui, M. (2011). Molecular Docking: A Powerful Approach for Structure-Based Drug Discovery. *Current Computer-Aided Drug Design.* **7 (2)**: 146–157.

16. Lopes, P. E. M., Guvench, O., and MacKerell, A. D. (2015). *Current Status of Protein Force Fields for Molecular Dynamics Simulations*. 47–71.

17. Fischer, E. (1894). Einfluss Der Configuration Auf Die Wirkung Der Enzyme. *Berichte der Dtsch. Chem. Gesellschaft.* **27 (3)**: 2985–2993. https://doi.org/10.1002/cber.18940270364.
18. Matsuzaki, Y., Uchikoga, N., Ohue, M., and Akiyama, Y. (2016). *Rigid-Docking Approaches to Explore Protein–Protein Interaction Space.* 33–55.
19. Kuntz, I. D., Blaney, J. M., Oatley, S. J., Langridge, R., and Ferrin, T. E. (1982). A Geometric Approach to Macromolecule-Ligand Interactions. *Journal of Molecular Biology.* **161 (2):** 269–288.
20. Ewing, T. J. A., Makino, S., Skillman, A. G., and Kuntz, I. D. (2001). DOCK 4.0: Search Strategies for Automated Molecular Docking of Flexible Molecule Databases. *Journal of Computer-Aided Molecular Design.* **15 (5)**: 411–428.
21. Leach, A. R., and Kuntz, I. D. (1992). Conformational Analysis of Flexible Ligands in Macromolecular Receptor Sites. *Journal of Computational Chemistry.* **13 (6)**: 730–748.
22. Shoichet, B. K., Stroud, R. M., Santi, D. V., Kuntz, I. D., and Perry, K. M. (1993). Structure-Based Discovery of Inhibitors of Thymidylate Synthase. *Science.* **259 (5100)**: 1445–1450.
23. Miller, M. D., Kearsley, S. K., Underwood, D. J., and Sheridan, R. P. (1994). FLOG: A System to Select ? Quasi-Flexible? Ligands Complementary to a Receptor of Known Three-Dimensional Structure. *Journal of Computer-Aided Molecular Design.* **8 (2)**: 153–174.
24. Gabb, H. A., Jackson, R. M., and Sternberg, M. J. E. (1997). Modelling Protein Docking Using Shape Complementarity, Electrostatics and Biochemical Information 1. Edited by J. Thornton. *Journal of Molecular Biology.* **272 (1)**: 106–120.
25. Anderson, A. C., O'Neil, R. H., Surti, T. S., and Stroud, R. M. (2001). Approaches to Solving the Rigid Receptor Problem by Identifying a Minimal Set of Flexible Residues during Ligand Docking11 PDB Coordinates Have Been Deposited with the RSCB with Accession ID: 1F28. *Chemistry & Biology.* **8 (5)**: 445–457.
26. Du, X., Li, Y., Xia, Y.-L., Ai, S.-M., Liang, J., Sang, P., Ji, X.-L., and Liu, S.-Q. (2016). Insights into Protein–Ligand Interactions: Mechanisms, Models, and Methods. *International Journal of Molecular Sciences.* **17 (2)**: 144.
27. Rarey, M., Kramer, B., Lengauer, T., and Klebe, G. A. (1996). Fast Flexible Docking Method Using an Incremental Construction Algorithm. *Journal of Molecular Biology.* **261 (3):** 470–489.

28. Morris, G. M., Goodsell, D. S., Halliday, R. S., Huey, R., Hart, W. E., Belew, R. K., and Olson, A. J. (1639). Automated Docking Using a Lamarckian Genetic Algorithm and an Empirical Binding Free Energy Function. *Journal of Computational Chemistry.* **19 (14)**: 1639–1662.

29. Andrusier, N., Mashiach, E., Nussinov, R., and Wolfson, H. J. (2008). Principles of Flexible Protein–Protein Docking. *Proteins: Structure, Function, and Bioinformatics.* **73 (2)**: 271–289.

30. Repasky, M. P., Shelley, M., and Friesner, R. A. (2007). Flexible Ligand Docking with Glide. *Current Protocols in Bioinformatics.* **18 (1)**: *Chapter 8.*

31. Chaudhary, M., and Tyagi, K. (2024). A Review on Molecular Docking and Its Application. *International Journal of Advanced Research.* **12 (03)**:1141–1153.

32. Astalakshmi, N., Gokul, T., Gowri Sankar, K. B., Nandhini, M, Hari Hara Sudhan, M. R., Gowtham, S., Latha, S. T., and Surendra Kumar, M. (2022). Overview on Molecular Docking: A Powerful Approach for Structure-Based Drug Discovery. *International Journal of Pharmaceutical Sciences Review and Research.* **77 (2)**: 146–157.

33. Totrov, M., and Abagyan, R. (1997). Flexible Protein-Ligand Docking by Global Energy Optimization in Internal Coordinates. *Proteins: Structure, Function, and Genetics.* **(29)**: 215–220.

34. McMartin, C., and Bohacek, R. S. (1997). QXP: Powerful, Rapid Computer Algorithms for Structure-Based Drug Design. *Journal of Computer-Aided Molecular Design.* **11 (4)**: 333–334.

35. Taylor, R. D., Jewsbury, P. J., and Essex, J. W. (2003). FDS: Flexible Ligand and Receptor Docking with a Continuum Solvent Model and Soft-core Energy Function. *Journal of Computational Chemistry.* **24 (13)**:1637–1656.

36. Tietze, S., and Apostolakis, J. (2007). GlamDock: Development and Validation of a New Docking Tool on Several Thousand Protein–Ligand Complexes. *Journal of Chemical Information and Modeling.* **47 (4)**: 1657–1672.

37. Liu, M., and Wang, S. (199). MCDOCK: A Monte Carlo Simulation Approach to the Molecular Docking Problem. *Journal of Computer-Aided Molecular Design.* **13 (5)**: 435–451.

38. Landaburu, J. (2016). PRODOCK: Software Package for Protein Modeling and Docking. *Journal of Computational Chemistry.* **16 (1)**: 1–23.

39. Meiler, J., and Baker, D. (2006). ROSETTALIGAND: Protein–Small Molecule Docking with Full Side-chain Flexibility. *Proteins: Structure, Function, and Bioinformatics.* **65 (3)**: 538–548.
40. Morley, S. D., and Afshar, M. (2004). Validation of an Empirical RNA-Ligand Scoring Function for Fast Flexible Docking Using RiboDock. *Journal of Computer-Aided Molecular Design.* **18 (3)**: 189–208.
41. DesJarlais, R. L., Sheridan, R. P., Seibel, G. L., Dixon, J. S., Kuntz, I. D., and Venkataraghavan, R. (1988). Using Shape Complementarity as an Initial Screen in Designing Ligands for a Receptor Binding Site of Known Three-Dimensional Structure. *Journal of Medicinal Chemistry.* **31 (4)**: 722–729.
42. Böhm, H. J. (1992). The Computer Program LUDI: A New Method for the de Novo Design of Enzyme Inhibitors. *Journal of Computer-Aided Molecular Design.* **6 (1)**: 61–78.
43. Bentham Science Publisher, B. S. P. (2006). EHiTS: An Innovative Approach to the Docking and Scoring Function Problems. *Current Protein & Peptide Science.* **7 (5)**: 421–435.
44. Venkatachalam, C. M., Jiang, X., Oldfield, T., and Waldman, M. (2003). LigandFit: A Novel Method for the Shape-Directed Rapid Docking of Ligands to Protein Active Sites. *Journal of Molecular Graphics and Modelling.* **21 (4)**: 289–307.
45. Hassan, N. M., Alhossary, A. A., Mu, Y., and Kwoh, C. K. (2017). Protein-Ligand Blind Docking Using QuickVina-W with Inter-Process Spatio-Temporal Integration. *Scientific Reports.* **7 (1)**: 1–13.
46. Hetényi, C., and van der Spoel, D. (2002). Efficient Docking of Peptides to Proteins without Prior Knowledge of the Binding Site. *Protein Science.* **11 (7)**: 1729–1737.
47. Hetényi, C., and van der Spoel, D. (2006). Blind Docking of Drug-sized Compounds to Proteins with up to a Thousand Residues. *FEBS Letters.* **580 (5)**: 1447–1450.
48. Adcock, S. A., and McCammon, J. A. (2006). Molecular Dynamics: Survey of Methods for Simulating the Activity of Proteins. *Chemical Reviews.* **106 (5)**: 1589–1615.
49. Boresch, S., Tettinger, F., Leitgeb, M., and Karplus, M. (2003). Absolute Binding Free Energies: A Quantitative Approach for Their Calculation. *Journal of Physical Chemistry B.* **107 (35)**: 9535–9551.
50. Mobley, D. L., Graves, A. P., Chodera, J. D., McReynolds, A. C., Shoichet, B. K., and Dill, K. A. (2007). Predicting Absolute Ligand Binding Free Energies to a Simple Model Site. *Journal of Molecular Biology.* **371 (4)**: 1118–1134.

51. Aldeghi, M., Heifetz, A., Bodkin, M. J., Knapp, S., and Biggin, P. C. (2016). Accurate Calculation of the Absolute Free Energy of Binding for Drug Molecules. *Chemical Science.* **7 (1)**: 207–218.
52. Bash, P. A., Singh, U. C., Brown, F. K., Langridge, R., and Kollman, P. A. (1987). Calculation of the Relative Change in Binding Free Energy of a Protein-Inhibitor. *Complex Science.* **235 (4788)**: 574–576.
53. Cournia, Z., Allen, B., and Sherman, W. (2017). Relative Binding Free Energy Calculations in Drug Discovery: Recent Advances and Practical Considerations. *Journal of Chemical Information and Modeling.* **57 (12)**: 2911–2937.
54. Hingerty, B. E., Ritchie, R. H., Ferrell, T. L., and Turner, J. E. (1985). Dielectric Effects in Biopolymers: The Theory of Ionic Saturation Revisited. *Biopolymers.* **24 (3)**: 427–439.
55. Gilson, M. K., and Honig, B. H. (1986). The Dielectric Constant of a Folded Protein. *Biopolymers.* **25 (11)**: 2097–2119.
56. Huang, S.-Y., Grinter, S. Z., and Zou, X. (2010). Scoring Functions and Their Evaluation Methods for Protein-Ligand Docking: Recent Advances and Future Directions. *Physical Chemistry Chemical Physics (PCCP).* **12 (40)**: 12899–12908.
57. Shoichet, B. K., Kuntz, I. D., and Bodian, D. L. (1992). Molecular Docking Using Shape Descriptors. *Journal of Computational Chemistry.* **13 (3)**: 380–397.
58. Moustakas, D. T., Lang, P. T., Pegg, S., Pettersen, E., Kuntz, I. D., Brooijmans, N., and Rizzo, R. C. (2006). Development and Validation of a Modular, Extensible Docking Program: DOCK 5. *Journal of Computer-Aided Molecular Design.* **20 (10)**: 601–619.
59. Allen, W. J., Balius, T. E., Mukherjee, S., Brozell, S. R., Moustakas, D. T., Lang, P. T., Case, D. A., Kuntz, I. D., and Rizzo, R. C. (2015). DOCK 6: Impact of New Features and Current Docking Performance. *Journal of Computational Chemistry.* **36 (15)**: 1132–1156.
60. Meng, E. C., Shoichet, B. K., and Kuntz, I. D. (1992). Automated Docking with Grid-Based Energy Evaluation. *Journal of Computational Chemistry.* **13 (4)**: 505–524.
61. Morris, G. M., Huey, R., Lindstrom, W., Sanner, M. F., Belew, R. K., Goodsell, D. S., and Olson, A. J. (2009). AutoDock4 and AutoDockTools4: Automated Docking with Selective Receptor Flexibility. *Journal of Computational Chemistry.* **30 (16)**: 2785–2791.
62. Huey, R., Morris, G. M., Olson, A. J., and Goodsell, D. S. (2007). A Semiempirical Free Energy Force Field with Charge-Based Desolvation. *Journal of Computational Chemistry.* **28 (6)**: 1145–1152.

63. Jones, G., Willett, P., and Glen, R. C. (1995). Molecular Recognition of Receptor Sites Using a Genetic Algorithm with a Description of Desolvation. *Journal of Molecular Biology.* **245 (1)**: 43–53.
64. Jones, G., Willett, P., Glen, R. C., Leach, A. R., and Taylor, R. (1997). Development and Validation of a Genetic Algorithm for Flexible Docking11 Edited by F. E. Cohen. *Journal of Molecular Biology.* **267 (3)**: 727–748.
65. Shin, W.-H., and Seok, C. (2012). GalaxyDock: Protein–Ligand Docking with Flexible Protein Side-Chains. *Journal of Chemical Information and Modeling.* **52 (12)**: 3225–3232.
66. Li, H., Peng, J., Sidorov, P., Leung, Y., Leung, K.-S., Wong, M.-H., Lu, G., and Ballester, P. J. (2019). Classical Scoring Functions for Docking Are Unable to Exploit Large Volumes of Structural and Interaction Data. *Bioinformatics.* **35 (20)**: 3989–3995.
67. Guedes, I. A., Pereira, F. S. S., and Dardenne, L. E. (2018). Empirical Scoring Functions for Structure-Based Virtual Screening: Applications, Critical Aspects, and Challenges. *Frontiers in Pharmacology.* **9**:1089.
68. Muegge, I., and Martin, Y. C. (1999). A General and Fast Scoring Function for Protein–Ligand Interactions: A Simplified Potential Approach. *Journal of Medicinal Chemistry.* **42 (5)**: 791–804.
69. Muegge, I., and Rarey, M. (2001). Small Molecule Docking and Scoring. In *Reviews in Computational Chemistry*. 1–60.
70. Muegge, I. (2000). A Knowledge-Based Scoring Function for Protein-Ligand Interactions: Probing the Reference State. *Perspectives in Drug Discovery and Design.* **20 (1)**: 99–114.
71. Neudert, G., and Klebe, G. (2011). DSX: A Knowledge-Based Scoring Function for the Assessment of Protein–Ligand Complexes. *Journal of Chemical Information and Modeling.* **51 (10)**: 2731–2745.
72. Velec, H. F. G., Gohlke, H., and Klebe, G. (2005). DrugScoreCSD Knowledge-Based Scoring Function Derived from Small Molecule Crystal Data with Superior Recognition Rate of Near-Native Ligand Poses and Better Affinity Prediction. *Journal of Medicinal Chemistry.* **48 (20)**: 6296–6303.
73. Gohlke, H., Hendlich, M., and Klebe, G. (2000). Knowledge-Based Scoring Function to Predict Protein-Ligand Interactions. *Journal of Molecular Biology.* **295 (2)**: 337–356.
74. Huang, S.-Y., and Zou, X. (2006). An Iterative Knowledge-Based Scoring Function to Predict Protein–Ligand Interactions: I. Derivation of Interaction Potentials. *Journal of Computational Chemistry.* **27 (15)**: 1866–1875.

75. Huang, S.-Y., and Zou, X. (2010). Inclusion of Solvation and Entropy in the Knowledge-Based Scoring Function for Protein–Ligand Interactions. *Journal of Chemical Information and Modeling.* **50 (2)**: 262–273.

76. Morris, G. M., Goodsell, D. S., Halliday, R. S., Huey, R., Hart, W. E., Belew, R. K., and Olson, A. J. (1998). Automated Docking Using a Lamarckian Genetic Algorithm and an Empirical Binding Free Energy Function. *Journal of Computational Chemistry.* **19 (14)**: 1639–1662.

77. El-Hachem, N., Haibe-Kains, B., Khalil, A., Kobeissy, F. H., and Nemer, G. (2017). *AutoDock and AutoDockTools for Protein-Ligand Docking: Beta-Site Amyloid Precursor Protein Cleaving Enzyme 1(BACE1) as a Case Study*, 391–403.

78. Moghadam, N. H., Salehzadeh, S., and Shahabadi, N. (2017). Spectroscopic and Molecular Docking Studies on the Interaction of Antiviral Drug Nevirapine with Calf Thymus DNA. *Nucleosides, Nucleotides, and Nucleic Acids.* **36(9)**: 553–570.

79. Eberhardt, J., Santos-Martins, D., Tillack, A. F., and Forli, S. (2021). AutoDock Vina 1.2.0: New Docking Methods, Expanded Force Field, and Python Bindings. *Journal of Chemical Information and Modeling.* **61 (8)**: 3891–3898.

80. Trott, O., and Olson, A. J. (2010). AutoDock Vina: Improving the Speed and Accuracy of Docking with a New Scoring Function, Efficient Optimization, and Multithreading. *Journal of Computational Chemistry.* **31 (2)**: 455–461.

81. Kramer, B., Rarey, M., and Lengauer, T. (199). Evaluation of the FLEXX Incremental Construction Algorithm for Protein-Ligand Docking. *Proteins: Structure, Function, and Genetics.* **37 (2)**: 228–241.

82. Bauer, A., Kovári, Z., and Keserű, G. M. (2004). Optimization of Virtual Screening Protocols: FlexX-Based Virtual Screening for COX-2 Inhibitors Reveals the Importance of Tailoring Screen Parameters. *Journal of Molecular Structure: THEOCHEM.* **676 (1–3)**: 1–5.

83. Friesner, R. A., Banks, J. L., Murphy, R. B., Halgren, T. A., Klicic, J. J., Mainz, D. T., Repasky, M. P., Knoll, E. H., Shelley, M., Perry, J. K., Shaw, D. E., Francis, P., and Shenkin, P. S. (2004). Glide: A New Approach for Rapid, Accurate Docking and Scoring. 1. Method and Assessment of Docking Accuracy. *Journal of Medicinal Chemistry.* **47 (7)**: 1739–1749.

84. Boittier, E. D., Tang, Y. Y., Buckley, M. E., Schuurs, Z. P., Richard, D. J., and Gandhi, N. S. (2020). Assessing Molecular Docking Tools to Guide Targeted Drug Discovery of CD38 Inhibitors. *International Journal of Molecular Sciences.* **21 (15)**: 5183.

85. Jones, G., Willett, P., Glen, R. C., Leach, A. R., and Taylor, R. (1997). Development and Validation of a Genetic Algorithm for Flexible Docking. *Journal of Molecular Biology.* **267 (3)**: 727–748.

86. Verdonk, M. L., Cole, J. C., Hartshorn, M. J., Murray, C. W., and Taylor, R. D. (2003). Improved Protein–Ligand Docking Using GOLD. *Proteins: Structure, Function, and Bioinformatics.* **52 (4)**: 609–623.

87. Sato, H., Shewchuk, L. M., and Tang, J. (2006). Prediction of Multiple Binding Modes of the CDK2 Inhibitors, Anilinopyrazoles, Using the Automated Docking Programs GOLD, FlexX, and LigandFit: An Evaluation of Performance. *Journal of Chemical Information and Modeling.* **46 (6)**: 2552–2562.

88. Pierce, B. G., Wiehe, K., Hwang, H., Kim, B.-H., Vreven, T., and Weng, Z. (2014). ZDOCK Server: Interactive Docking Prediction of Protein–Protein Complexes and Symmetric Multimers. *Bioinformatics.* **30 (12)**: 1771–1773.

89. Grosdidier, A., Zoete, V., and Michielin, O. (2011). SwissDock, a Protein-Small Molecule Docking Web Service Based on EADock DSS. *Nucleic Acids Research.* **39**: W270–W277.

90. Bitencourt-Ferreira, G., and de Azevedo, W. F. (2019). *Docking with SwissDock.* 189–202.

91. Tapia-Abellán, A., Angosto-Bazarra, D., Martínez-Banaclocha, H., de Torre-Minguela, C., Cerón-Carrasco, J. P., Pérez-Sánchez, H., Arostegui, J. I., and Pelegrin, P. (2019). MCC950 Closes the Active Conformation of NLRP3 to an Inactive State. *Nature Chemical Biology.* **15 (6)**: 560–564.

92. Bitencourt-Ferreira, G., and de Azevedo, W. F. (2019). *Molecular Docking Simulations with ArgusLab.* 203–220.

93. Tanguenyongwatana, P., and Jongkon, N. (2016). Molecular Docking Study of Tyrosinase Inhibitors Using ArgusLab 4.0.1: A Comparative Study. *Thai Journal of Pharmaceutical Sciences.* **40 (1)**: 21–25.

94. Guterres, H., and Im, W. (2020). Improving Protein-Ligand Docking Results with High-Throughput Molecular Dynamics Simulations. *Journal of Chemical Information and Modeling.* **60 (4)**: 2189–2198.

95. O'Boyle, N. M., Banck, M., James, C. A., Morley, C., Vandermeersch, T., and Hutchison, G. R. (2011). Open Babel: An Open Chemical Toolbox. *Journal of Cheminformatics.* **3 (1)**: 33.

96. Wishart, D. S., Knox, C., Guo, A. C., Shrivastava, S., Hassanali, M., Stothard, P., Chang, Z., and Woolsey, J. (2006). DrugBank: A Comprehensive Resource for in Silico Drug Discovery and Exploration. *Nucleic Acids Research.* **34**: D668–72.

97. Li, Q., Cheng, T., Wang, Y., and Bryant, S. H. (2010). PubChem as a Public Resource for Drug Discovery. *Drug Discovery Today*. **15 (23–24)**: 1052–1057.

98. Irwin, J. J., Sterling, T., Mysinger, M. M., Bolstad, E. S., and Coleman, R. G. (2012). ZINC: A Free Tool to Discover Chemistry for Biology. *Journal of Chemical Information and Modeling.* **52 (7)**: 1757–1768.

99. Gaulton, A., Bellis, L. J., Bento, A. P., Chambers, J., Davies, M., Hersey, A., Light, Y., McGlinchey, S., Michalovich, D., Al-Lazikani, B., and Overington, J. P. (2012). ChEMBL: A Large-Scale Bioactivity Database for Drug Discovery. *Nucleic Acids Research.* **40**: D1100–7.

100. Pence, H. E., and Williams, A. (2010). ChemSpider: An Online Chemical Information Resource. *Journal of Chemical Education.* **87 (11)**: 1123–1124.

101. Laskowski, R. A. (1995). SURFNET: A Program for Visualizing Molecular Surfaces, Cavities, and Intermolecular Interactions. *Journal of Molecular Graphics.* **13 (5)**: 323–330.

102. Goodford, P. J. (1985). A Computational Procedure for Determining Energetically Favorable Binding Sites on Biologically Important Macromolecules. *Journal of Medicinal Chemistry.* **28 (7)**: 849–857.

103. Narang, P., Bhushan, K., Bose, S., and Jayaram, B. (2006). Protein Structure Evaluation Using an All-Atom Energy-Based Empirical Scoring Function. *Journal of biomolecular Structure & Dynamics.* **23 (4)**: 385–406.

104. Yang, J., Roy, A., and Zhang, Y. (2013). Protein-Ligand Binding Site Recognition Using Complementary Binding-Specific Substructure Comparison and Sequence Profile Alignment. *Bioinformatics.* **29 (20)**: 2588–2595.

105. Jiménez, J., Doerr, S., Martínez-Rosell, G., Rose, A. S., and De Fabritiis, G. (2017). DeepSite: Protein-Binding Site Predictor Using 3D-Convolutional Neural Networks. *Bioinformatics.* **33 (19)**: 3036–3042.

106. Binkowski, T. A., Naghibzadeh, S., and Liang, J. (2003). CASTp: Computed Atlas of Surface Topography of Proteins. *Nucleic Acids Research.* **31 (13)**: 3352–3355.

107. Kandel, J., Tayara, H., and Chong, K. T. (2021). PUResNet: Prediction of Protein-Ligand Binding Sites Using Deep Residual Neural Network. *Journal of Cheminformatics.* **13 (1)**: 65.

108. Sugio, S., Kashima, A., Mochizuki, S., Noda, M., and Kobayashi, K. (199). Crystal Structure of Human Serum Albumin at 2.5 Å Resolution. *Protein Engineering, Design and Selection.* **12 (6)**: 439–446.

109. Sudlow, G., Birkett, D. J., and Wade, D. N. (1975). The Characterization of Two Specific Drug Binding Sites on Human Serum Albumin. *Molecular Pharmacology.* **11 (6)**: 824–832.

110. Blake, C. C., Koenig, D. F., Mair, G. A., North, A. C., Phillips, D. C., and Sarma, V. R. (1965). Structure of Hen Egg-White Lysozyme. A Three-Dimensional Fourier Synthesis at 2 Angstrom Resolution. *Nature.* **206 (4986)**: 757–761.

111. Strynadka, N. C. J., and James, M. N. G. (1991). Lysozyme Revisited: Crystallographic Evidence for Distortion of an N-Acetylmuramic Acid Residue Bound in Site D. *Journal of Molecular Biology.* **220 (2)**: 401–424.

112. Perutz, M. F., Rossmann, M. G., Cullis, A. F., Muirhead, H., Will, G., and North, A. C. T. (1960). Structure of Hæmoglobin: A Three-Dimensional Fourier Synthesis at 5.5-Å. Resolution, Obtained by X-Ray Analysis. *Nature.* **185 (4711)**: 416–422.

113. Perutz, M. F. (1989). Myoglobin and Haemoglobin: Role of Distal Residues in Reactions with Haem Ligands. *Trends in Biochemical Sciences.* **14 (2)**: 42–44.

114. Salehi, B., Fokou, P. V. T., Sharifi-Rad, M., Zucca, P., Pezzani, R., Martins, N., and Sharifi-Rad, J. (2019). The Therapeutic Potential of Naringenin: A Review of Clinical Trials. *Pharmaceuticals.* **12 (1)**: 11.

115. Chen, R., Qi, Q.-L., Wang, M.-T., and Li, Q.-Y. (2016). Therapeutic Potential of Naringin: An Overview. *Pharmaceutical Biology.* **54 (12)**: 3203–3210.

116. Park, C. M., and Song, Y.-S. (2019). Luteolin and Luteolin-7-O-Glucoside Protect against Acute Liver Injury through Regulation of Inflammatory Mediators and Antioxidative Enzymes in GalN/LPS-Induced Hepatic ICR Mice. *Nutrition Research and Practice.* **13 (6)**: 473.

117. Mani, R., and Natesan, V. (2018). Chrysin: Sources, Beneficial Pharmacological Activities, and Molecular Mechanism of Action. *Phytochemistry.* **145**: 187–196.

118. Anand David, A., Arulmoli, R., and Parasuraman, S. (2016). Overviews of Biological Importance of Quercetin: A Bioactive Flavonoid. *Pharmacognosy Reviews.* **10 (20)**: 84.

119. Annunziata, F., Pinna, C., Dallavalle, S., Tamborini, L., and Pinto, A. (2020). An Overview of Coumarin as a Versatile and Readily Accessible Scaffold with Broad-Ranging Biological Activities. *International Journal of Molecular Sciences.* **21 (13)**: 4618.

139. Lyndem, S., Gazi, R., Belwal, V. K., Bhatta, A., Jana, M., and Roy, A. S. (2022). Binding of Bioactive Esculin and Esculetin with Hen Egg White Lysozyme: Spectroscopic and Computational Methods to Comprehensively Elucidate the Binding Affinities, Interacting Forces, and Conformational Alterations at Molecular Level. *Journal of Molecular Liquids.* **360**, 119423.

140. Lyndem, S., Hazarika, U., Bhatta, A., Prakash, V., Jha, A. N., and Singha Roy, A. (2023). In Vitro Interactions of Esculin and Esculetin with Bovine Hemoglobin Alter Its Structure and Inhibit Aggregation: Insights from Spectroscopic and Computational Studies. *New Journal of Chemistry.* **47 (30)**: 14447–14468.

141. Das, S., Santra, S., Rohman, M. A., Ray, M., Jana, M., and Singha Roy, A. (2019). An Insight into the Binding of 6-Hydroxyflavone with Hen Egg White Lysozyme: A Combined Approach of Multi-Spectroscopic and Computational Studies. *Journal of Biomolecular Structure and Dynamics. Taylor & Francis.* **37(15)**: 4019–4034.

142. Lu, N., Li, J., Tian, R., and Peng, Y.-Y. (2014). Binding of Human Serum Albumin to Single-Walled Carbon Nanotubes Activated Neutrophils to Increase Production of Hypochlorous Acid, the Oxidant Capable of Degrading Nanotubes. *Chemical Research in Toxicology.* **27 (6)**: 1070–1077.

143. Wang, Y. Q., Zhang, H. M., and Cao, J. (2014). Binding of Hydroxylated Single-Walled Carbon Nanotubes to Two Hemoproteins, Hemoglobin and Myoglobin. *Journal of Photochemistry and Photobiology B: Biology.* **141**, 26–35.

144. Yamada, Y.(2021). Nucleic Acid Drugs—Current Status, Issues, and Expectations for Exosomes. *Cancers.* **13 (19)**: 5002.

145. Pray, L. A. (2008). Discovery of DNA Structure and Function: Watson and Crick. *Nature Education.* **1 (1)**: 100.

146. ROSENBERG, B., VAN CAMP, L., and KRIGAS, T. (1965). Inhibition of Cell Division in Escherichia Coli by Electrolysis Products from a Platinum Electrode. *Nature.* **205 (4972)**: 698–699.

147. Ang, W. H., and Dyson, P. J. (2006). Classical and Non-Classical Ruthenium-Based Anticancer Drugs: Towards Targeted Chemotherapy. *ChemInform.* **37 (49)**: 4003–4018.

148. Bruijnincx, P. C., and Sadler, P. J. (2008). New Trends for Metal Complexes with Anticancer Activity. *Current Opinion in Chemical Biology.* **12 (2)**: 197–206.

149. Markman, M. (2003). Toxicities of the Platinum Antineoplastic Agents. *Expert Opinion on Drug Safety.* **2 (6)**: 597–607.

150. Tang, B., Shen, F., Wan, D., Guo, B.-H., Wang, Y.-J., Yi, Q.-Y., and Liu, Y.-J. (2017). DNA-Binding, Molecular Docking Studies, and Biological Activity Studies of Ruthenium-Polypyridyl Complexes. *RSC Advances.* **7 (56)**: 34945–34958.

151. Singh, A. K., Pandey, D. S., Xu, Q., and Braunstein, P. (2014). Recent Advances in Supramolecular and Biological Aspects of Arene Ruthenium(II) Complexes. *Coordination Chemistry Reviews.* **270–271**:31–56.

152. Yan, Y. K., Melchart, M., Habtemariam, A., and Sadler, P. J. (2005). Organometallic Chemistry, Biology and Medicine: Ruthenium Arene Anticancer Complexes. *Chemical Communications.* **38**: 4764.

153. Chaviara, A. T., Christidis, P. C., Papageorgiou, A., Chrysogelou, E., Hadjipavlou-Litina, D. J., and Bolos, C. A. (2005). In Vivo Anticancer, Anti-Inflammatory, and Toxicity Studies of Mixed-Ligand Cu(II) Complexes of Dien and Its Schiff Dibases with Heterocyclic Aldehydes and 2-Amino-2-Thiazoline. Crystal Structure of [Cu(Dien)(Br)(2a-2tzn)](Br)(H2O). *Journal of Inorganic Biochemistry.* **99 (11)**: 2102–2109.

154. Fujimori, T., Yamada, S., Yasui, H., Sakurai, H., In, Y., and Ishida, T. (2005). Orally Active Antioxidative Copper(II) Aspirinate: Synthesis, Structure Characterization, Superoxide Scavenging Activity, and in Vitro and in Vivo Antioxidative Evaluations. *JBIC Journal of Biological Inorganic Chemistry.* **10 (8)**: 831–841.

155. Yin, F., Zhu, J., Hao, J., Cui, J., and Yang, J. (2013). Synthesis, Characterization, Crystal Structure, and Antitumor Activities of a Novel Demethylcantharidato Bridged Copper(II) Phenanthroline Complex. *Science China Chemistry.* **56 (4)**: 481–489.

156. Jeyalakshmi, K., Selvakumaran, N., Bhuvanesh, N. S. P., Sreekanth, A., and Karvembu, R. (2014). DNA/Protein Binding and Cytotoxicity Studies of Copper(II) Complexes Containing N,N′,N″-Trisubstituted Guanidine Ligands. *RSC Advances.* **4 (33)**: 17179–17195.

157. Sundaravadivel, E., Reddy, G. R., Manoj, D., Rajendran, S., Kandaswamy, M., and Janakiraman, M. (2018). DNA Binding and Cleavage Studies of Copper(II) Complex Containing N_2O_2 Schiff Base Ligand. *Inorganica Chimica Acta.* **482**: 170–178.

158. Patel, N. J., Bhatt, B. S., and Patel, M. N. (2019). Heteroleptic N,N-Donor Pyrazole-Based Pt(II) and Pd(II) Complexes: DNA Binding, Molecular Docking, and Cytotoxicity Studies. *Inorganica Chimica Acta.* **498 (II)**: 119130.

159. Quraishi, S., Saha, D., Kumari, K., Jha, A. N., Roy, A. S. (2024). Non-Covalent Binding Interaction of Bioactive Coumarin Esculetin with Calf Thymus DNA and Yeast Transfer RNA: A Detailed Investigation to Decipher the Binding Affinities, Binding Location, Interacting Forces and Structural Alterations at a Molecular Level. *International Journal of Biological Macromolecules.* **257**: 128568.

160. Lewis, E. A., Munde, M., Wang, S., Rettig, M., Le, V., Machha, V., and Wilson, W. D. (2011). Complexity in the Binding of Minor Groove Agents: Netropsin Has Two Thermodynamically Different DNA Binding Modes at a Single Site. *Nucleic Acids Research.* **39 (22)**: 9649–9658.

161. Hampshire, A. J., and Fox, K. R. (2008). The Effects of Local DNA Sequence on the Interaction of Ligands with Their Preferred Binding Sites. *Biochimie.* **90 (7)**: 988–998.

162. Neidle, S. (2001). DNA Minor-Groove Recognition by Small Molecules (up to 2000). *Natural Product Reports.* **18 (3)**: 291–309.

163. Wilson, W. D., Tanious, F. A., Mathis, A., Tevis, D., Hall, J. E., and Boykin, D. W. (2008). Antiparasitic Compounds That Target DNA. *Biochimie.* **90 (7)**: 999–1014.

164. Dolenc, J., Borštnik, U., Hodoščcek, M., Koller, J., and Janežič, D. (2005). An Ab Initio QM/MM Study of the Conformational Stability of Complexes Formed by Netropsin and DNA. The Importance of van Der Waals Interactions and Hydrogen Bonding. *Journal of Molecular Structure: THEOCHEM.* **718 (1–3)**: 77–85.

165. Tse, W. C., and Boger, D. L. (2004). Sequence-Selective DNA Recognition: Natural Products and Nature's Lessons. *Chemistry & Biology.* **11 (12)**: 1607–1617.

166. Waring, M. J., and Bailly, C. (1994). DNA Recognition by Intercalators and Hybrid Molecules. *Journal of Molecular Recognition.* **7 (2)**: 109–122.

167. Hannon, M. J. (2007). Supramolecular DNA Recognition. *Chem Soc Rev.* **36 (2)**: 280–295.

168. Rohs, R. (2005). Molecular Flexibility in Ab Initio Drug Docking to DNA: Binding-Site and Binding-Mode Transitions in All-Atom Monte Carlo Simulations. *Nucleic Acids Research.* **33 (22)**: 7048–7057.

169. Quraishi, S., Nudrat, S., Kumari, K., Marboh, E. W. M., Aguan, K., and Singha Roy, A. (2024). Elucidation of Inhibitory Effects of Bioactive Anthraquinones towards the Formation of DNA Advanced Glycation End Products (DNA-AGEs). *International Journal of Biological Macromolecules.* **269**: 131810.

170. Agudelo, D., Bourassa, P., Bérubé, G., and Tajmir-Riahi, H.-A. (2014). Intercalation of Antitumor Drug Doxorubicin and Its Analogue by DNA Duplex: Structural Features and Biological Implications. *International Journal of Biological Macromolecules.* **66**: 144–150.

171. Zhou, X., Zhang, G., and Wang, L. (2014). Probing the Binding Mode of Psoralen to Calf Thymus DNA. *International Journal of Biological Macromolecules.* **67**: 228–237.

172. Tang, B., Shen, F., Wan, D., Guo, B. H., Wang, Y. J., Yi, Q. Y., and Liu, Y. J. (2017). DNA-Binding, Molecular Docking Studies, and Biological Activity Studies of Ruthenium(II) Polypyridyl Complexes. *RSC Advances.* **7 (56)**: 34945–34958.

173. Mahmood, K., Akhter, Z., Perveen, F., Aisha, N., Bibi, M., Ismail, H., Tabassum, N., Yousuf, S., Ashraf, A. R., and Qayyum, M. A. (2023). Synthesis, DNA Binding, and Biological Evaluation of Benzimidazole Schiff Base Ligands and Their Metal(II) Complexes. *RSC Advances.* **13 (18)**: 11982–11999.

174. Suntharalingam, K., Mendoza, O., Duarte, A. A., Mann, D. J., and Vilar, R. (2013). A Platinum Complex That Binds Non-Covalently to DNA and Induces Cell Death via a Different Mechanism than Cisplatin. *Metallomics,* **5 (5)**: 514–523.

175. Páll, S., Zhmurov, A., Bauer, P., Abraham, M., Lundborg, M., Gray, A., Hess, B., and Lindahl, E. (2020). Heterogeneous Parallelization and Acceleration of Molecular Dynamics Simulations in GROMACS. *Journal of Chemical Physics.* **153 (13)**: 134110.

176. Sahihi, M., and Borhan, G. (2017). The Effects of Single-Walled Carbon Nanotubes (SWCNTs) on the Structure and Function of Human Serum Albumin (HSA): Molecular Docking and Molecular Dynamics Simulation Studies. *Structural Chemistry.* **28 (6)**: 1815–1822.

Chapter 5

Computational Drug Development Studies Against SARS-CoV-2 Spike (S) Protein: A Novel Case Study with Peptidomimetic Molecules

Prabuddha Bhattacharya,[a] Ritam Maharana,[a,b,*] and Sourish Das[a,b,*]

[a]*Department of Chemistry, Mrinalini Datta Mahavidyapith, Kolkata, West Bengal, India*

[b]*Department of Biochemistry, West Bengal State University, Kolkata, West Bengal, India*

b.prabuddha3@gmail.com

COVID-19 emerged as a highly transmissible disease following its initial outbreak in December 2019, caused by the novel coronavirus SARS-CoV-2. Originating in Wuhan, China, this pathogen rapidly escalated into a global health emergency. Among the various genomic products of SARS-CoV-2, the spike (S) protein plays a pivotal role in the initiation of infection. Consequently, the spike protein represents a critical therapeutic target for combating the disease.

* These authors contributed equally to this work.

Computational Methods for Molecular Modelling and Docking
Edited by Arup Kumar Ghosh and Suban Kumar Sahoo

ISBN 978-981-5352-25-2 (Hardcover), 978-1-042-00204-7 (eBook)
www.jennystanford.com

DOI: 10.1201/9781042002047-5

This chapter aims to outline some of the basic underlying protocols in computer-aided drug discovery through concise discussions of drug development against the SARS-CoV-2 spike (S) protein. To further exemplify the discussed methods, a structure-based drug design (SBDD) approach has been followed to screen a large library of peptidomimetic molecules and identify novel hits that may be taken for further clinical consideration.

5.1 Introduction

The COVID-19 pandemic, caused by the SARS-CoV-2 virus, continues to spread globally, even five years after the initial outbreak. Despite mass vaccination efforts, daily case numbers remain high, although they are lower than during the peak of the pandemic. As of October 20, 2024, the World Health Organization (WHO) reported over 77.6 million confirmed cases and more than 7 million deaths across 229 countries [1]. The virus was first identified in Wuhan, China, in December 2019, and was initially named the "Novel Coronavirus" by the WHO. Later, after phylogenetic analysis, it was renamed SARS-CoV-2. This virus is the third zoonotic coronavirus, with the previous two originating from animals like bats and civet cats. The MERS-CoV outbreak, which started in Saudi Arabia, had around 2,500 cases and 800 deaths (as of January 2020) [2]. SARS-CoV-2 spreads primarily through respiratory droplets and fomites when infected individuals are in close, unprotected contact with others. Common symptoms include dry cough, fever, headache, and shortness of breath, with severe cases leading to organ failure and secondary infections. This outbreak is the third major coronavirus epidemic, following the SARS-CoV and MERS-CoV outbreaks [3].

SARS-CoV-2, a positive-sense single-stranded RNA (+ssRNA) virus from the Coronaviridae family, has a genome of approximately 30 kb. It belongs to the beta-coronavirus genus and exhibits an elliptical structure with a diameter of 60–140 nm [4]. The viral genome contains two major open reading frames (ORFs), ORF1a and ORF1b, which are translated into polyproteins pp1a and pp1b. These polyproteins are cleaved by two cysteine proteases—papain-like protease (PL^{pro}, nsp3) and 3C-like protease ($3CL^{pro}$, nsp5)—to produce 16 non-structural proteins involved in viral replication [5]. The RNA-dependent RNA polymerase, comprising nsp12, is central to the replication-transcription complex. The genome also encodes four major structural proteins: Spike (S), nucleocapsid (N), membrane (M), and envelope (E), with the S protein consisting of S1 (containing the receptor-binding domain) and S2 subunits. In addition, SARS-CoV-2 produces several accessory proteins (ORF3, ORF6, ORF7a, ORF7b, ORF8, ORF9b) that play crucial roles in viral pathogenesis and host interaction [6].

Drug development to combat the disease involved targeting various key biomolecular pathways involved in the virus's replication and entry into host cells (Figure 5.1). Some of the most prominent targets, along with the corresponding therapeutic approaches, are given in the following section [7].

(a) **S Protein:** Drugs, including monoclonal antibodies, block the S protein's interaction with the ACE2 receptor, preventing viral entry into cells.
(b) **ACE2 Receptor:** Though not commonly targeted directly, some strategies aim to modulate ACE2 activity to reduce viral entry.
(c) **Main Protease (M^{pro}):** Inhibitors of M^{pro} (also known as $3CL^{pro}$), like nirmatrelvir (Paxlovid), block viral replication by preventing the processing of viral polyproteins.
(d) **RNA-Dependent RNA Polymerase (RdRp):** Drugs like remdesivir and molnupiravir inhibit RdRp, preventing viral RNA replication.
(e) **Nucleocapsid Protein (N):** Research is ongoing into inhibitors of the N protein, which is involved in viral genome packaging.
(f) **Viral RNA and Translation Machinery:** Strategies targeting viral RNA or translation machinery aim to prevent viral protein synthesis.
(g) **Host Cell Factors and Immune Modulation:** Drugs like corticosteroids (e.g., dexamethasone) reduce inflammation and modulate immune responses.
(h) **Viral Fusion Inhibitors:** These drugs block the fusion of the viral envelope with the host cell membrane, preventing infection.
(i) **Papain-like Protease (PLpro):** Inhibitors target PLpro to block the release of early non-structural proteins and prevent the virus from suppressing the host's innate immune response.

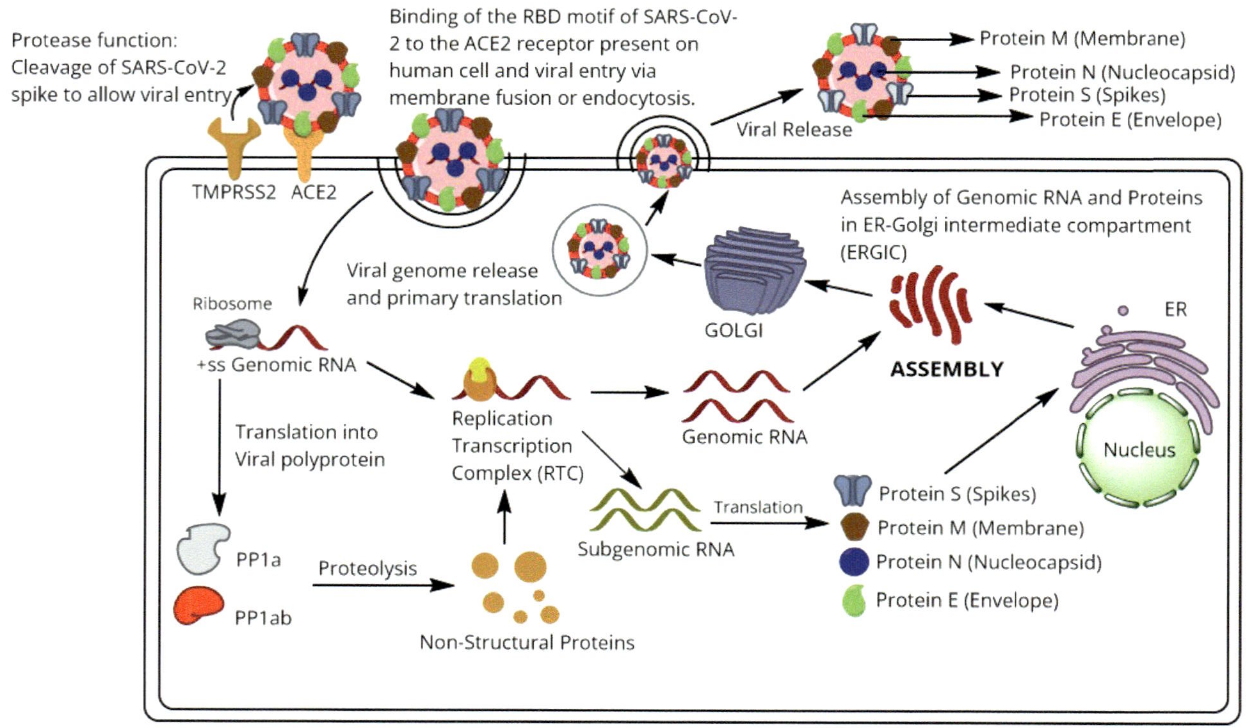

Figure 5.1 SARS-CoV-2 infection cycle.

While several drugs have been developed and approved for emergency use, ongoing research on these and other molecular targets continues to be critical in the fight against COVID-19, especially with the emergence of new variants that may alter the virus's behavior.

At this juncture, it is worth acknowledging that computer-aided drug design (CADD) plays a seminal role in modern drug discovery by leveraging computational techniques to streamline the identification, optimization, and development of therapeutic compounds [8]. It has emerged as a highly multidisciplinary domain encompassing biophysical chemistry, structural bioinformatics, biomolecular chemistry, molecular/quantum mechanics, statistical thermodynamics, and computer science. It enables virtual screening of vast chemical libraries, predicting the binding affinity and specificity of potential drug candidates for target proteins, which significantly accelerates the early stages of drug discovery. Simulating molecular interactions reduces the need for extensive *in vitro* and *in vivo* testing, minimizing time, costs, and potential toxicity. Furthermore, CADD aids in the design of molecules with desired properties, such as improved bioavailability, selectivity, and reduced side effects, making it an indispensable tool for addressing complex diseases. In silico approaches, including virtual screening, ADMET predictions, molecular dynamics simulations, and free energy calculations, have emerged as indispensable tools in the identification and optimization of candidate molecules, offering a cost-effective and time-efficient alternative to traditional experimental drug discovery methods. These computational techniques enable high-throughput screening of vast chemical libraries, providing insights into the binding affinity and specificity of small molecules for the target protein.

In this chapter, we focus on computational methods for identifying small organic molecules that may potentially target the SARS-CoV-2 S protein. Towards this end, this chapter has been primarily divided into three parts as follows: (i) A brief discussion on the structural biology of the SARS-CoV-2 S protein, (ii) Literature survey on the in silico drug development strategies targeting the S protein, (iii) A novel case study through screening of peptidomimetic molecules.

5.2 Structural Biology of the SARS-CoV-2 S Protein

Before going into the in silico drug design studies, it is pertinent to briefly discuss the relevant structural biology of the S protein. The S protein of SARS-CoV-2 is a class-I transmembrane glycoprotein responsible for viral entry into host cells [9]. It is a critical feature of all human coronaviruses and other viruses like HIV, influenza, paramyxoviruses, and Ebola virus [10]. The SARS-CoV-2 S protein, weighing 180–200 kDa, consists of three domains: an extracellular *N*-terminus, a transmembrane domain, and a short intracellular C-terminus. The S protein enables viral fusion with the host cell membrane, and its polysaccharide molecules help the virus evade immune detection during entry. The S protein comprises 1,273 amino acids and has two subunits: S1, which includes a receptor-binding domain (RBD), and S2, which contains fusion-related elements. The S1 and S2 subunits of the SARS-CoV-2 S protein are essential in viral entry and fusion with host cells. The S1 subunit contains the RBD, which binds to the ACE2 receptor on host cells, specifically near aminopeptidase N. The RBD in SARS-CoV-2 contains several conserved residues, although differences from SARS-CoV reduce antibody cross-reactivity, making it a challenging therapeutic target due to its mutability [11].

The S2 subunit comprises a fusion peptide, heptapeptide repeat sequences (HR1 and HR2), a transmembrane domain, and a cytoplasmic domain. This subunit facilitates viral fusion with host cell membranes. The fusion peptide anchors the virus by interacting with the lipid bilayer, while HR1 and HR2 form a six-helix bundle that drives viral fusion and entry. Due to the conserved structure of HR1–HR2 interactions across coronaviruses, the S2 subunit, particularly the HR region, presents a target for antiviral drug development. Peptides derived from HR2 can bind HR1, thereby inhibiting viral infection, highlighting a potential therapeutic avenue for coronavirus infections. The trimeric S protein forms a crown-like structure around the viral particle. In its native state, the S protein is inactive, but upon host-cell entry, it is cleaved by proteases, notably TMPRSS2, to activate the fusion process. The structure of the trimeric S protein has been determined at the atomic level using cryo-electron microscopy [12].

The infection cycle of SARS-CoV-2 begins (Figure 5.1) with the binding of the RBD of the S protein's S1 subunit to the host cell receptor ACE2, which is expressed in respiratory and intestinal epithelial cells [13]. Following receptor engagement, the virus enters the host cell via endocytosis, facilitated by the S2 subunit of the S protein and activated by cleavage by the transmembrane protease serine-type 2 (TMPRSS2), along with other proteases such as trypsin, factor Xa, and plasmin. This cleavage enables membrane fusion and the release of the viral genome into the host cell. Inside the cytoplasm, the viral RNA is translated into two polyproteins, pp1a and pp1b, which are processed by the viral proteases $3CL^{pro}$ (main protease) and PL^{pro} to generate 16 non-structural proteins (NSPs), including NSP12, the RdRp. The RdRp, in conjunction with other NSPs, forms the replication-transcription complex that produces both genomic and subgenomic RNA. The subgenomic RNA is translated into structural proteins—Spike (S), envelope (E), membrane (M), and nucleocapsid (N)—which, along with the genomic RNA, are transported to the Endoplasmic Reticulum-Golgi intermediate compartment (ERGIC) for virion assembly. Finally, newly formed virions are transported via Golgi vesicles and released from the host cell through exocytosis, completing the viral replication cycle [14].

5.3 In Silico Drug Development Strategies Targeting the S Protein

Following the identification of the S protein as a viable medicinal target, numerous studies have used conventional computational tools to identify potent pharmacophoric scaffolds against the target. In this section, we provide a brief overview of notable research outputs in this domain.

In 2020, multiple research groups rapidly shared their findings on molecules that target the S protein of SARS-CoV-2. Maurya and coworkers investigated several Ayurvedic natural products for their antiviral potential against SARS-CoV-2, focusing on structure-based drug design aimed at the S glycoprotein and ACE2 receptor [15]. A literature search was conducted to identify natural products with immunomodulatory properties, demonstrated antiviral activity against RNA viruses, and relevant to symptoms associated with SARS-CoV-2 infection. Pharmacokinetic analysis indicated that these natural products had favorable characteristics, including low molecular weights and adherence to Lipinski's Rule of Five, suggesting they would have good absorption and high oral bioavailability. The drug-likeness of these compounds was evaluated using the SwissADME webserver, and their ADME toxicity was assessed via the admetSAR webserver. Molecular docking studies using Molegro Virtual Docker (MVD-3.0.0) identified five potential ligand-binding sites on the S glycoprotein and ACE2, with one site considered the active one. Curcumin and nimbin exhibited the strongest interactions, as revealed through molecular docking, with the S protein (PDB ID: 6VXX), with MolDock scores of −141.36 and −148.621 kcal/mol, respectively. Both compounds also showed significant binding affinity for the ACE2 receptor (PDB ID: 1R42), with binding scores of −142.647 kcal/mol for curcumin and −140.108 kcal/mol for nimbin, outperforming other natural products and controls.

Kulkarni et al. performed computational studies to evaluate the potential of essential oil components as inhibitors of the SARS-CoV-2 S protein [16]. They identified several compounds - anethole, cinnamaldehyde, carvacrol, geraniol, cinnamyl acetate, L-4-terpineol, thymol, and pulegone, derived from plants in the *Lamiaceae, Lauraceae, Myrtaceae, Apiaceae, Geraniaceae*, and *Fabaceae* families, all of which are known for their antiviral properties. The RBD of the S1 subunit of the SARS-CoV-2 S protein (PDB ID: 6M0J) was used as the target for the potential inhibitory ligands. The essential oil components were selected for their known antimicrobial, antiviral, antifungal, and antibacterial activities. Major constituents of essential oils, such as thyme, cinnamon, clove, star anise, basil, holy basil, eucalyptus, geranium, oregano, and ajwain, were chosen as ligands due to their established efficacy against various pathogens. As part of their in silico pipeline, active sites on the protein were identified using the SCFBio webserver at IIT-D, and docking studies were performed in PyRx with AutoDock Vina. The compound thymol showed the strongest binding affinity with a value of −5.4 kcal/mol. Density Functional Theory (DFT) was applied to analyze the relevant electronic properties of the molecules, offering insights into their structure-activity relationships. The phytochemicals were optimized using the B3LYP functional with a 6-31G (d) basis set in

Gaussian 16 to estimate their molecular properties. The energies of the Highest Occupied Molecular Orbital (HOMO) and the Lowest Unoccupied Molecular Orbital (LUMO) were calculated, and electron density maps for the HOMO and LUMO regions were generated and analyzed. The energy gap between HOMO and LUMO was calculated, with α-terpinene showing the smallest gap (ΔE = 2.58 eV) and *o*-cymene exhibiting the largest energy difference (ΔE = 12.57 eV) among the compounds.

Lu and coworkers developed an assay to examine SARS-CoV-2 S protein-mediated cell-cell fusion, revealing that SARS-CoV-2 has stronger plasma membrane fusion capabilities than SARS-CoV [17]. They reported the X-ray crystal structure of the six-helical bundle (6-HB) core in the HR1 and HR2 domains of the SARS-CoV-2 S2 subunit. Subsequently, they synthesized a series of lipopeptides derived from EK1 (a pan-coronavirus fusion inhibitor) and identified EK1C4 as the potent inhibitor of SARS-CoV-2 S protein-mediated membrane fusion and pseudovirus infection, showing IC_{50} values of 1.3 and 15.8 nM—about 241 and 149 times more effective than the original EK1 peptide. EK1C4 also proved effective against membrane fusion and infection from other tested human coronavirus pseudoviruses, including SARS-CoV and MERS-CoV, and reduced replication of five live human coronaviruses, including SARS-CoV-2. The interaction model of EK1C4 and the HR1 domain of SARS-CoV-2 was predicted using SWISS-MODEL and AutoDock 4 software. Structural simulation and docking of EK1C4 with the HR1 domain of SARS-CoV-2 showed that the cholesterol group could anchor to the HR1 trimer via hydrophobic interactions, enhancing binding affinity and facilitating the binding of the peptide to its target. The 6-HB fusion core structure formed by HR1 and HR2 in the S2 subunit of human coronaviruses plays a crucial role in viral membrane fusion, making it a target for drug development.

Souza et al. reported eight synthetic peptides (MoCBP3-PepI, MoCBP3-PepII, MoCBP3-PepIII, RcAlb-PepI, RcAlb-PepII, RcAlb-PepIII, PepGAT, and PepKAA) as potential inhibitors of the SARS-CoV-2 S (S) protein [18]. The findings revealed that all peptides interacted with the S protein, with Mo-CBP3-PepII and PepKAA showing the strongest binding affinities. The initial docking experiments tested the peptides against the S protein in both its open (PDB ID: 6VYB) and closed (PDB ID: 6VXX) states. Afterward, the complexes formed between the S protein and the peptides were docked with the human ACE2 receptor (PDB ID: 1R42). The ClusPro 2.0 webserver was used for all molecular docking simulations. Among the peptides interacting with the S protein's S1 domain, Mo-CBP3-PepII showed the lowest binding energy, at −814.4 kJ/mol. Six key systems—combinations of the S protein (both open and closed states) with ACE2 and the two peptides (Mo-CBP3-PepII and PepKAA)—were minimized and equilibrated using GROMACS 2018.4. The system's topology was generated using the OPLS all-atom force field. Structural analysis was carried out using the PDBsum database to examine the interface regions from the docking tests. MD simulations of these eight antimicrobial peptides in complex with the SARS-CoV-2 S protein were performed. The interaction of Mo-CBP3-PepII with the S protein caused RMSD values to increase up to 1.258 Å in the open state and 1.331 Å in the closed state. Similarly, PepKAA induced conformational changes in the S protein's S2 domain, with RMSD values shifting up to 1.262 Å and 1.342 Å in the open and closed states, respectively. These results suggest that Mo-CBP3-PepII and PepKAA can alter the conformation of the S protein, potentially reducing its binding affinity to ACE2. LigPlot software was used to visualize 2D representations of hydrogen bonds and hydrophobic interactions, while PyMOL was employed to generate 3D structural models. Refinement and validation of the structures were performed using Galaxy Refine Complex (a tool from the GalaxyWEB molecular interaction webserver). The 3D structures were validated for steric conflicts using the MolProbity and PROCHECK webservers.

Padmanabhan and coworkers published a study on a computational analysis involving virtual screening, molecular docking, MD simulations, and MM-GBSA calculations using ligands derived from DrugBank and PubChem libraries [19]. After removing the ACE2 receptor from the complex, they performed docking of the S protein (PDB: 6LZG) and ligands using AutoDock Vina. To filter out the suitable ligands, they applied Lipinski's Rule of Five. The selected ligands included 1407 FDA-approved drug molecules from DrugBank and 6942 pharmacologically annotated compounds from PubChem. Molecular docking was performed with AutoDock Vina, using blind docking to scan the protein surface and identify ligands binding at the ACE2–S-protein interaction site. Bisoxatin exhibited the best binding free energy of −7.4 kcal/mol among all the compounds tested. For the molecular dynamics simulations, the

Desmond module was used. The docked poses were subjected to 100 ns atomistic MD simulations. The resulting trajectory was analyzed using both Desmond and Maestro from the Schrödinger Suite. The root mean square deviation (RMSD) of the complex was 2.5 Å, and MM-GBSA calculations, performed with the Prime MM-GBSA script in Schrödinger Suite 2019-2, estimated the binding free energy of the protein–ligand interaction to be –30.52 kcal/mol, confirming the stability and strength of the Bisoxatin-SARS-CoV-2 S protein complex. Their analysis focused on the RBD of the SARS-CoV-2 S protein and ultimately identified Bisoxatin as a potential repurposable drug. The compound exhibited strong binding at the RBD-ACE2 interface, suggesting its potential to inhibit the interaction between ACE2 and the S protein.

In 2021, Pandey et al. reported the use of in silico methods to assess the binding potential of 10 naturally occurring compounds (both flavonoids and non-flavonoids) for drug development [20]. Their molecular dynamics simulations revealed that kaempferol, quercetin, and fisetin strongly bind to the human ACE2-S protein complex, particularly at the interface. Further analyses, including free energy calculations, validated these findings. As part of their computational workflow, all compounds adhered to Lipinski's Rule of Five, suggesting they could be candidates for anti-COVID drugs, though further *in vitro* validation would be needed. AutoDock Vina was used to assess the binding of curcumin and hydroxychloroquine to the SARS-CoV-2 S protein (PDB ID: 6VYB), with a blind docking of the ligands to ascertain the probable binding site further. The ligand, kaempferol, showed binding affinity of –7.4 kcal/mol with S1 domain, while quercetin and fisetin, showed binding affinities of –8.5 kcal/mol upon interacting with the S2 domain. Molecular dynamics simulations were performed using GROMACS 2018.1 with the CHARMM36 force field and CGenFF-derived ligand parameters. Each system was simulated for 100 ns, with key analyses focusing on RMSD, radius of gyration, and binding free energy, calculated via the MM-PBSA method using snapshots from the last 20 ns to ensure convergence. The RMSD time evolution plot showed that the fisetin-humanACE2-S complex reached equilibrium around 40 ns and remained stable until the end of the simulation at 100 ns. In contrast, the kaempferol complex stabilized at around 17–20 Å. The RMSD distribution showed splits near 15 Å, 20 Å, and 30 Å. The radius of gyration (Rg) distribution revealed that fisetin's Rg values ranged from 42.0 to 43.0 Å, while kaempferol and quercetin exhibited higher Rg population densities around 42.0–44.0 Å and 47.0–48.0 Å, respectively. Using MD trajectories, the binding free energies of the ligands with the S protein were calculated via MM-PBSA. Among the compounds, quercetin showed the lowest binding free energy (−22.17 ± 3.04 kcal/mol), indicating the strongest affinity for the human ACE2-S protein, followed by fisetin (−21.11 ± 3.49 kcal/mol), and kaempferol, which had the weakest affinity (−15.07 ± 2.42 kcal/mol). Additionally, ADME profiling was conducted using the online pkCSM webserver.

In 2022, Gangarapu et al. studied the properties of phytoconstituents from the Siddha formulation Kabasura Kudineer and a novel herbal preparation, JACOM, which are traditionally used to treat viral fevers [21]. Molecular docking studies of 37 phytochemicals against the SARS-CoV-2 S protein (PDB ID: 6VSB) identified strong binding affinities for chrysoeriol and luteolin from Kabasura Kudineer, and quercetin from JACOM. In addition, in silico predictions of pharmacokinetics and toxicity suggested that these compounds have good oral bioavailability and low toxicity. Based on these findings, the researchers proposed a new formulation called "SNACK-V," which included nine phytochemicals from six different herbs. In their computational studies, they used Cresset Flare Docking software to analyze 37 phytochemicals from Kabasura Kudineer Chooranam and JACOM, docking them to the SARS-CoV-2 S protein. Ligands were optimized in Cresset Flare before docking. Among the tested phytocompounds, all exhibited effective binding with the SARS-CoV-2 S protein, with ligand fitness (LF) rank scores ranging from −5.75 to −11.03. The highest docking LF rank scores were observed for chrysoeriol and luteolin from Kabasura Kudineer Chooranam (−11.478 and −11.392, respectively), and quercetin from JACOM (−11.159). Other compounds like 5-hydroxy-7,8-dimethoxyflavanone, cirsimaritin, and scutellarein showed moderate binding affinity with LF scores of −9.035, −9.228, and −10.27, respectively. The remaining compounds had lower binding affinities. The LF rank score reflects the binding strength of the protein–ligand complex. The pkCSM webserver was used for pharmacokinetic predictions.

Kar et al. investigated 17 selected phytocompounds by docking them into the binding pocket of the RBD of the SARS-CoV-2 S protein, leveraging existing knowledge of the active site residues that interact with the human ACE2 receptor [22]. Phytochemicals from *Justicia adhatoda, Ocimum sanctum,* and *Swertia chirata* were screened from Dr. Duke's Phytochemical and Ethnobotanical Databases, with their structures retrieved from PubChem. Their 3D structures were generated using Open Babel and optimized with the PRODRG webserver. Molecular docking was performed with AutoDock Vina on the active sites of the RBD (PDB ID: 6LZG), M^{pro} (PDB ID: 6LU7), and RdRp (PDB ID: 7BV2), with results validated using the DINC webserver and receptor-ligand interactions analyzed in PyMOL. The compounds anisotine, adhatodine, beta-carotene, eugenol, mangiferin, and beta-amyrin showed binding affinities of 7.8 ± 0.01 kcal/mol, 7.0 ± 0.02 kcal/mol, 7.2 ± 0.01 kcal/mol, 7.3 ± 0.03 kcal/mol, 7.5 ± 0.03 kcal/mol, and 7.1 ± 0.01 kcal/mol, respectively, with anisotine displaying the highest potential in terms of binding energy. The physicochemical properties of these 17 phytochemicals were analyzed using SwissADME and pkCSM webservers, and toxicity parameters were evaluated using ProTox-II and vNN-ADMET. Molecular dynamics simulations were performed to validate the docking results and assess the stability of receptor-ligand complexes over 120 ns, using GROMACS 2019 with GROMOS96 43a1 force field parameters. The average RMSD values for the anisotine-RBD SARS-CoV-2 S protein, anisotine-SARS-CoV-2 M^{pro}, and amarogentin-SARS-CoV-2 RdRp complexes were 1.8 Å, 1.3 Å, and 1.6 Å, respectively. Further analysis revealed that the average RMSF values for these complexes were 1.3 Å, 2.3 Å, and 1.9 Å, respectively. Interaction stability was analyzed using LigPlot, and binding free energies were calculated using the MM-PBSA method, with the 'g_mmpbsa' script in GROMACS. The final binding energy scores for the anisotine-RBD SARS-CoV-2 S protein, anisotine-SARS-CoV-2 M^{pro}, and amarogentin-SARS-CoV-2 RdRp complexes were 39.39 ± 1.29 kcal/mol, 42.44 ± 1.27 kcal/mol, and 37.38 ± 0.79 kcal/mol, respectively, suggesting significant conformational stability of the complexes after 120 ns of MD simulation. Detailed analysis of the anisotine and amarogentin complexes confirmed their stability, showing no changes in hydrogen bonds or hydrophobic interactions, which supported the docking results.

An interesting study reported by Mondal and coworkers explored the "Ex_3Lx_6L" motif (abbreviated as E-L-L) located in the HR2 domain of S proteins in both human and nonhuman coronaviruses [23]. This motif plays a key role in stabilizing the post-fusion six-helix bundle (6-HB) structure, which is essential for viral entry via membrane fusion. Mutations in this motif decrease the fusogenicity of the S protein but do not affect its stability or localization to the membrane. The amino acid sequences of multiple coronavirus S proteins were obtained from the NCBI database, and sequence alignment of the HR1 and HR2 regions was carried out using ClustalW. The conservation of the "E-L-L" motif across coronavirus HR2 regions was assessed, and HR1–HR2 peptide complexes, both wild-type (WT) and mutant variants of HR2, were subjected to 100 ns molecular dynamics simulations using GROMACS 2019 with CHARMM36 force field. Mutations were incorporated using Chimera. MM-GBSA analyses were performed using gmx_MMPBSA software, with 200 snapshots from the last 10 ns of the MD simulations. The results indicated that the WT bis-peptide had the most stable ΔG (−35.02 ± 3.83 kcal/mol). Mutants such as E1182A (−30.44 ± 2.32 kcal/mol) and E1182K (−33.96 ± 3.58 kcal/mol) showed minor shifts in stability, while the double mutant (E1182K–L1186G) exhibited moderate increases in RMSD, Rg, and RMSF. The triple mutant (E1182K–L1186G–L1193G) showed destabilization of −16.80 ± 3.01 kcal/mol, highlighting the importance of the "E-L-L" motif for structural integrity. The MM-GBSA calculations further reinforced that the full "E-L-L" motif plays a critical role in stabilizing the HR1–HR2 complex. Through the method of molecular docking of the 6-HB (PDB ID: 6LXT – single pair of HR1–HR2 helices) with 25 FDA-approved or under-trial molecules, posaconazole, an anti-fungal drug, was identified as a lead molecule that binds to the "E-L-L" motif and may play a role in preventing the formation of the 6-HB and thereby inhibiting SARS-CoV-2 infection in cells. The 3D structures of ligands were sourced from Chem 3D Pro or databases like PubChem and DrugBank. Docking studies were performed using AutoDock 4.2, and posaconazole's binding energy was found to be −6.51 kcal/mol. The protein–ligand interactions were analyzed using LigPlot v2.2.4. A multidimensional in silico screening approach was employed, focusing on criteria such as binding energy, cluster population, and the interaction with the "E-L-L" motif. Mutations in the "E-L-L" motif conferred resistance, confirming the drug's specificity. Despite its potent ability to block viral entry mediated by the S protein, mutations in the "E-L-L" motif conferred complete resistance to posaconazole, underscoring its specificity. The findings demonstrate that posaconazole acts early in infection by selectively blocking membrane fusion and the release of the viral genome into host cells. It is also effective against major SARS-CoV-2 variants, including β, κ, δ, and o. Posaconazole inhibited SARS-CoV-2 pseudovirus entry in a dose-dependent manner, with 90% inhibition at 10 μM and an IC_{50} of 3.37 μM. It showed no activity against VSV G pseudovirus or other viruses such as influenza A/H1N1 and HSV1, indicating its selective inhibition of SARS-CoV-2. Additionally, posaconazole blocked 6-HB formation, reduced syncytium formation in fusion assays, and effectively inhibited entry of pseudoviruses from multiple SARS-CoV-2 variants (D614G, κ, δ, and o), confirming its ability to block the membrane fusion.

5.4 A Novel Case Study Through Screening of Peptidomimetic Molecules

As discussed earlier, through multiple sequence alignment and subsequent MD simulations, Mondal and coworkers identified that the E1182-L1186-L1193 motif (Figure 5.2) in the HR2 domain of SARS-CoV-2 S plays a vital role in the 6-helix bundle (HB) formation and in the fusion-mediated viral entry [23]. Viral fusion refers to the process by which the viral membrane fuses with the host cell membrane, allowing the viral genome to enter the host cytoplasm. This process is mediated by the S protein and other host cell factors. The formation of a six-helix bundle (6-HB) is crucial for viral fusion [17, 24]. Following cleavage, the fusion peptide in the S2 subunit is exposed and undergoes a conformational change, enabling its insertion into the host cell membrane. This interaction brings the viral and host cell membranes into proximity, and the HR1 and HR2 regions form a six-helix bundle, facilitating membrane fusion. The formation of 6-HB pulls the viral membrane tightly against the host membrane, allowing the two membranes to fuse and the viral genome to enter the host cytoplasm. Hence, Mondal and coworkers explored the possibility of inhibiting viral fusion using organic small molecules. Inspired by the same work, we decided to screen a large library of peptidomimetic molecules as potential fusion inhibitors against the specific "E-L-L" motif as our target.

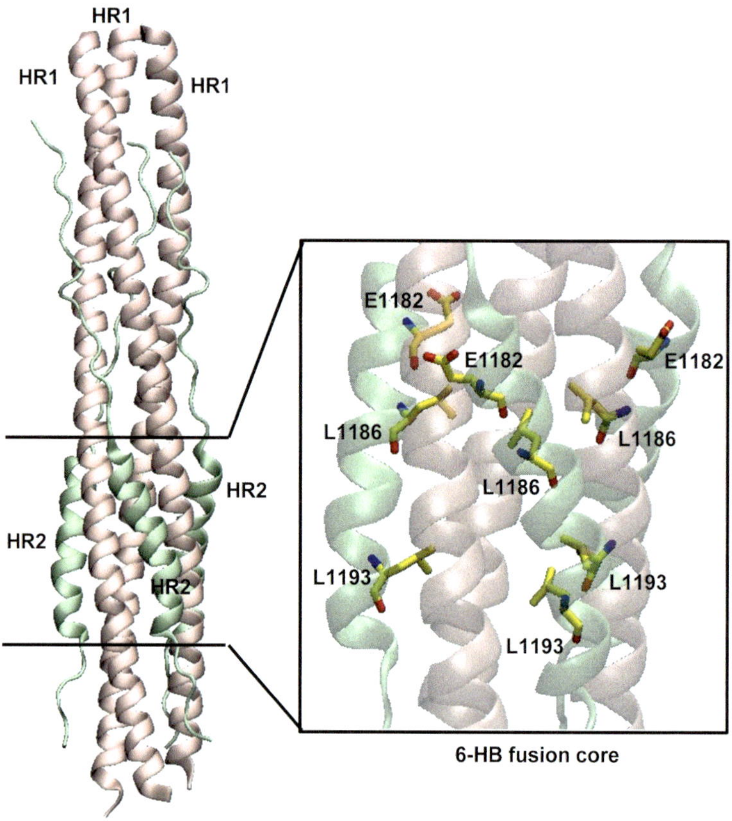

Figure 5.2 The structure of the SARS-CoV-2 6-HB is displayed in a cartoon representation, with HR1 shown in pink and HR2 in lime green colors, respectively. The inset shows the fusion core of 6-HB, where the "E-L-L" motif is highlighted.

Peptidomimetic molecules are small protein-like molecules that mimic the structure of real peptides or bioactive proteins [25]. The term 'peptidomimetic' describes the modification of peptide sequence with improved biological properties and the replacement of the peptide backbone with a molecular scaffold. Peptidomimetic molecules are designed in a way that they become biologically active, metabolically stable, and highly selective. As compared to other simple organic molecules, peptides and peptidomimetic molecules have more stereogenic centers and biological relevance. Peptidomimetic molecules can be produced in various ways. Some of them include (i) modifying the peptide backbone to introduce non-natural features while retaining or enhancing biological activity, (ii) replacing the side chains of amino acids on the α-carbon with substitutions on the nitrogen atom of the peptide backbone, and (iii) constraining peptides or peptidomimetics into cyclic structures to improve stability and binding affinity.

Peptidomimetic molecules offer significant advantages over natural peptide-based drugs, including improved stability, enhanced bioavailability, and easier synthesis. Unlike natural peptides, which are prone to enzymatic degradation, peptidomimetics are designed to resist proteolysis, increasing their therapeutic half-life. They also have better oral bioavailability and can more efficiently cross biological barriers. Additionally, their synthetic nature allows for cost-effective production and optimization for greater potency and selectivity, making them an alternative to natural peptides in drug development [26].

These reasons prompted us to explore a diverse range of peptidomimetic molecules as potential inhibitors of the 6-HB formation.

The computational methodology involved creating a 3D ligand library of 8000 peptidomimetic molecules from different structural databases. The ligand structures were procured in SDF format and subsequently optimized to generate their energy-minimized and hydrogen-added 3D conformation using MMFF94 [27] force field implemented through Open Babel [28]. The protonation state of each ligand was ensured at the physiological pH of 7.4. The 3D structure of the post-fusion core of the SARS-CoV-2 S2 subunit was obtained from the RCSB PDB database, having PDB ID 6LXT. UCSF Chimera 1.16 [29] was used to remove the water and other heteroatoms, followed by the addition of hydrogen atoms and furnishing the Kollman charges [30]. One of the chains (a single pair of HR1–HR2 helices) from the homo-trimer (three pairs of HR1–HR2 fusion-core helices) was used for docking. The underlying concept of this approach is consistent with that of Mondal and coworkers, wherein the strong binding of the ligand to one of the protomers interferes with the binding of the other two chains, thereby disrupting the formation of the 6-HB structure [23]. Batch docking was performed using AutoDock Vina version 1.2.5 [31], employing the following parameters: grid center coordinates set to –2.335, 1.035, and

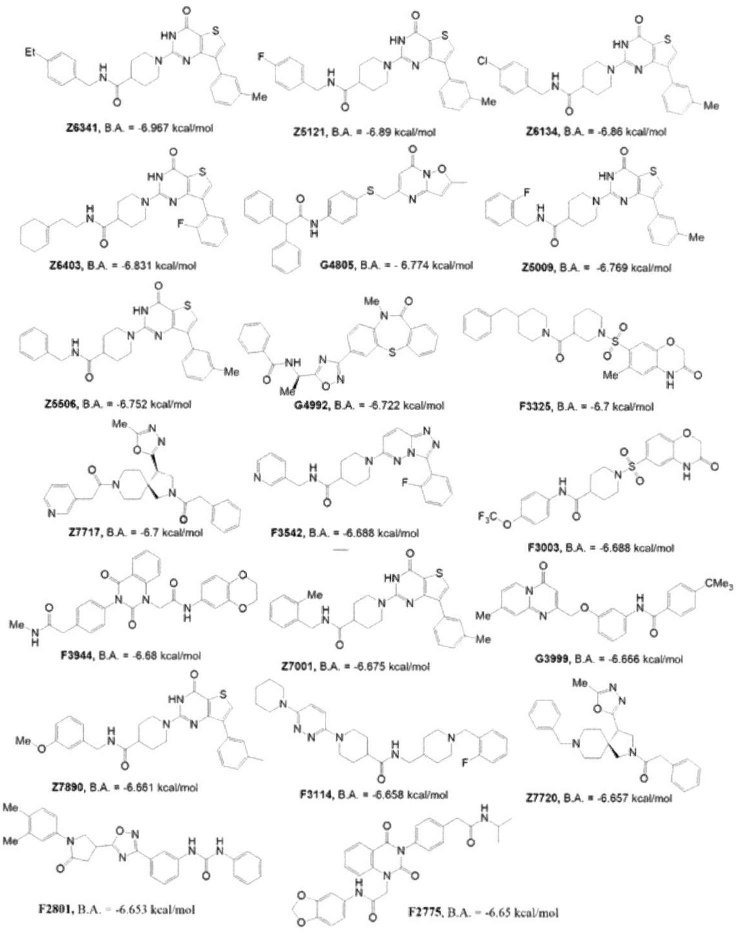

Figure 5.3 Structures of the top 20 ligands ranked by binding affinity (B.A.) values calculated using AutoDock Vina.

−28.732; grid dimensions defined as 18 Å, 14 Å, and 30 Å; and exhaustiveness level of 16. Along with the peptidomimetic molecules, posaconazole (reported by Mondal and coworkers) was also docked as a ligand. A single seed was used to dock all the ligands reported in this work. The binding affinity ranged from −6.97 kcal/mol to 3.718 kcal/mol. The molecules were ranked based on their binding affinities with ligand **Z6341,** showing the highest binding affinity. It was notable that posaconazole exhibited a binding affinity of −6.24 kcal/mol and ranked 231st. Thus, our docking studies revealed a library of more than 200 peptidomimetic molecules that might show better inhibition of 6-HB than the reported drug posaconazole. Figure 5.3 shows the structure of the top 20 ligands. The protein–ligand interaction profile (obtained using LigPlot v2.2.4) [32] shown in Table 5.1 depicts that all the ligands interacted primarily through hydrophobic interactions, which is understandable since the three parallel HR1 helices from three separate protomers create a trimeric coiled-coil hydrophobic core, around which the HR2 helices fold back and intertwine in an antiparallel orientation.

The top 20 candidates were studied for their ADMET (Absorption, Distribution, Metabolism, Excretion, and Toxicity) properties using the pkCSM and SwissADME web webservers (Table 5.2) [33–35]. These studies are crucial in computational drug design as they provide essential insights into the pharmacokinetic and safety profiles of potential drug candidates. By predicting how a compound behaves in the body, ADMET predictions help to identify drug candidates early in the design process, minimizing the risk of failure in later-stage development.

Hydrogen-bond donors and acceptors can alter passive diffusion rates across cell membranes. LogP, or the logarithm of the partition coefficient, is called the lipophilicity of a drug. Lipinski's rule of five defines a molecule's oral bioavailability [36]. This is based on four physiological properties, including molecular mass less than 500 Da, no more than 5 HBD, no more than 10 HBA, and a partition coefficient not greater than 5. Gastrointestinal (GI) absorption determines if the drug reaches the bloodstream. The GI tract is the primary route for absorbing drugs and toxicants, and the small intestine is where most absorption happens. P-glycoprotein (P-gp) substrates are molecules that can be actively effluxed from cells by the ATP-binding cassette transporter P-gp, significantly affecting drug absorption, distribution, and the therapeutic efficacy of a given drug target. The BBB permeance refers to the ability of a drug to cross the blood-brain barrier (BBB), allowing it to reach the central nervous system (CNS). Cytochrome P450 enzymes, including CYP1A2, metabolize several important drugs, pharmaceuticals, and chemicals. To increase the efficacy of a given drug molecule, we should use CYP1A2 inhibitors with it if it is not a CYP1A2 inhibitor, but only in limited amounts. The Ames test is a widely employed test to assess the mutagenic potential of a drug using bacteria. A positive test indicates that the compound has mutagenic abilities and, therefore, might act as a carcinogen. The maximum recommended tolerated dose (MRTD) provides an estimate of the toxic dose threshold of chemicals in the human body.

All 20 top-ranked ligands (except three) did not exhibit any Lipinski violations and had reasonably acceptable maximum tolerated doses (log mg/kg/day). Thus, most of the screened candidates showed favorable ADMET properties for subsequent clinical considerations.

Table 5.1 Protein–ligand interaction data for the corresponding best-docked pose

		Interaction Profile	
Ligand ID	**Binding Affinity (kcal/mol)**	**Residues Involved in Hydrogen Bonding, along with Donor-Acceptor Distance**	**Residues Involved in Hydrophobic Interactions**
Z6341	−6.967	–	F927, I934, S1196, I931, L938, R1185, V1189, **E1182**, **L1186**, **L1193**
Z5121	−6.89	–	L945, L938, I934, I931, R1185, V1189, **E1182, L1186, L1193**
Z6134	−6.86	–	F927, L938, R1185, I934, V1189, **E1182, L1186, L1193**
Z6403	−6.831	–	I931, I934, L938, R1185, V1189, **E1182, L1186, L1193**
G4805	−6.774	S 1196 (3.00 Å, 3.06 Å)	L945, N1192, V1189, R1185, **E1182, L1186, L1193**
Z5009	−6.769	–	I931, I934, L938, L945, V1189. R1185, **E1182, L1186, L1193**
Z5506	−6.752		I931, I934, L938, L945, V1189, R1185, **E1182, L1186, L1193**

Ligand ID	Binding Affinity (kcal/ mol)	Interaction Profile	
		Residues Involved in Hydrogen Bonding, along with Donor-Acceptor Distance	Residues Involved in Hydrophobic Interactions
G4992	-6.722	-	R1185, L938, V1189, I934, I945, I931, **E1182, L1186, L1193**
F3325	-6.7	-	L938, L945, I1179, V1189, R1185, **E1182, L1186**
Z7717	-6.7	-	L938, V1189, L945, R1185, I1179, **E1182, L1186**
F3542	-6.688	-	I934, L938, R1185, V1189, **E1182, L1186, L1193**
F3003	-6.688	-	L945, I1179, R1185, V1189 **E1182, L1186, L1193**
F3944	-6.68	-	I934, L938, L945, R1185, S1196, N1192, Q1195, V1189, **E1182, L1186, L1193**
Z7001	-6.675	-	V1189, I931, I934, L938, R1185, L945, **E1182, L1186, L1193**
G3999	-6.666	-	L938, I934, L945, I1179, V1189, R1185, **E1182, L1186, L1193**
Z7890	-6.661	-	I931, I934, L938, L945, V1189, S1196, R1185, **E1182, L1186, L1193**
F3114	-6.658	-	F927, I931, I934, L945, L938, V1189, I1179, R1185, **E1182, L1186, L1193**

(Continued)

Table 5.1 (*Continued*)

		Interaction Profile	
Ligand ID	**Binding Affinity (kcal/mol)**	**Residues Involved in Hydrogen Bonding, along with Donor-Acceptor Distance**	**Residues Involved in Hydrophobic Interactions**
Z7720	-6.657	R1185 (2.98 Å)	L945, L938, I1179, V1189, **E1182, L1186**
F2801	-6.653	–	L938, L945, R1185, V1189, **E1182, L1186, L1193**
F2775	-6.65	–	**E1182, L1186, L1193,** L945, L938, S1196, I934, V1189, N1192, R1185

Generally, docking provides an initial static model of the protein–ligand interaction, but MD simulations allow for the exploration of conformational flexibility and the influence of solvent, temperature, and pressure, which can impact binding interactions. MD simulations help identify key molecular interactions, stability of binding poses, and potential conformational changes that may not be captured by docking alone, providing a more accurate and realistic assessment of the protein–ligand interaction. Hence, a 150 ns all-atom MD simulation was performed for the five top-ranked ligands (in complex with the protein) to refine and validate the conformational stability, binding affinity, and interaction dynamics of these protein–ligand complexes over time under physiological conditions. For the purpose of further comparison, apo-proteins (a single pair of HR1–HR2 fusion-core helices and three pairs of HR1–HR2 fusion-core helices) were also made to undergo MD simulation under identical conditions.

Table 5.2 ADMET prediction for the top 20 candidates

Ligand ID	HBD	HBA	LogP	Lipinski Violation	GI Absorption	P-gp Substrate	BBB Permeant	CYP1A2 Inhibitor	AMES Toxicity	Max. Tolerated Dose (log mg/kg/day)
Z6341	2	3	4.77	No	High	Yes	No	No	No	0.355
Z5121	2	4	4.47	No	High	Yes	No	No	Yes	0.508
Z6134	2	3	4.70	No	High	Yes	No	No	Yes	0.52
Z6403	2	4	4.70	No	High	Yes	No	No	No	0.265
G4805	1	4	4.71	No	High	No	No	No	No	0.246
Z5009	2	4	4.44	No	High	Yes	No	No	Yes	0.548
Z5506	2	3	4.03	No	High	Yes	No	No	Yes	0.519
G4992	1	5	3.89	No	High	Yes	No	Yes	Yes	0.059
F3325	1	6	2.89	Yes (MW>500)	High	Yes	No	No	No	−0.455
Z7717	0	6	2.43	No	High	Yes	No	No	No	−0.584
F3542	1	6	2.59	No	High	Yes	No	No	No	−0.092
F3003	2	10	2.43	No	High	Yes	No	Yes	No	−0.55
F3944	2	6	2.26	Yes (MW>500)	High	Yes	No	No	No	0.546

(*Continued*)

Table 5.2 (*Continued*)

Ligand ID	HBD	HBA	LogP	Lipinski Violation	GI Absorption	P-gp Substrate	BBB Permeant	CYP1A2 Inhibitor	AMES Toxicity	Max. Tolerated Dose (log mg/kg/day)
Z7001	2	3	4.33	No	High	Yes	No	No	Yes	0.514
G3999	1	4	4.24	No	High	Yes	No	No	Yes	−0.178
Z7890	2	4	4.11	No	High	Yes	No	No	Yes	0.499
F3114	1	5	3.72	No	High	Yes	Yes	No	No	−1.008
Z7720	0	5	3.37	No	High	Yes	Yes	No	No	−0.216
F2801	2	5	3.96	No	High	Yes	No	No	No	0.093
F2775	2	6	2.95	Yes (MW>500)	High	Yes	No	No	No	0.452

HBD: Hydrogen bond donor, HBA: Hydrogen bond acceptor, LogP: Logarithm of partition coefficient, GI absorption: Gastrointestinal absorption, P-gp substrate: P-glycoprotein substrate, BBB Permeant: Blood brain barrier permeant.

For all five ligands, MD simulation studies were done for the following cases:

(a) Complex between ligand and a single pair of HR1–HR2 fusion-core helices: In this case, the input structure was the one obtained as the best-docked pose as discussed earlier. The objective was to understand the time-dependent stability of the single-pair HR1–HR2 domain in complex with ligands.

(b) Complex between ligand and the three pairs of HR1–HR2 fusion-core helices: For this second approach, the other two pairs of HR1–HR2 fusion-core helices (initially excluded from the docking) were added to the best-docked pose to manually construct the 6-HB fusion core in the complex with the ligand. The aim was to study the effect of these ligands on the overall structural integrity of the 6-HB.

Thus, the overall goal of the simulation studies was to optimally identify the ligand(s) that showed the most favorable binding to the single pair of HR1–HR2 fusion-core helices while inducing maximum distortion of 6-HB.

GROMACS 2023.3 was used for the MD simulations in this study [37]. Protein topology was prepared using the CHARMM36 force field, while the CHARMM-compatible ligand topology and parameters were generated using the CGenFF webserver [38–39]. To maintain isotonic conditions representative of physiological environments, each system was neutralized with Na^+ and Cl^- ions at a concentration of 0.15 M, followed by solvation with the TIP3P water model [40]. Energy minimization was conducted using the steepest descent algorithm with a maximum of one million steps or until the forces were reduced below 10.0 kJ/mol. Each system was equilibrated through a 1 ns NVT ensemble followed by a 1 ns NPT ensemble. Solvent molecules were allowed unrestricted movement to achieve solvent equilibrium. Long-range electrostatic interactions were calculated with the Particle Mesh Ewald (PME) method, employing a van der Waals and electrostatic cutoff of 1.2 nm and a Fourier spacing of 1.2 nm [41]. Subsequently, each equilibrated system underwent a 150 ns MD simulation with a 2-fs time step. System temperature was set to 310 K using the V-rescale weak coupling method, while pressure was maintained at 1 atm using the Parrinello-Rahman barostat [42]. Following periodic boundary condition (PBC) corrections and least-squares fitting to the protein backbone, various structural parameters, such as root mean square deviation (RMSD) and radius of gyration (Rg), and hydrogen bonds, were analyzed using GROMACS' built-in functions. Cluster analysis was done with a 0.35 nm cutoff with the entire trajectory at 500 frame intervals. The representative cluster was considered to visualize the system(s) of interest.

The free energies of binding for the protein–ligand complexes were obtained using gmx_MMPBSA 1.6.3 [43–44]. Frames from the entire simulation trajectory of 150 ns, as well as the last 50 ns (both with an interval of 1 ns), were considered for the Molecular Mechanics/Poisson-Boltzmann Surface Area (MM-PBSA) calculations. The internal and external dielectric constants of 1 and 80, respectively, were used, while the iPB model 2 was selected for the analyses. The binding energy ($\Delta G_{binding}$) of a protein–ligand system is given by the following equation: $\Delta G_{binding} = G_{complex} - (G_{receptor} + G_{ligand})$, where $G_{receptor}$ is the binding energy of the free receptor, and G_{ligand} is the energy of the unbounded ligand. $\Delta G_{binding}$ can also be represented as $\Delta G_{binding} = \Delta H - T\Delta S$, where ΔH corresponds to the enthalpy of binding and $-T\Delta S$ to the conformational entropy after ligand binding. When the entropic term is not considered, the computed value is known as the **effective free energy**, which is usually good enough for comparing relative binding free energies for similar systems [44]. The lower value of binding energy (higher negative value) of a ligand indicates the higher binding affinity of that ligand toward the target protein.

Table 5.3 Average values of some of the geometrical parameters obtained from post-MD analysis

One pair of HR1–HR2 fusion-core helices	
System	**Average RMSD (nm)**
Apo-protein	0.73 ± 0.18
Protein (in complex with Z6341)	0.62 ± 0.13
Protein (in complex with Z5121)	0.58 ± 0.15
Protein (in complex with Z6134)	0.63 ± 0.12
Protein (in complex with Z6403)	0.67 ± 0.13
Protein (in complex with G4805)	0.66 ± 0.13

One pair of HR1–HR2 fusion-core helices	
System	**Average Rg (nm)**
Apo-protein	2.93 ± 0.1
Protein (in complex with Z6341)	2.97 ± 0.08
Protein (in complex with Z5121)	3.02 ± 0.06
Protein (in complex with Z6134)	3.06 ± 0.09
Protein (in complex with Z6403)	2.97 ± 0.07
Protein (in complex with G4805)	3.01 ± 0.05
Ligand	**Average RMSD (nm)**
Z6341	0.23 ± 0.08
Z5121	0.37 ± 0.06
Z6134	0.40 ± 0.09
Z6403	0.35 ± 0.05
G4805	0.31 ± 0.05
Three pairs of HR1–HR2 fusion-core helices (6-HB)	
System	**Average RMSD (nm)**
Apo-protein	0.15 ± 0.02
Protein (in complex with Z6341)	0.19 ± 0.03
Protein (in complex with Z5121)	0.18 ± 0.02
Protein (in complex with Z6134)	0.22 ± 0.03
Protein (in complex with Z6403)	0.18 ± 0.02
Protein (in complex with G4805)	0.16 ± 0.02
System	**Average Rg (nm)**
Apo-protein	3.20 ± 0.01
Protein (in complex with Z6341)	3.21 ± 0.01
Protein (in complex with Z5121)	3.22 ± 0.01
Protein (in complex with Z6134)	3.22 ± 0.01
Protein (in complex with Z6403)	3.22 ± 0.01
Protein (in complex with G4805)	3.21 ± 0.01

In MD simulations, the root-mean-square deviation (RMSD) and radius of gyration (Rg) of the protein backbone are crucial indicators of structural stability and compactness, respectively. RMSD measures the average positional deviation of backbone atoms from a reference structure, providing insight into conformational stability over time; a stable RMSD suggests that the protein maintains its native or near-native structure throughout the simulation. Rg, on the other hand, represents the distribution of atoms relative to the protein's center of mass, indicating its level of compactness. A stable Rg reflects a consistent protein folding state, while fluctuations may suggest unfolding, flexibility, or conformational changes. Together, RMSD and Rg offer a comprehensive view of the protein's structural behavior under simulated conditions, with stable values typically interpreted as evidence of structural integrity and functional relevance.

As evident from the average values (Table 5.3) and the RMSD time evolution plots (Figure 5.4 A,B), the average RMSD of the single pair of HR1–HR2 helices (protein) backbone reduced after binding of all 5 ligands, which indicated that the ligand-binding conferred stability to the protein. The same was further substantiated through the RMSD clustered histogram. The radius of gyration for the HR1–HR2 helices increased (Figure 5.4 C,D) upon binding, which might indicate that the protein underwent some conformational change, possibly through unfolding. The increase was most prominent for **Z6134**. The corresponding RMSF plots (Figure 5.4 G,H) showed that none of the ligands induced any major structural flexibility to the target "E-L-L" motif, as suggested by reasonably similar values of the per-residue RMSF for the concerned region. Interestingly, for the HR1 helix, the binding of **Z6134** enhanced the structural fluctuations for most of the residues in the range 962–982, possibly through some allosteric effect. In the case of HR2, **Z6341** imparted the least structural mobility to the residues in the range 1161–1181. **Z6341**, **Z6134**, and **G4805** showed lower average RMSF for E1182, I1186, and L1193. The positional fluctuation of the ligands (Figure 5.4 E,F) with respect to the initial docked pose was found to be minimal for **Z6341.** The representative binding poses (Figure 5.5 A–E) of the most populated clusters obtained through cluster analysis for the corresponding protein–ligand complex showed that, in the case of ligands **Z6341** and **Z6134**, the ligand molecules remained relatively closer to the target "E-L-L" motif.

Of particular interest to us was the effect of the ligand on the 6-HB. The average RMSD of the 6-HB protein backbone was increased (Table 5.3) after binding to all the ligands, with the maximum distortion observed for **Z6134**, followed by **Z6341**. The same was also validated through the RMSD clustered histogram (Figure 5.6 A,B).

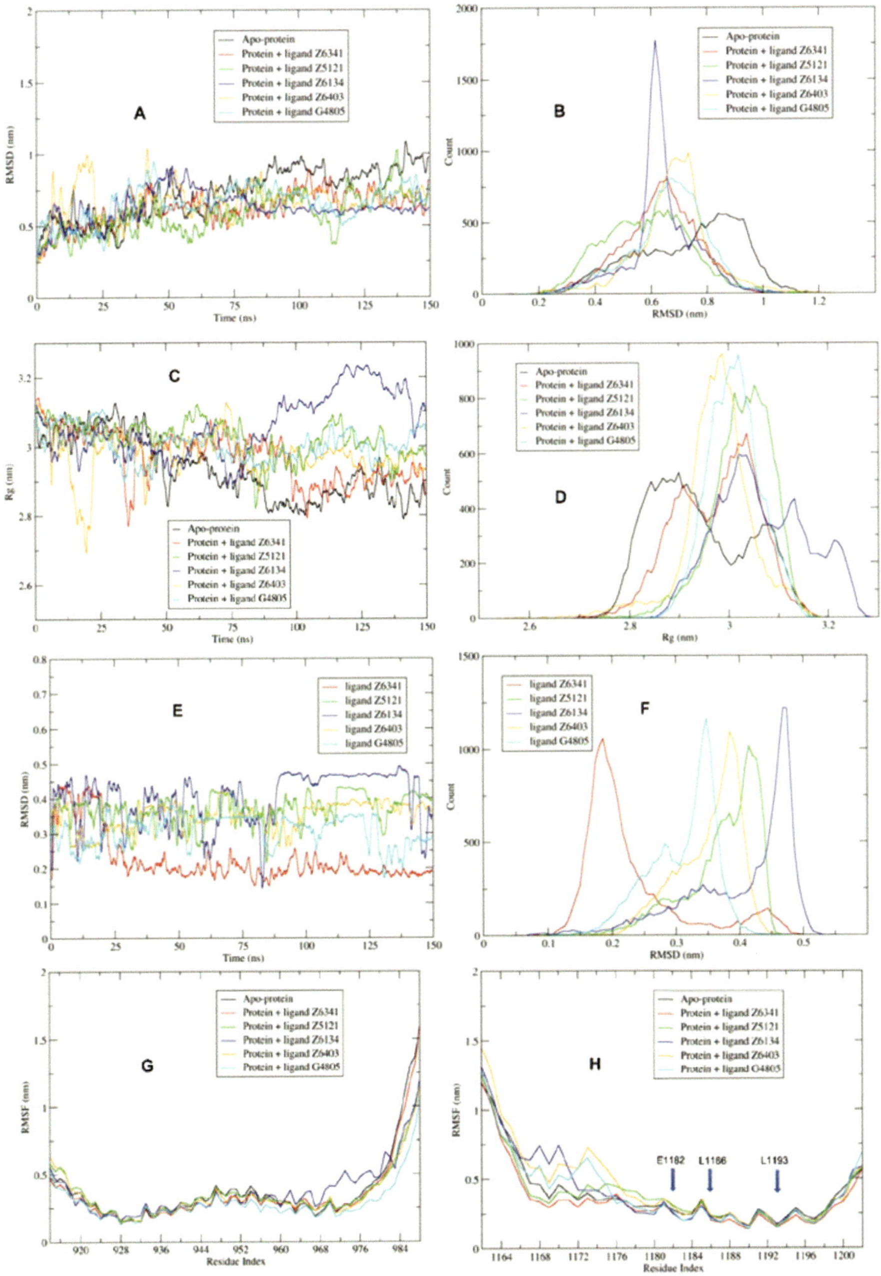

Figure 5.4 (A) Backbone RMSD of the apo-protein (single pair of HR1–HR2 helices) and the same protein in complex with the 5 ligands; (B) Clustered histogram corresponding to the plot in Figure 5.4 A; (C) Rg of the apo-protein (single pair of HR1–HR2 helices) and the same protein in complex with the 5 ligands; (D) Clustered histogram corresponding to the plot in Figure 5.4 C; (E) Ligand heavy atoms RMSD for the 5 ligands; (F) Clustered histogram corresponding to the plot in Figure 5.4 E; (G) RMSF (per residue) of the HR1 helix for the apo-protein (single pair of HR1–HR2 helices) and the same protein in complex with the 5 ligands; and (H) RMSF (per residue) of the HR2 helix for the apo-protein (single pair of HR1–HR2 helices) and the same protein in complex with the 5 ligands.

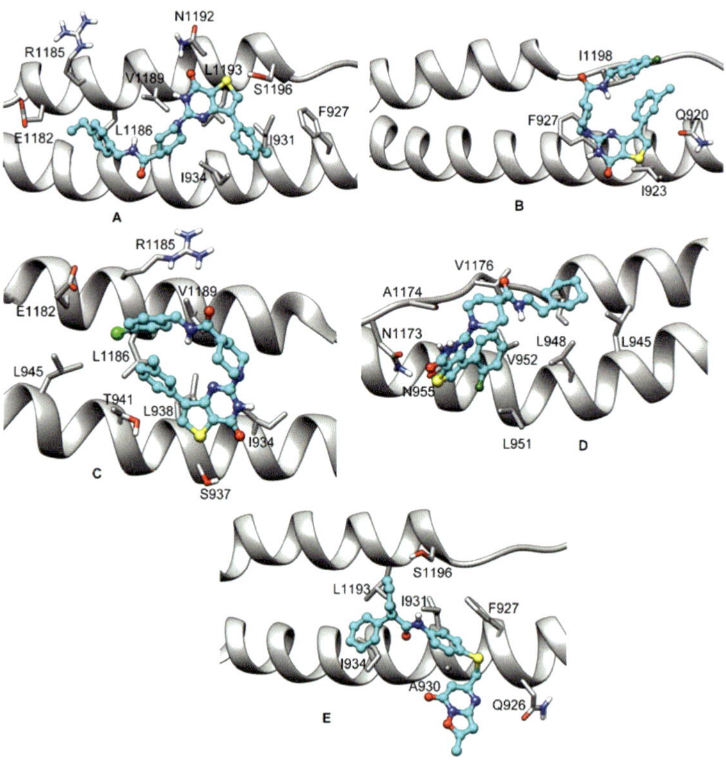

Figure 5.5 Representative structure of the protein (single pair of HR1–HR2 helices)-ligand complex obtained through cluster analysis for ligands (A) Z6341, (B) Z5121, (C) Z6134, (D) Z6403, and (E) G4805.

The radius of gyration increased to a smaller extent after binding to the ligands (Figure 5.6 C,D). At this point, it is worth noting that in the present study, we assume a stoichiometry of one ligand molecule per fusion core in the six-helix bundle (6-HB). Introducing a greater number of ligand molecules relative to the 6-HB could potentially lead to increased structural distortions within the 6-HB. However, this hypothesis requires comprehensive experimental validation in future investigations.

Principal Component Analysis (PCA) in MD simulations of proteins is a valuable tool for investigating the large-scale motions and structural flexibility of proteins [45]. PCA identifies the dominant principal components (PCs) of protein dynamics by analyzing the covariance matrix of atomic positions over time. Since the first and

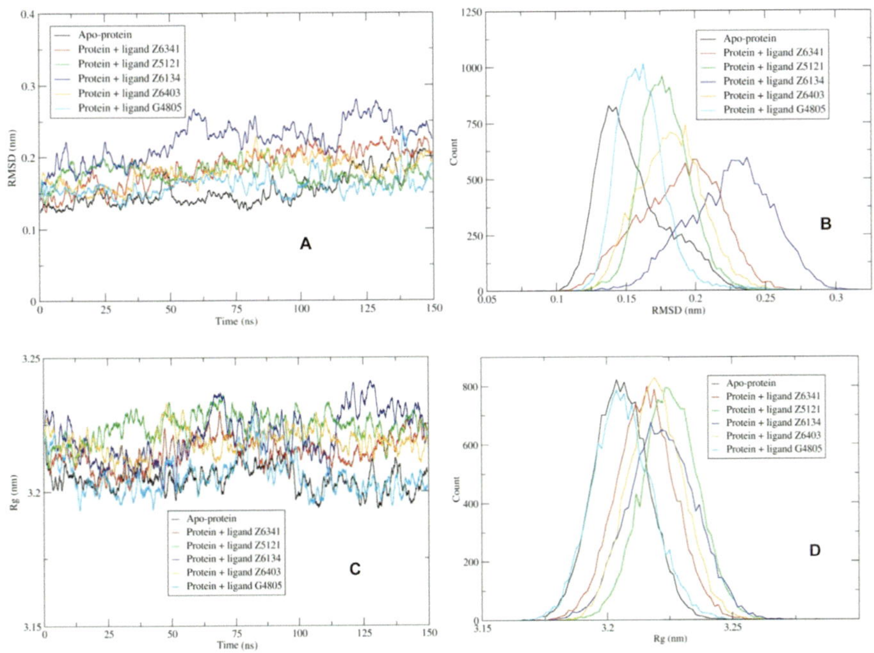

Figure 5.6 (A) Backbone RMSD of the apo-protein (6-HB: three pairs of HR1–HR2 helices) and the same protein in complex with the 5 ligands; (B) Clustered histogram corresponding to the plot in Figure 5.6 A; (C) Rg of the apo-protein (6-HB: three pairs of HR1–HR2 helices) and the same protein in complex with the 5 ligands; and (D) Clustered histogram corresponding to the plot in Figure 5.6 C.

second principal components (PC1 and PC2) captured the most significant molecular motions in the protein based on distance, these two eigenvector projections for the apo-6-HB and protein–ligand complexes were used in the PCA analysis. The backbone of the apo-protein showed a trace of the covariance matrix of 16.26 nm^2. Upon binding to the ligands, the protein showed a trace of the covariance matrix as 18.09 nm^2, 16.73 nm^2, 18.21 nm^2, 13.36 nm^2, and 14.77 nm^2, corresponding to the ligands **Z6341**, **Z5121**, **Z6134**, **Z6403**, and **G4805** (Figure 5.7 A–F). An increase in trace values in the covariance matrix, as compared to the apo-protein, suggested increased conformational flexibility. Thus, it was evident that the bound protein corresponding to **Z6134** and **Z6341** occupied the greater global motion and spanned a larger conformational space among the complexes.

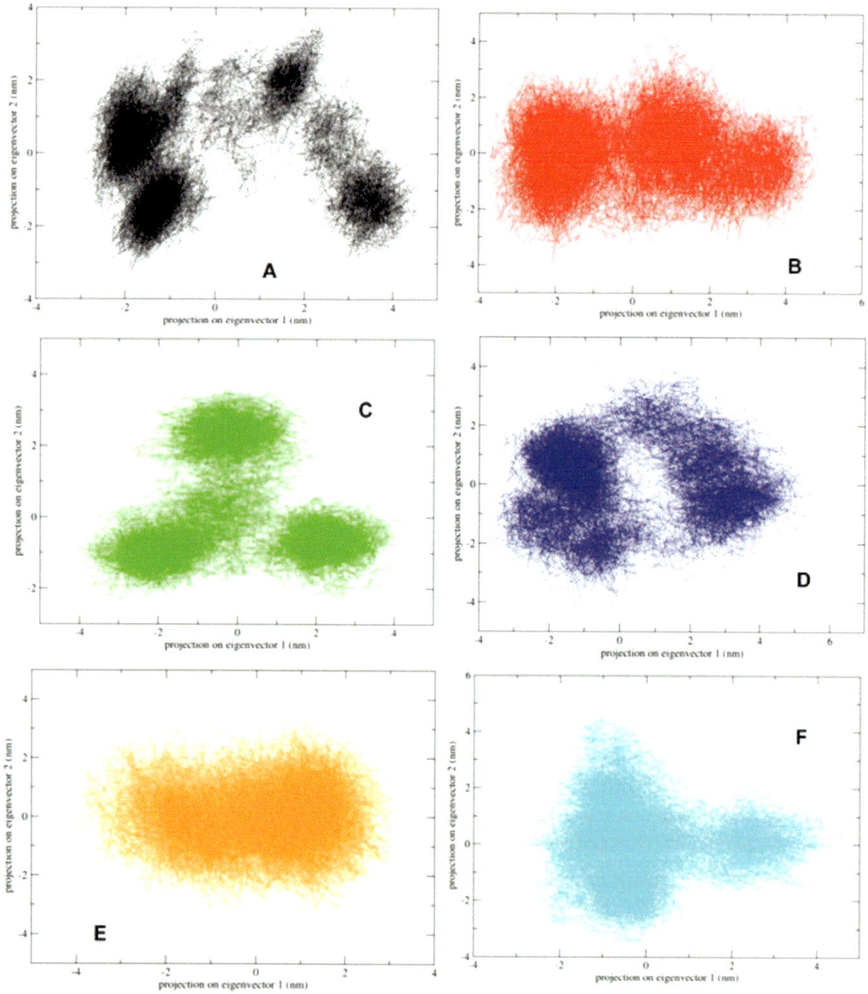

Figure 5.7 Projection of the motion of the protein in phase space along the PC1 and PC2 for the apo-protein (A) and the protein in complex with the ligands Z6341 (B), Z5121 (C), Z6134 (D), Z6403 (E), and G4805 (F).

Subsequently, we determined the effective free energy of binding between the ligand molecules and the single pair of HR1–HR2 domains using the gmx_MMPBSA tool. We considered the entire 150 ns trajectory and the last 50 ns (both with frame intervals of 100) for the calculations. Ligand **Z6341** showed the highest effective free energy (ΔH) of binding (Table 5.4), followed by **Z6134**. Decomposition analysis revealed that the ligand for the ligand

Table 5.4 Energy components for the binding obtained using MM-PBSA calculations

Energy Components (kcal/mol)	Z6341		Z5121		Z6134		Z6403		G4805	
	150 ns	Last 50 ns	150 ns	Last 50 ns	150 ns	Last 50 ns	150 ns	Last 50 ns	150 ns	Last 50 ns
ΔVDWAALS	−25.07 ± 0.77	−25.30 ± 0.58	−22.99 ± 0.95	−23.38 ± 0.95	−24.78 ± 1.64	−27.45 ± 0.78	−21.49 ± 0.09	−21.77 ± 1.01	−22.28 ± 0.45	−24.01 ± 1.15
ΔEEL	−2.16 ± 1.89	−1.91 ± 1.19	−4.98 ± 1.05	−3.71 ± 1.12	−4.28 ± 1.78	−6.02 ± 1.79	−4.56 ± 0.72	−2.28 ± 1.63	−4.06 ± 0.65	−1.48 ± 0.14
ΔEPB	7.31 ± 0.52	6.46 ± 0.15	9.29 ± 0.28	8.33 ± 0.12	9.78 ± 0.09	13.42 ± 0.54	8.68 ± 0.36	7.31 ± 0.72	8.96 ± 0.08	6.52 ± 0.31
ΔENPOLAR	−3.70 ± 0.15	−3.75 ± 0.03	−3.09 ± 0.18	−2.96 ± 0.17	−3.29 ± 0.18	−3.38 ± 0.21	−3.15 ± 0.09	−3.16 ± 0.03	−3.27 ± 0.01	−3.43 ± 0.04
ΔGGAS	−27.23 ± 2.19	−27.22 ± 1.51	−27.97 ± 1.66	−27.10 ± 5.36	−29.06 ± 2.57	−33.46 ± 2.06	−26.05 ± 1.21	−24.05 ± 2.17	−26.35 ± 1.22	−25.50 ± 1.43
ΔGSOLV	3.61 ± 0.54	2.71 ± 0.15	6.20 ± 0.33	5.36 ± 0.21	6.50 ± 0.20	10.04 ± 0.58	5.54 ± 0.37	4.15 ± 0.72	5.70 ± 0.08	3.09 ± 0.32
ΔTOTAL	**−23.61 ± 2.25**	**−24.50 ± 1.51**	**−21.77 ± 1.70**	**−21.73 ± 1.67**	**−22.56 ± 2.58**	**−23.42 ± 2.14**	**−20.51 ± 1.27**	**−19.90 ± 2.28**	**−20.65 ± 1.22**	**−22.40 ± 1.46**

ΔVDWAALS = van der Waals contribution; ΔEEL = electrostatic contribution; ΔEPB = polar contribution to the solvation free energy; ΔENPOLAR = non-polar contribution to the solvation free energy; ΔGGAS = ΔVDWAALS + ΔEEL; ΔGSOLV = ΔEPB + ENPOLAR; ΔTOTAL = ΔGGAS + ΔGSOLV.

Z6341, stabilizing interactions were obtained with I931, I934, L938, L945, L1186, V1189, and L1193 (Figure 5.8 A,B). The **Z6134**, F927, I931, I934, L938, R1185, L1186, V1189, and L1193 had a stabilizing effect, while E1182 had a destabilizing effect on the complex (Figure 5.8 C,D).

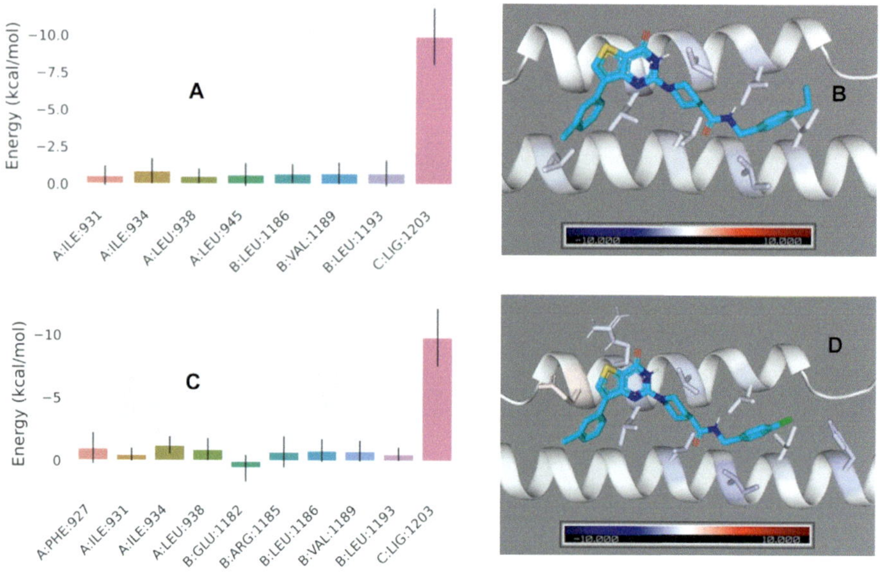

Figure 5.8 Per residue contribution in the effective binding free energy for protein (single pair of HR1–HR2 helices)-ligand complex (above the threshold of 0.50 kcal/mol) corresponding to Z6341 (A and B) and Z6134 (C and D).

5.5 Conclusion

The SARS-CoV-2 spike (S) protein is a key structural component of the virus, playing a crucial role in viral entry by interacting with the angiotensin-converting enzyme 2 (ACE2) in human host cells. As a surface-exposed protein, it has become a prominent focus for vaccine and drug development targeting SARS-CoV-2. In this chapter, we focus on computational drug discovery approaches for identifying molecular scaffolds that bind to the S protein and may thus exhibit therapeutic potential against COVID-19. The basic structural biology of the S protein has been discussed for a better understanding of the macromolecular target that would facilitate the structure-based drug design (SBDD) approach. A selective coverage of the literature in this domain has been presented to give an overview of the workflow followed, which mostly involved protein and ligand preparation, molecular docking, prediction of the pharmacokinetic properties, drug-likeness, and toxicity, followed by molecular dynamics simulation and solvation-based free energy calculations for the protein–ligand binding. To exemplify the pipeline, an experimental approach to screen peptidomimetic molecules as a potential 6-helix bundle fusion inhibitor, by targeting the conserved "E-L-L" motif in the HR2 domain, has been discussed. Virtual screening of 8000 molecules, ADMET predictions, followed by 150 ns atomistic MD simulations and subsequent MM-PBSA calculations, were strategically carried out, leading to the identification of two peptidomimetic molecules (**Z6341** and **Z6134**) that may emerge as potent fusion inhibitors. However, considering the experimental limits of in silico drug discovery, extensive wet-lab validation is necessary for further clinical considerations.

In the end, it must be understood that continued research on drug molecules against COVID-19, particularly targeting the S protein, remains vital due to the emergence of new variants that can evade existing treatments and vaccines. Furthermore, identifying novel effective inhibitors can enhance therapeutic strategies, address vaccine limitations, and better prepare for future viral outbreaks.

References

1. WHO SARS-CoV-2. 2024. https://covid19.who.int/[accessed: 20th October 2024].
2. https://www.emro.who.int/pandemic-epidemic-diseases/mers-cov/mers-situation-update-january-2020.html [accessed: 20th October 2024].
3. https://www.who.int/news-room/commentaries/detail/transmission-of-sars-cov-2-implications-for-infection-prevention-precautions [accessed: 20th October 2024].
4. Chen, Y., Liu, Q., Guo, D. (2020). Emerging coronaviruses: Genome structure, replication, and pathogenesis, *J. Med. Virol.*, **92**, pp. 418–423.
5. Morse, J. S., Lalonde, T., Xu, S., Liu, W. R. (2020). Learning from the past: Possible urgent prevention and treatment options for severe acute respiratory infections caused by 2019-nCoV, *ChemBioChem.*, **21**, pp. 730–738.

6. Wu, A., Peng, Y., Huang, B., Ding, X., Wang, X., Niu, P., Meng, J., Zhu, Z., Zhang, Z., Wang, J., Sheng, J., Quan, L., Xia, Z., Tan, W., Cheng, G., Jiang, T. (2020). Genome composition and divergence of the novel coronavirus (2019-nCoV) originating in China, *Cell Host Microbe*, **27**, pp. 325–328.

7. Su, H., Xu, Y., Jiang, H. (2021). Drug discovery and development targeting the life cycle of SARS-CoV-2, *Fundam. Res.*, **1**, pp. 151–165.

8. Niazi, S. K., Mariam, Z. (2023). Computer-aided drug design and drug discovery: A prospective analysis, *Pharmaceuticals (Basel)*, **17**, pp. 22.

9. Bosch, B. J., van der Zee, R., de Haan, C. A., Rottier, P. J. (2003). The coronavirus spike protein is a class I virus fusion protein: structural and functional characterization of the fusion core complex, *J. Virol.*, **77**, pp. 8801–8811.

10. Weissenhorn, W., Dessen, A., Calder, L. J., Harrison, S. C., Skehel, J. J., Wiley, D. C. (1999). Structural basis for membranc fusion by enveloped viruses. *Mol. Membr. Biol.*, **16**, pp. 3–9.

11. Chen, Y., Guo, Y., Pan, Y., Zhao, Z. J. (2020). Structure analysis of the receptor binding of 2019-nCoV, *Biochem. Biophys. Res. Commun.*, **525**, pp. 135–140.

12. Heurich, A., Hofmann-Winkler, H., Gierer, S., Liepold, T., Jahn, O., Pöhlmann, S. (2014). TMPRSS2 and ADAM17 cleave ACE2 differentially, and only proteolysis by TMPRSS2 augments entry driven by the severe acute respiratory syndrome coronavirus spike protein, *J. Virol.*, **88**, pp. 1293–1307.

13. V'kovski, P., Kratzel, A., Steiner, S., Stalder, H., Thiel, V. (2021). Coronavirus biology and replication: Implications for SARS-CoV-2, *Nat. Rev. Microbiol.*, **19**, pp. 155–170.

14. Malone, B., Urakova, N., Snijder, E. J., Campbell, E. A. (2022). Structures and functions of coronavirus replication-transcription complexes and their relevance for SARS-CoV-2 drug design, *Nat. Rev. Mol. Cell. Biol.*, **23**, pp. 21–39.

15. Maurya, V. K., Kumar, S., Prasad, A. K., Bhatt, M. L. B., Saxena, S. K. (2020). Structure-based drug designing for potential antiviral activity of selected natural products from Ayurveda against SARS-CoV-2 spike glycoprotein and its cellular receptor, *Virusdisease*, **31**, pp. 179–193.

16. Kulkarni, S. A., Nagarajan, S. K., Ramesh, V., Palaniyandi, V., Selvam, S. P., Madhavan, T. (2020). Computational evaluation of major components from plant essential oils as potent inhibitors of SARS-CoV-2 spike protein, *J. Mol. Struct.*, **1221**, pp. 128823.

17. Xia, S., Liu, M., Wang, C., Xu, W., Lan, Q., Feng, S., Qi, F., Bao, L., Du, L., Liu, S., Qin, C., Sun, F., Shi, Z., Zhu, Y., Jiang, S., Lu, L. (2020). Inhibition of SARS-CoV-2 (previously 2019-nCoV) infection by a highly potent pan-coronavirus fusion inhibitor targeting its spike protein that harbors a high capacity to mediate membrane fusion, *Cell Res.*, **30**, pp. 343–355.

18. Souza, P. F. N., Lopes, F. E. S., Amaral, J. L., Freitas, C. D. T., Oliveira, J. T. A. (2020). A molecular docking study revealed that synthetic peptides induced conformational changes in the structure of SARS-CoV-2 spike glycoprotein, disrupting the interaction with human ACE2 receptor, *Int. J. Biol. Macromol.*, **164**, pp. 66–76.

19. Unni, S., Aouti, S., Thiyagarajan, S., Padmanabhan, B. (2020). Identification of a repurposed drug as an inhibitor of the spike protein of human coronavirus SARS-CoV-2 by computational methods, *J. Biosci.*, **45**, pp. 130.

20. Pandey, P., Rane, J. S., Chatterjee, A., Kumar, A., Khan, R., Prakash, A., Ray, S. (2021). Targeting SARS-CoV-2 spike protein of COVID-19 with naturally occurring phytochemicals: An in silico study for drug development, *J. Biomol. Struct. Dyn.*, **39**, pp. 6306–6316.

21. Kiran, G., Karthik, L., Shree Devi, M. S., Sathiyarajeswaran, P., Kanakavalli, K., Kumar, K. M., Ramesh Kumar, D. (2022). In Silico computational screening of Kabasura Kudineer – Official Siddha Formulation and JACOM against SARS-CoV-2 spike protein, *J. Ayurveda Integr. Med.*, **13**, pp. 100324.

22. Kar, P., Kumar, V., Vellingiri, B., Sen, A., Jaishee, N., Anandraj, A., Malhotra, H., Bhattacharyya, S., Mukhopadhyay, S., Kinoshita, M., Govindasamy, V., Roy, A., Naidoo, D., Subramaniam, M. D. (2022). Anisotine and amarogentin as promising inhibitory candidates against SARS-CoV-2 proteins: A computational investigation, *J. Biomol. Struct. Dyn.*, **40**, pp. 4532–4542.

23. Jana, I. D., Bhattacharya, P., Mayilsamy, K., Banerjee, S., Bhattacharje, G., Das, S., Aditya, S., Ghosh, A., McGill, A. R., Srikrishnan, S., Das, A. K., Basak, A., Mohapatra, S. S., Chandran, B., Bhimsaria, D., Mohapatra, S., Roy, A., Mondal, A. (2022). Targeting an evolutionarily conserved "E-L-L" motif in spike protein to identify a small molecule fusion inhibitor against SARS-CoV-2. *PNAS nexus*, **1**, pgac198.

24. Jackson, C. B., Farzan, M., Chen, B., Choe, H. (2022). Mechanisms of SARS-CoV-2 entry into cells, *Nat. Rev. Mol. Cell Biol.*, **23**, pp. 3–20.

25. Lenci, E., Trabocchi, A. (2020). Peptidomimetic toolbox for drug discovery, *Chem. Soc. Rev.*, **49**, pp. 3262–3277.

26. Qvit, N., Rubin, S. J. S., Urban, T. J., Mochly-Rosen, D., Gross, E. R. (2017). Peptidomimetic therapeutics: Scientific approaches and opportunities, *Drug Discov. Today*, **22**, pp. 454–462.
27. Halgren, T. A. (1996). Merck molecular force field. I. Basis, form, scope, parameterization, and performance of MMFF94, *J. Comput. Chem.*, **17**, pp. 490–519.
28. O'Boyle, N. M., Banck, M., James, C. A., Morley, C., Vandermeersch, T., Hutchison, G. R. (2011). Open Babel: An open chemical toolbox, *J. Cheminform.*, **3**, 33.
29. Pettersen, E. F., Goddard, T. D., Huang, C. C., Couch, G. S., Greenblatt, D. M., Meng, E. C., Ferrin, T. E. (2004). UCSF Chimera: A visualization system for exploratory research and analysis, *J. Comput. Chem.*, **25**, pp. 1605–1612.
30. Singh, U. C., Kollman, P. A. (1984). An approach to computing electrostatic charges for molecules, *J. Comput. Chem.*, **5**, pp. 129–145.
31. Eberhardt, J., Santos-Martins, D., Tillack, A. F., Forli, S. (2021). AutoDock Vina 1.2.0: New docking methods, expanded force field, and Python bindings, *J. Chem. Inf. Model.*, **61**, pp. 3891–3898.
32. Laskowski, R. A., Swindells, M. B. (2011). LigPlot+: Multiple ligand-protein interaction diagrams for drug discovery, *J. Chem. Inf. Model.*, **51**, pp. 2778–2786.
33. Daoud, N. E., Borah, P., Deb, P. K., Venugopala, K. N., Hourani, W., Alzweiri, M., Bardaweel, S. K., Tiwari, V. (2021). ADMET profiling in drug discovery and development: Perspectives of in silico, in vitro and integrated approaches, *Curr. Drug Metab.*, **22**, pp. 503–522.
34. Pires, D. E., Blundell, T. L., Ascher, D. B. (2015). pkCSM: Predicting small-molecule pharmacokinetic and toxicity properties using graph-based signatures, *J. Med. Chem.*, **58**, pp. 4066–4072.
35. Daina, A., Michielin, O., Zoete, V. (2017). SwissADME: A free web tool to evaluate pharmacokinetics, drug-likeness, and medicinal chemistry friendliness of small molecules, *Sci. Rep.*, **7**, pp. 42717.
36. Lipinski, C. A. (2004). Lead- and drug-like compounds: the rule-of-five revolution, *Drug Discov. Today Technol.*, **1**, pp. 337–341.
37. Source Code: https://doi.org/10.5281/zenodo.10017699.
38. Huang, J., MacKerell, A. D. (2013). CHARMM36 all-atom additive protein force field: Validation based on comparison to NMR data, *J. Comput. Chem.*, **34**, pp. 2135–2145.
39. Vanommeslaeghe, K., Hatcher, E., Acharya, C., Kundu, S., Zhong, S., Shim, J., Darian, E., Guvench, O., Lopes, P., Vorobyov, I., Mackerell, A.

D. Jr. (2010). CHARMM general force field: A force field for drug-like molecules compatible with the CHARMM all-atom additive biological force fields, *J. Comput. Chem.*, **31**, pp. 671–690.

40. Mark, P., Nilsson, L. (2001). Structure and dynamics of the TIP3P, SPC, and SPC/E water models at 298 K, *J. Phys. Chem. A,* **105**, pp. 9954–9960.

41. Darden, T., York, D., Pedersen, L. (1993). Particle mesh Ewald: An N·log(N) method for Ewald sums in large systems, *J. Chem. Phys.*, **98**, pp. 10089–10092.

42. Parrinello, M., Rahman, A. (1981). Polymorphic transitions in single crystals: A new molecular dynamics method, *J. Appl. Phys.,* **52**, pp. 7182– 7190.

43. Valdés-Tresanco, M. S., Valdes-Tresanco, M. E., Valiente, P. A., Moreno, E. (2021). gmx_MMPBSA: A new tool to perform end-state free energy calculations with GROMACS, *J. Chem. Theory Comput.*, **17**, pp. 6281–6291.

44. Wang, E., Sun, H., Wang, J., Wang, Z., Liu, H., Zhang, J. Z. H., Hou, T. (2019). End-point binding free energy calculation with MM/PBSA and MM/GBSA: Strategies and applications in drug design, *Chem. Rev.*, **119**, pp. 9478–9508.

45. Maisuradze, G. G., Liwo, A., Scheraga, H. A. (2009). Principal component analysis for protein folding dynamics, *J. Mol. Biol.*, **385**, pp. 312–329.

Chapter 6

In silico Strategies to Combat COVID-19 Using Docking Simulations

Dipanjan Guha[a] and Angshuman Bagchi[a,b]
[a]*BIF Centre, University of Kalyani, Kalyani, Nadia, West Bengal, India*
[b]*Department of Biochemistry & Biophysics, University of Kalyani, Kalyani, Nadia, West Bengal, India*
angshumanb@gmail.com, angshu@klyuniv.ac.in

The binding interactions between a receptor and a ligand are predicted by the technique of molecular docking simulations, which is one of the important tools in structural bioinformatics studies. By simulating interactions between the drug and the protein, docking helps identify compounds with high affinity and specificity, crucial for therapeutic efficacy. The process utilizes scoring functions to rank potential drug candidates by assessing their binding energy and stability. During the SARS-CoV-2 pandemic, the importance and usefulness of molecular docking simulations were shown as an inevitable scientific tool for the drug discovery process. Among the vast collection of natural products, flavonoids are well-studied

Computational Methods for Molecular Modelling and Docking
Edited by Arup Kumar Ghosh and Suban Kumar Sahoo

ISBN 978-981-5352-25-2 (Hardcover), 978-1-042-00204-7 (eBook)
www.jennystanford.com

DOI: 10.1201/9781042002047-6

polyphenolic compounds that can reduce viral infection by modulating processes such as viral entry into hosts, viral replication, and protein translation. A few of the flavonoids have anti-SARS-CoV-1 activities. This chapter aims to describe the basic needs and software available for molecular docking and to present our primary data regarding the prediction of potential flavonoid candidates, which can be proven effective to stall SARS-CoV-2 propagation by deactivating NSP15, an endoribonuclease, and NSP16, a 2'-O-methyltransferase. Apigenin, Herbacetin, and Juglanin are those flavonoids that can inhibit the activity of NSP15 and NSP16 proteins in COVID-19. These natural inhibitors could hinder the functions of non-structural proteins that are important for viral life-cycle propagation.

6.1 Introduction

Molecular docking is one of the pivotal tools in rational drug design processes. The technique enables the detection of the interaction pattern between the target and the receptor, which is typically a protein [1]. By simulating these interactions, docking helps to identify compounds with the potential to bind effectively to the target receptor, which is essential for therapeutic efficacy. This approach accelerates the drug development process by screening vast libraries of compounds in silico, predicting the best candidates for further experimental validation. Molecular docking is invaluable in rational drug design, optimizing lead compounds, and understanding the molecular mechanisms of diseases [2, 3]. It reduces the need for expensive and time-intensive bench lab assays and plays a significant role in structure-based drug development, making it a key element of precision medicine [4, 5]. The importance of molecular docking in combating unknown pandemics lies in its ability to swiftly predict how potential therapeutic compounds interact with novel viral proteins [6, 7]. When a new virus emerges, molecular docking can rapidly screen vast libraries of drug candidates, identifying those most likely to inhibit viral replication or block critical pathways. This computational approach allows researchers to prioritize compounds for experimental validation, significantly speeding up the drug discovery process during urgent situations [8]. By focusing on viral structures or common viral protein targets, molecular docking provides a strategic advantage in designing treatments or vaccines to mitigate the spread of future pandemics.

The newly identified coronavirus type 2 (SARS-CoV-2) from a seafood market in Wuhan, China, caused the pandemic commonly referred to as COVID-19. The virus soon spread across the globe and led to severe devastation with the loss of millions of lives. Its containment attempts failed [9, 10]. Later, the World Health Organization (WHO) considered the situation and proclaimed it a pandemic. As of April 2024, this had resulted in 70,47,53,890 cases and 70,10,681 confirmed deaths, as reported by Worldometer (https://www.worldometers.info/coronavirus/), which makes it the deadliest pandemic in history. The virus's extremely rapid growth contributes to disease severity, and most infections are asymptomatic. Therefore, given the health risks of rapidly transmitting the virus to healthy individuals, researchers need to develop therapeutics to combat SARS-CoV-2 and prevent viral spread [11]. Advances in clinical diagnostic techniques have enabled rapid diagnosis and the identification of potential drug targets in less time. The use of combinatorial chemistry, artificial intelligence (AI), and deep learning approaches helps to find novel therapeutics in less time [12].

The target of several vaccines is the non-structural protein (NSP) in SARS-CoV-2. This particular protein is essential to carry out the biochemical processes necessary for the virus to live inside the host. Understanding the importance of NSPs is crucial for developing effective treatments and interventions. In SARS-CoV-2, NSP1 to NSP16 are reported to be found. Those are involved in controlling viral replication, transcription, immune evasion, viral protein processing, and structural and non-structural protein interactions [13, 14]. One of the important proteins, NSP5, which is also known as 3CLpro or main protease, has an important role in the replication of coronaviruses [14–16]. NSP15 or endoribonuclease of SARS-CoV-2, an enzyme needed for the coronavirus life cycle. This protein has a unique function in the proofreading and repair of the virus's RNA genome [17, 18]. NSP16 is also called 2'-O-methyltransferase, associated with the inclusion of variations of viral RNA. It works in conjunction with another protein called NSP10 to form a complex responsible for the insertion of a methyl group at the 2' position of the viral RNA; this is referred to as the viral cap [19, 20]. Targeting and inhibiting these non-structural proteins can affect the lifecycle of SARS-CoV-2.

This chapter aims to explore several key topics related to drug discovery against SARS-CoV-2. It will focus on molecular docking as a tool for identifying potential drugs, particularly to fight against SARS-CoV-2. Additionally, the literature will delve into the efficacy of flavonoids as antivirals targeting SARS-CoV-2. It will also cover the various databases available for searching the structures of proteins, ligands, or drug-like molecules. Furthermore, the chapter will discuss resources commonly used for molecular docking. Importantly, the aim is to present our primary data, highlighting the prediction of potential flavonoid candidates (which were effective against SARS-CoV-1) on important NSPs of SARS-CoV-2.

6.2 Molecular Docking for Searching Therapeutics Against SARS-CoV-2

6.2.1 Significance of Molecular Docking for the Drug Discovery Against COVID-19

Molecular docking has become an essential procedure to combat SARS-CoV-2, significantly contributing to the rapid identification and development of potential therapeutic agents [1–5]. As a computational method, it predicts the binding modes of small-molecule ligands and target receptor proteins, offering valuable insights into how drugs can inhibit viral mechanisms. In the context of the COVID-19 pandemic, molecular docking has been instrumental in understanding how the virus functions and how various compounds might impede its replication, binding, or entry into host cells. Key contributions of molecular docking during the COVID-19 drug discovery process include:

6.2.1.1 Rapid screening of potential drug candidates

The urgency for the discovery of potential therapeutics for the COVID-19 pandemic demanded the swift identification of antiviral agents. Molecular docking allows researchers to virtually screen vast libraries of existing drugs, natural compounds, and synthetic molecules against the different proteins of SARS-CoV-2 [21, 22].

This in silico screening dramatically accelerates the drug discovery process, narrowing down the candidates for *in vitro* and *in vivo* testing. For example, early in the pandemic, molecular docking studies helped identify drugs, such as remdesivir, favipiravir, ribavirin, and galidesivir, that showed promising interactions with viral proteins, and some of them advanced into clinical trials [23–25].

6.2.1.2 Drug repurposing

One of the most significant contributions of molecular docking to COVID-19 drug discovery has been its role in drug repurposing [26, 27]. Given the time constraints of developing novel drugs, repurposing existing FDA-approved drugs offered a faster route to finding effective treatments. Molecular docking enabled the rapid evaluation of known drugs, such as hydroxychloroquine, lopinavir, and ritonavir, for their potential to interact with SARS-CoV-2 proteins [28, 29]. Although not all repurposed drugs proved effective in clinical settings, docking studies were essential in guiding early research and focusing experimental efforts on the most promising compounds.

6.2.1.3 Understanding viral mechanisms

Molecular docking has also provided important details pertaining to the structures of SARS-CoV-2. By docking how potential drugs bind to viral proteins, researchers have gained a deeper understanding of the virus's mechanisms. For instance, docking studies of the spike protein of SARS-CoV-2, which uses the ACE2 receptor of human beings for entry into host cells, revealed key interactions that could be targeted by inhibitors [30, 31]. Similarly, studies involving Mpro have identified potential inhibitors that can block viral replication, a critical step in halting disease progression [32].

6.2.1.4 Design of novel inhibitors

Beyond repurposing drugs, molecular docking has facilitated the design of novel inhibitors tailored to combat SARS-CoV-2. Using docking simulations, researchers can modify the structure of molecules to optimize binding affinity to proteins [3, 33]. This rational drug design approach has led to the development of new compounds with higher specificity and potency against the virus.

These compounds can then be further refined through additional computational techniques, such as molecular dynamics simulations, to ensure their efficacy and stability [34].

6.2.1.5 Guiding rational drug design

Docking simulations have been instrumental in the rational design of novel antiviral drugs. By predicting how small molecules interact with viral proteins, docking has enabled researchers to optimize their structures, thereby enhancing their specificity and binding affinity to SARS-CoV-2 targets [35, 36].

6.2.1.6 Identification of allosteric sites

In addition to active site targeting, molecular docking has helped identify potential allosteric sites on viral proteins [37]. These alternative binding sites offer new therapeutic strategies, enabling the development of drugs that modulate protein function without directly competing with natural substrates.

6.2.1.7 Cost-effective approach

Compared with traditional bench lab methods, molecular docking is less expensive [3]. It allows researchers to test thousands of compounds in silico, significantly reducing the need for expensive and time-intensive bench work, which was crucial during the pandemic when resources were limited.

6.2.1.8 Prediction of drug-protein interaction

Molecular docking predicts not only the likelihood of a drug binding to a viral protein but also the nature of the interaction, such as hydrogen bonds, van der Waals forces, and electrostatic interactions [38]. This detailed information is valuable for optimizing drug efficacy and selectivity.

6.2.2 Limitations of Molecular Docking for Drug Design Against COVID-19

Despite its contributions, the method comes with several limitations that must be considered. These limitations can affect the accuracy, reliability, and overall success of the docking predictions, potentially leading to false positives or negatives in drug discovery.

6.2.2.1 Protein flexibility

One of the most significant limitations of molecular docking is its inability to accurately account for the flexibility of proteins [39]. The bindings of ligands to protein receptors lead to the induction of conformational fluctuations in them. Most docking algorithms treat the protein as a rigid structure or allow only limited flexibility in the active site, which may not reflect the actual binding environment [40]. This simplification can lead to incorrect predictions about how a drug interacts with the target protein, resulting in potential false negatives or missing viable drug candidates.

6.2.2.2 Scoring function inaccuracies

Docking relies on scoring functions to determine the binding efficacies of ligands for their target protein receptors. These scoring functions are often based on simplified models of molecular interactions (such as hydrogen bonding, van der Waals forces, and hydrophobic interactions) and may fail in capturing the full complexity of binding events [41]. Consequently, predicted binding affinities may not always correlate with experimental results, leading to false positives or to the overlooking of potentially active compounds [42].

6.2.2.3 Limited sampling of ligand conformations

While docking programs generate multiple conformations of a ligand to find the best fit within the protein's active site, they often explore only a limited subset of possible conformations. This restricted sampling can lead to inaccurate predictions of the optimal binding pose [43]. Complex ligands with high flexibility may not be adequately modeled, thereby reducing the reliability of docking results.

6.2.2.4 Neglect of solvent effects

Molecular docking typically does not consider the explicit role of water molecules or other solvent effects in the protein-ligand interaction. However, water can play a crucial role in mediating binding, either by stabilizing interactions or by participating in hydrogen bonds. Neglecting the solvent environment can result in an inaccurate prediction of binding affinities, which is especially relevant for highly polar or charged molecules [3, 44].

6.2.2.5 Focus on a single binding site

Docking studies often focus on a single target or binding site, ignoring the possibility of allosteric binding sites or secondary interactions that could also influence drug efficacy [45]. For COVID-19, most docking efforts have focused on key viral proteins like the spike protein and main proteases, but other less obvious sites may also be therapeutically relevant. This focus on a single site could lead to missing the opportunities for identifying drugs that act through alternative mechanisms.

6.2.2.6 Viral mutations

SARS-CoV-2, like other viruses, mutates over time, leading to variations in its protein structures. Docking studies are typically conducted using static crystal structures of viral proteins. However, mutations, especially in the spike protein, can significantly alter the binding site, potentially rendering previously identified drug candidates less effective [46]. Docking studies performed on one variant may not be applicable to newly emerging strains unless the method is regularly updated with new structural data.

6.2.2.7 Lack of integration with other biological factors

Molecular docking focuses primarily on the interaction between a ligand and its corresponding target protein receptor, but successful drug discovery requires understanding additional biological factors, such as cell permeability, metabolic stability, and toxicity [47]. Docking results alone do not account for these factors, meaning that a drug candidate predicted to bind well to a target protein may still fail in later stages of drug development due to poor pharmacokinetics or off-target effects.

6.2.2.8 Dependence on structural data

Molecular docking depends heavily on the availability and quality of structural data, such as crystal, cryo-electron microscopy, or NMR structures [48]. For some proteins, especially those that are difficult to crystallize or express, reliable structures may not be available, limiting the applicability of docking studies. Additionally, low-resolution structures can introduce inaccuracies into the docking process [49].

6.3 Flavonoids Against Viral Diseases Emphasizing COVID-19

Flavonoids are important molecules that inhibit various viral diseases, including SARS-CoV-1 and SARS-CoV-2. Some polyphenolic secondary metabolites inside fruits, vegetables, seeds, and tea are called flavonoids [50]. There are more than 6,000 structurally identified flavonoids, most of which are found in plants. During the stress response, plants use these flavonoids to defend against biological pathogens such as insects and pests [51]. Flavonoids are known to have lesser systemic toxicity and can be synergized with conventional drugs. Flavonoids are known to be "pleiotropic" compounds. Their functional group(s) are found to get involved in modulating multiple pathways [52, 53]. These are the important features that make flavonoids a promising agent for targeting the virus life cycle. Chemically, flavonoids are hydroxylated phenolic molecules, which are synthesized through the phenylpropanoid pathway. Their pharmacological properties include antimicrobial, antiviral, antioxidant, and anti-inflammatory effects. Flavonoids have been studied against various DNA and RNA viruses [54]. Apigenin has shown activity against picornaviruses (RNA viruses) by inhibiting protein synthesis through the suppression of IRES viral activity [55]. Bafandeh et al. also highlighted that natural products, such as plant extracts, essential oils, and other natural substances, can be used as candidate drugs for SARS-CoV-2 [56]. They found, through bioinformatics analyses, that flavonoids are one of the promising compounds. Kumar et al. studied the FDA-approved library to perform virtual screening experiments against several different protein targets of SARS-CoV-2, like NSP3, NSP9, NSP12, and NSP15, and reported inhibitor(s) against this virus [57]. Previous work from this group has identified and reported interactions between antiviral compounds with NSP1 [58]. In this chapter, we have presented evidence for the efficacy of these flavonoids, which have been shown to inhibit SARS-CoV-1. We are reporting the potency of flavonoids Apigenin, Herbacetin, and Juglanin against important non-structural proteins of COVID-19, i.e., NSP5, NSP15, and NSP16.

6.4 Databases to Search the Structure of Protein and Ligand or Drug-Like Molecules

6.4.1 Databases to Search the Structure of Protein Molecules

Several databases provide access to protein structures, enabling researchers to find detailed structural information for drug discovery, molecular docking, and other applications. Here are some of the most prominent databases for searching protein structures, along with short descriptions:

6.4.1.1 Protein Data Bank (PDB)

It is the most comprehensive storehouse of 3D protein structures. It provides information regarding structural details of biomolecules-proteins, nucleic acids, and complex assemblies derived from X-ray crystallography, NMR, and cryo-electron microscopy [59].

6.4.1.2 UniProt

UniProt is a comprehensive database that contains information about the amino acid sequences of proteins and their functional information. While primarily focused on protein sequences, UniProt links to structural data from PDB and provides information on protein function, domain organization, and post-translational modifications [60].

6.4.1.3 Swiss-model repository

This repository provides homology-modeled protein structures. If an experimentally determined structure is not available, Swiss-Model can generate models based on homologous sequences. It is useful for proteins that lack resolved structures [61].

6.4.1.4 AlphaFold protein structure database

Created by DeepMind, this database provides predicted 3D structures of proteins using the AlphaFold AI system. It has expanded structural coverage to many proteins that lack experimental structures, making it a valuable resource for exploring protein conformations [62].

6.4.1.5 Protein Data Bank in Europe (PDBe)

Part of the Worldwide Protein Data Bank (wwPDB), PDBe provides detailed structural data of proteins and offers search and visualization tools for analyzing molecular structures. It also supports advanced searches based on protein features and interactions [63].

6.4.1.6 Protein Data Bank Japan (PDBj)

PDBj is the Japanese branch of the wwPDB, providing access to the same set of protein structures as PDB but with unique tools and additional resources for analyzing and visualizing macromolecular structures [64].

6.4.1.7 ModBase

ModBase is a database of comparative protein structure models, created using homology modeling. It allows researchers to search for modeled structures when experimental structures are not available [65].

6.4.1.8 Structural Classification of Proteins (SCOP)

SCOP provides a classification of protein structures based on similarities in their folds and architectures. It groups proteins into evolutionary families and super-families, helping researchers explore structural and functional relationships [66].

6.4.1.9 Protein Structure Classification Database (CATH)

CATH classifies protein structures into evolutionary super-families and folds, similar to SCOP. It organizes proteins by their structural and functional similarities and provides detailed information on structure-function relationships [67].

6.4.1.10 Molecular Modeling Database (MMDB)

Hosted by NCBI, MMDB contains 3D macromolecular structures, similar to those in the PDB, and integrates structural information with sequence and functional annotation through the Entrez search system [69].

These databases offer diverse resources for accessing and analyzing protein structures, each with unique tools and features to assist researchers in exploring structural biology.

6.4.2 Databases to Search the Structure of Ligand or Drug-Like Molecules

Several databases specialize in providing structures, properties, and information about ligand molecules, which are essential for drug discovery, molecular docking, and other research applications. Here are some of the most popular databases for searching ligand structures:

6.4.2.1 PubChem

PubChem is one of the largest open-source chemistry databases, offering detailed information on chemical compounds, including ligand structures, biological activities, and properties. It includes 3D structures of ligands and tools for searching, visualizing, and downloading molecular data [69].

6.4.2.2 ZINC database

ZINC is a database that contains details of commercially available compounds, including 3D structures, physicochemical properties, and vendor information. It is widely used for virtual screening in drug discovery and contains millions of ligand molecules in ready-to-dock formats [70].

6.4.2.3 DrugBank

DrugBank is a comprehensive resource combining detailed drug data with chemical, pharmacological, and pharmaceutical information. It provides 2D and 3D structures of candidate drugs, along with their targets, mechanisms of action, and clinical data [71].

6.4.2.4 ChEMBL

ChEMBL is a database of bioactive compounds having drug-like properties. It provides chemical structures, bioactivity data, and information about ligand interactions with protein targets, making it an excellent resource for drug discovery [72].

6.4.2.5 BindingDB

This is a database that provides information pertaining to the binding affinities between receptors and ligands. It contains structures of ligands and is widely used for studying drug-target interactions and virtual screening [73].

6.4.2.6 Ligand Expo

Ligand Expo, part of the Protein Data Bank (PDB), contains information on small molecules, ligands, and heteroatoms that have been observed in 3D structures of macromolecules in the PDB. It provides 3D structures, as well as information on how these ligands interact with proteins [74].

6.4.2.7 ChemSpider

ChemSpider is a database of chemical structures to get access to millions of structures from multiple data sources. It offers detailed information on molecular properties, 2D/3D structures, and spectral data, making it useful for exploring chemical compounds [75].

6.4.2.8 Selleckchem

Selleckchem is a commercial database of small bioactive molecules, primarily focused on inhibitors, agonists, and antagonists used in biological and pharmacological research. It provides structural information and data for thousands of ligands used in drug discovery [76].

6.4.2.9 MolPort

MolPort is a database of chemical compounds from suppliers worldwide. It is used primarily for sourcing and searching for small molecules and provides structural information, chemical properties, and vendor details. This database was frequently used during COVID-19 [77].

6.4.2.10 eMolecules

eMolecules is a chemical search engine that allows researchers to search for chemical compounds, including ligands, with access to vendor information, pricing, and 3D structures. It is useful for acquiring and analyzing small molecules for drug discovery projects [78].

6.4.2.11 Human Metabolome Database (HMDB)

The HMDB provides the necessary information about metabolites, which are small molecules found in the human body, including their chemical structure, biological activities, and interactions with proteins. It is particularly valuable for studying metabolic pathways and disease-related ligands [79].

6.4.2.12 InterBioScreen

InterBioScreen is a database of natural compounds and synthetic ligands that offer structural information on compounds of biological interest. It is widely used in virtual screening for natural product-based drug discovery [80].

These databases provide extensive resources for finding ligand structures, making them valuable tools for researchers involved in drug design, molecular docking, and bioactivity studies.

6.5 Strategy to Select Ligand or Drug-Like Molecules

6.5.1 Energy Minimization and Structure Optimization

The 3D structural details of ligand molecules (flavone derivatives) were fetched from the PubChem database (https://pubchem.ncbi.nlm.nih.gov/). We have selected such groups of flavones as Rhoifolin, Pectolinarin, Juglanin, Herbacetin, and Apigenin, which were proven to be effective against SARS-CoV-1. Hydrogen molecules were added to the downloaded 3D coordinates of those ligands, and applying the MMFF force field, ligands were energy minimized for 1000 cycles to 0.01 kcal/mol using Discovery Studio 2.5. During the process of energy minimization, structure deformation was observed for Rhoifolin and Pectolinarin; we therefore selected Apigenin, Herbacetin, and Juglanin for further studies.

6.5.2 ADMET Property and Toxicity Prediction

Physicochemical, pharmacokinetics, and drug likeliness of the ligand molecules were predicted through the Swiss ADME web portal (http://www.swissadme.ch/). The Swiss ADME web portal was used to predict various molecular properties of ligand molecules, including molecular weight, the number of heavy non-aromatic and aromatic heavy atoms, fraction of C sp3, the number of rotatable bonds, hydrogen bond acceptors, hydrogen bond donors, molar refractivity, and topological polar surface area (TPSA). Additionally, pharmacokinetic properties such as GI absorption, BBB permeability, and whether the molecules are P-gp substrates were also predicted. The drug-likeness of the molecules was evaluated using several filters, including the Lipinski, Ghose, Veber, Egan, and Muegge filters, as well as the Bioavailability Score, all through the Swiss ADME web portal. Furthermore, the toxic nature of the molecules was analyzed using the server ProTox-II (https://tox-new.charite.de/protox_II/).

6.6 Strategy to Prepare Protein Molecules

The 3D X-ray crystallographic structures of NSPs were downloaded from the RCSB Protein Data Bank (PDB) (https://www.rcsb.org/), the details of which are presented in Table 6.1. The corresponding amino acid sequences of the proteins, in FASTA format, were obtained from UniProt (https://www.uniprot.org/). Then, sequence alignment was performed using each NSPs to detect whether the 3D structures of all amino acids are available or not. Those proteins were selected for further experiments, where most of the amino acids have 3D coordinates. Those selected proteins were cleaned using Discovery Studio 2.5. Thereafter, the CHARMm force field was applied to each protein molecule, and energy minimization was done using Discovery Studio 2.5 to clear the steric clashes. Thereafter, the Ramachandran plot (https://saves.mbi.ucla.edu/) was constructed for each of the proteins to check their stereochemical fitness. The faulty structures were further refined using the tool ModLoop (https://modbase.compbio.ucsf.edu/modloop/), and thereafter, energy minimization was performed using the aforementioned protocol. Again, the Ramachandran plots were generated to reconfirm the stereochemical fitness of the faulty proteins.

Table 6.1 The NSPs of COVID-19 with their corresponding PDB IDs

Sl. No.	Protein Name	PDB ID
1	NSP1	7K3N
2	NSP5	7KYU
3	NSP7 & NSP8	7JLT
4	NSP9	7BWQ
5	NSP10	7MC5
6	NSP12	6M71
7	NSP14	7QGI
8	NSP15	6WLC
9	NSP16	6W4H

6.7 Resources Available for Molecular Docking

Widely used offline and online resources for molecular docking that researcher uses for structure-based drug discovery and virtual screening are discussed in brief:

6.7.1 Offline Resources for Molecular Docking

6.7.1.1 AutoDock

AutoDock is one of the most popular and widely used open-source molecular docking programs. It supports both flexible-ligand and rigid/flexible-protein docking. AutoDock uses a Lamarckian Genetic Algorithm for docking simulations and allows detailed modeling of ligand flexibility [81]. The software is compatible with Windows, Linux, and Mac operating systems.

6.7.1.2 AutoDock Vina

AutoDock Vina is a faster and more accurate version of AutoDock. It has an improved search algorithm and scoring function, offering significantly faster docking calculations with comparable or better accuracy. Vina can be integrated into local pipelines for virtual screening [82]. The software is compatible with Windows, Linux, and Mac operating systems.

6.7.1.3 LeDock

LeDock is an offline, fast molecular docking software designed for protein-ligand docking. It uses a combination of grid-based sampling and energy scoring functions to evaluate ligand-protein interactions and predict binding poses [83]. The software is compatible with Windows, Linux, and Mac operating systems.

6.7.1.4 Schrödinger Glide

Schrödinger's Glide is a highly precise docking program, known for producing accurate binding poses and excellent scoring of ligand-protein interactions. It is part of the Schrödinger molecular modeling suite, which integrates many tools for computational drug design [84]. The software is popular on Windows and Linux platforms.

6.7.1.5 Molecular Operating Environment (MOE)

MOE is a comprehensive computational chemistry software suite that includes tools for molecular docking, visualization, molecular dynamics, and QSAR modeling. It provides flexible ligand docking with multiple scoring functions and optimization algorithms [85]. The software is compatible with Windows, Linux, and Mac operating systems.

6.7.1.6 Genetic Optimization for Ligand Docking (GOLD)

GOLD is a high-accuracy docking tool based on a genetic algorithm to model flexible ligand docking. It is well-regarded for accurately predicting ligand binding modes and is widely accepted and used by pharma companies [86]. The software is compatible with Windows and Linux operating systems.

6.7.1.7 rDock

rDock is an open-source molecular docking program designed for high-throughput virtual screening of small molecules. It is lightweight and can be run locally for docking compounds to proteins or nucleic acids [87]. The software is compatible with Linux and Mac operating systems.

6.7.1.8 FlexX

FlexX is a molecular docking program that focuses on flexible ligand docking with a highly efficient search algorithm. It is known for its fast docking speeds while maintaining accuracy in predicting binding poses and affinities [88]. The software is compatible with Windows and Linux operating systems.

6.7.1.9 Surflex-Dock

Surflex-Dock is a docking program that generates ligand conformations and evaluates their fit within the protein target receptor. It uses a unique molecular similarity algorithm and provides high accuracy in predicting ligand binding [89]. The software is compatible with Windows and Linux operating systems.

These software tools allow researchers to perform molecular docking simulations locally, giving them the flexibility and control to optimize docking workflows without relying on cloud or web-based services.

6.7.2 Online Resources for Molecular Docking

6.7.2.1 SwissDock

SwissDock is a webserver (http://www.swissdock.ch/) that performs protein-ligand docking using the EADock DSS engine. It offers both blind and targeted docking, allowing users to specify the binding site or search the entire protein surface [90].

6.7.2.2 DockThor

DockThor is an online molecular docking platform (https://dockthor.lncc.br/v2/) that supports both protein-ligand and protein-protein docking. It uses a grid-based algorithm to predict the best binding poses of small-molecule ligands to their corresponding target protein receptors and provides detailed output reports [91].

6.7.2.3 Mcule

Mcule is an online platform (https://mcule.com/) designed for virtual drug discovery. It provides molecular docking tools along with databases of small molecules for virtual screening. Mcule's docking engine allows users to dock ligands to target proteins and includes tools for structure-based drug design [92].

6.7.2.4 Click2Drug

Click2Drug is an online drug discovery platform (https://www.click2drug.org/) that integrates molecular docking, QSAR, and ADMET prediction tools. It provides a comprehensive environment for ligand-based and structure-based drug design, including protein-ligand docking [93].

6.7.2.5 MolProbity docking server

MolProbity offers docking tools (http://molprobity.biochem.duke.edu/) as part of a larger suite for validating and improving macromolecular models. It provides high-accuracy docking predictions and focuses on improving structural accuracy in protein-ligand complexes [94].

6.7.2.6 GNINA

GNINA is an open-source deep learning-based molecular docking platform (https://gnina.github.io/). Its web version allows users to dock ligands into protein targets using traditional docking methods and deep learning models to improve accuracy [95].

6.7.2.7 GalaxyWeb (GalaxyDock)

GalaxyWeb offers GalaxyDock, a protein-ligand docking server designed for high-throughput screening and flexible docking (http://galaxy.seoklab.org/). It provides a simple interface and allows users to upload their protein and ligand files for docking studies [96].

6.7.2.8 VirtualFlow

VirtualFlow is an online docking platform (https://virtual-flow.org/) designed for large-scale virtual screening. It provides a pipeline for docking millions of small molecules against protein targets using a distributed computing framework [97].

6.7.2.9 GRAMM-X

GRAMM-X is a protein-protein docking web server (https://gramm.compbio.ku.edu/) that uses a fast Fourier transform (FFT) docking algorithm to predict the possible binding modes between two proteins. It is designed to provide a rough estimate of interaction interfaces by searching for complementary shapes and electrostatic potentials between proteins [98].

6.7.2.10 ClusPro

ClusPro is a highly popular and automated protein-protein docking server (https://cluspro.org/tut_dock.php). It works by first generating thousands of potential docking conformations (poses) and then clustering them based on structural similarity. The most populated clusters are selected as the most likely binding conformations. It provides options to refine docking models using energy minimization [99].

6.7.2.11 PatchDock

PatchDock employs molecular geometry to identify docking poses based on the complementarity between ligand and target receptor shapes (https://bioinfo3d.cs.tau.ac.il/PatchDock/). It decomposes the surfaces of the docking molecules into patches, matches complementary patches, and filters the solutions based on steric constraints. PatchDock is often used for protein-protein, protein-ligand, and protein-DNA docking [100].

6.7.2.12 HADDOCK

HADDOCK is a flexible docking platform specifically designed for protein-protein and protein-ligand docking (https://wenmr.science.uu.nl/haddock2.4/). It accounts for molecular flexibility and experimental data such as NMR or cross-linking data during docking calculations [101].

6.7.2.13 CB-Dock

CB-Dock is a web-based docking tool that performs flexible docking simulations by determining the presence of potential binding sites (http://cao.labshare.cn/cb-dock/). It offers quick and efficient docking, requiring minimal user input [102].

These online molecular docking platforms are accessible, easy to use, and suitable for use in a number of conditions, offering flexibility and powerful capabilities without requiring users to install software locally. During our study, AutoDock4.2 was used to predict the interactions between NSPs and selected flavonoids.

6.8 Efficacy of Selected Flavonoids Against SARS-CoV-2

6.8.1 Physicochemical, Pharmacokinetic, and Drug-Like Properties of Flavones

The physicochemical, pharmacokinetic, and drug-like properties of flavones, as shown in Table 6.2, were studied to find out the potentialities of Apigenin, Herbacetin, and Juglanin as drug candidates against SARS-CoV-2. Table 6.2 shows that the molecular weights of all compounds range between 270.24 g/mol and 418.35 g/mol. Results from Table 6.2 show that the number of rotatable bonds in Juglanin is 4, which is higher than the other 2 ligands. This might help Juglanin interact with proteins effectively [103]. Moreover, Juglanin has 10 hydrogen bond acceptors and 6 hydrogen bond donors, which might make this ligand more interactive during protein-drug interactions. A study of molar refractivity can shed light on the total polarizability of a substance [104]. Table 6.2 shows that the molar refractivity of Apigenin is 73.99, which is the lowest, and for Juglanin, it is 102.17, which is the highest among the chosen three ligand molecules. The TPSA of Juglanin is also the highest. A ligand molecule's polarity can be predicted by studying the value of TPSA, the number of atoms as hydrogen bond acceptors and donors [105]. In general, the TPSA value should be less than 140 Å^2 [106]. Here, we found the values of TPSA for Apigenin, Herbacetin, and Juglanin to be 90.90 Å^2, 131.36 Å^2, and 170.05 Å^2, respectively.

From the results, it can be concluded that the values of GI absorptions of Apigenin and Herbacetin are high. BBB permeability is low for all three ligands. All flavones are predicted not to be the substrate of P-glycoproteins or P-gp pumps, which could predict that the ligands can be retained inside the system. All ligands were also tested using the drug-like filters, which include Lipinski, Ghose, Veber, Egan, and Muegge. Apigenin and Herbacetin met all the aforementioned filters; whereas, a few deviations were observed for Juglanin due to the TPSA value.

Table 6.2 Physicochemical, pharmacokinetics, and drug-like properties of the ligands

Properties		Apigenin	Herbacetin	Juglanin
Physicochemical	Formula	C15H10O5	C15H10O7	C20H18O10
	Molecular weight	270.24 g/mol	302.24 g/mol	418.35 g/mol
	Number of heavy atoms	20	22	30
	Number of aromatic heavy atoms	16	16	16
	Fraction Csp3	0	0	0.25
	Number of rotatable bonds	1	1	4
	Number of H-bond acceptors	5	7	10
	Number of H-bond donors	3	5	6
	Molar Refractivity	73.99	78.03	102.17
	TPSA	90.90 Å^2	131.36 Å^2	170.05 Å^2
Pharmacokinetics	GI absorption	High	High	Low
	BBB permeant	No	No	No
	P-gp substrate	No	No	No
Drug likeliness	Lipinski	Yes	Yes	Yes, 1 violation: NH or OH>5
	Ghose	Yes	Yes	Yes
	Veber	Yes	Yes	No, 1 violation: TPSA>140
	Egan	Yes	Yes	No, 1 violation: TPSA>131.6
	Muegge	Yes	Yes	No, 2 violations: TPSA>150, H-don>5
	Bioavailability score	0.55	0.55	0.55

Table 6.3 Toxicity prediction of the ligands

Toxicity Against	Apigenin		Herbacetin		Juglanin	
	Prediction	Probability	Prediction	Probability	Prediction	Probability
Hepatotoxicity	Inactive	0.68	Inactive	0.69	Inactive	0.8
Carcinogenicity	Inactive	0.62	Active	0.68	Inactive	0.79
Immunotoxicity	Inactive	0.99	Inactive	0.92	Inactive	0.62
Mutagenicity	Inactive	0.57	Active	0.51	Inactive	0.73
Cytotoxicity	Inactive	0.87	Inactive	0.99	Inactive	0.72
Aryl Hydrocarbon Receptor (AhR)	Active	1	Active	0.91	Inactive	0.85
Androgen Receptor (AR)	Inactive	0.99	Inactive	0.99	Inactive	0.92
Androgen Receptor Ligand Binding Domain (AR-LBD)	Inactive	1	Inactive	0.97	Inactive	0.98
Aromatase	Active	0.61	Inactive	0.91	Inactive	0.98
Estrogen Receptor Alpha (ER)	Active	1	Active	0.87	Inactive	0.85

(*Continued*)

Table 6.3 *(Continued)*

Toxicity Against	Apigenin		Herbacetin		Juglanin	
	Prediction	Probability	Prediction	Probability	Prediction	Probability
Estrogen Receptor Ligand Binding Domain (ER-LBD)	Active	1	Active	0.95	Inactive	0.99
Peroxisome Proliferator Activated Receptor Gamma (PPAR-Gamma)	Active	1	Inactive	0.98	Inactive	0.93
Nuclear factor (erythroid-derived 2)-like 2/antioxidant responsive element (nrf2/ARE)	Inactive	0.99	Inactive	0.99	Inactive	0.91
Heat shock factor response element (HSE)	Inactive	0.99	Inactive	0.99	Inactive	0.91

Toxicity Against	Apigenin		Herbacetin		Juglanin	
	Prediction	Probability	Prediction	Probability	Prediction	Probability
Mitochondrial Membrane Potential (MMP)	Active	1	Active	1	Inactive	0.89
Phosphoprotein (Tumor Suppressor) p53	Active	1	Inactive	0.97	Active	0.55
ATPase family AAA domain-containing protein 5 (ATAD5)	Active	0.96	Inactive	0.99	Inactive	0.96

6.8.2 Toxicity Prediction of the Molecules

Toxicity values of the selected flavonoids were tested. LD_{50} values of Apigenin, Herbacetin, and Juglanin were found to be 2500 mg/kg, 3919 mg/kg, and 5000 mg/kg, respectively. These values show that the selected flavonoids have very little toxicity. The toxicity values of known poisons like potassium cyanate or EtBr were 54 mg/kg and 100 mg/kg, respectively, as obtained from the same database. In comparison to them, the selected flavonoids can safely be administered to the patients at higher doses (according to the need). Table 6.3 presents the probability values for the flavonoids to exert toxicity across different body systems/pathways. Apigenin was found to be highly active against cancer, MMPs, p53, and cell-surface receptors, whereas Herbacetin may be mutagenic and could induce carcinogenicity. The flavonoid Juglanin has anti-p53 activity. In this study, we have thoroughly screened these flavonoids against NSP5, NSP15, and NSP16 of SARS-CoV-2 to predict their efficacies against the disease.

6.8.3 Interaction of Flavones with the Target Protein

The structures of Apigenin, Herbacetin, and Juglanin are presented in Figure 6.1 A–C, respectively, and select NSP5, NSP15, and NSP16 for further studies. The 3D structures of energy minimized NSP5, NSP15, and NSP16 proteins are shown in Figure 6.2 (upper panel A–C), respectively. It is important to study the Ramachandran plot for all proteins, as it predicts whether all amino acids in the protein are within the allowed region or not [107]. After the preparation of the proteins, Ramachandran plots were drawn. Figure 6.2 (upper panel D and F) shows that no residues were present in the disallowed region for NSP5 and NSP16, respectively, in their corresponding Ramachandran plots. On the other hand, ASN29 of NSP15 was in the disallowed region. The protein structure was refined, and the Ramachandran plot was again drawn. Figure 6.2 (upper panel E) shows that all the amino acid residues of NSP15 satisfy the criterion of stability as determined from the Ramachandran plot. The 3D coordinates of NSP5, NSP15, and NSP16 are considered for docking experiments.

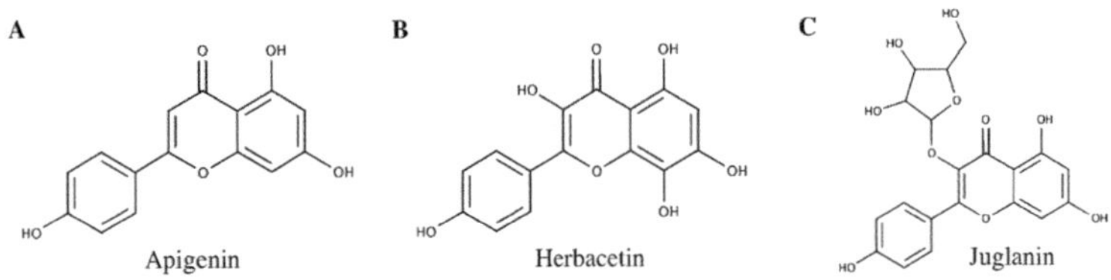

Figure 6.1 Structures of the flavonoids Apigenin (A), Herbacetin (B), and Juglanin (C).

Protein-ligand docking was done using AutoDock 4.2, which is a widely used software to analyze and predict the binding interactions of ligands with a protein in terms of free energy [108]. Changes in free energy are important parameters in the study of protein-ligand interactions [109]. Positive value of change in free energy of binding signifies less strong binding of the ligand with the protein molecule, whereas the negative value of change in free energy signifies stronger binding [110]. Figure 6.2 (middle panel) shows the free energy of binding of the selected flavonoids with NSP5, NSP15, and NSP16. Figure 6.2 (middle panel A) shows that the bindings of Apigenin, Herbacetin, and Juglanin with NSP5 are not energetically favorable, as the free energy of binding is positive in all instances. However, when Apigenin, Herbacetin, and Juglanin interact with NSP15 and NSP16 at the binding pocket, free energies of binding, shown in Figure 6.2 (middle panel A and B, respectively), were found to be negative, which signifies that all the selected flavonoids can interact with NSP15 and NSP16 very strongly. We therefore analyzed the interaction patterns of flavonoids with NSP15 and NSP16.

Hydrogen bonding is an important intermolecular interaction that takes place when a hydrogen atom covalently linked to a highly electronegative atom undergoes an attractive interaction with a lone pair of electrons on another electronegative atom. The strength of hydrogen bonds is stronger than other types of intermolecular forces [111]. Hydrogen bonding is of great significance in various fields of chemistry, biology, and materials science. Its unique strength and ability to stabilize structures, influence physical properties, and impact chemical reactivity make it a fundamental force in understanding molecular interactions and the behavior of substances [112]. Figure 6.2 (lower panel A and B) shows the binding patterns of Juglanin with NSP15. From Figure 6.2 (lower

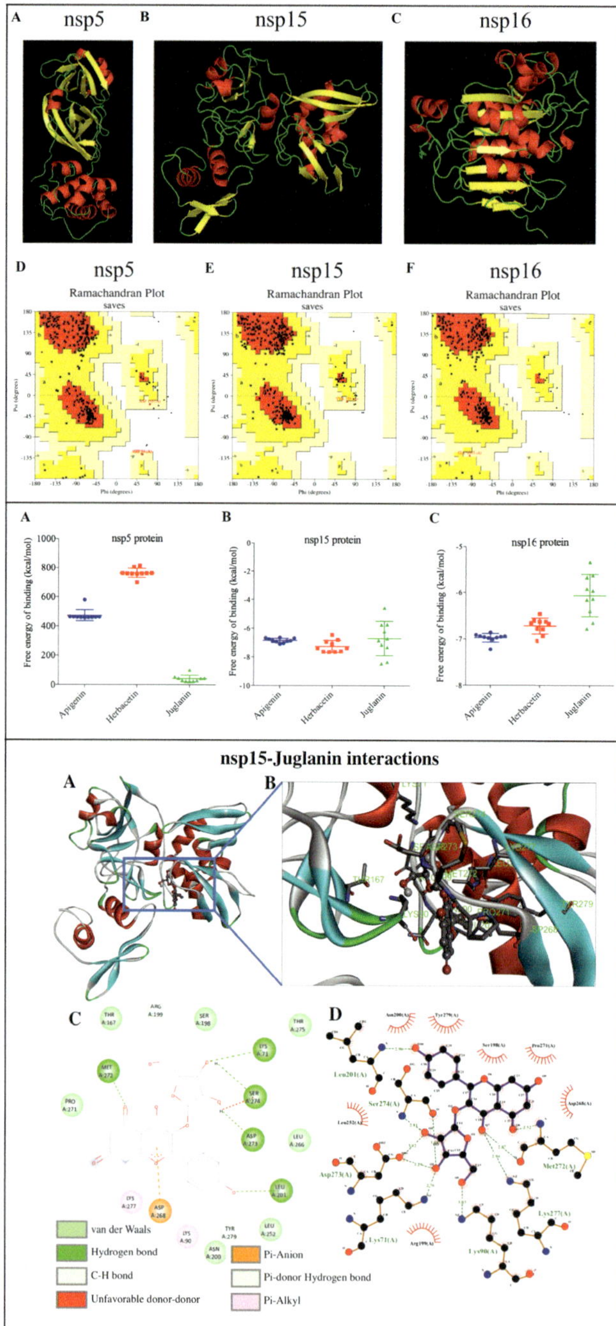

Figure 6.2 Upper panel shows the energy-minimized structures of NSP5 (A), NSP15 (B), and NSP16 (C), and their corresponding Ramachandran plots

in (D), (E), and (F), respectively. The middle panel shows the free energy of binding of NSP5 (A), NSP15 (B), and NSP16 (C) with Apigenin, Herbacetin, and Juglanin. The lower panel shows the binding of Juglanin to NSP15 (A and B) and the interaction studied using Discovery Studio Visualizer (v21.1.0.20298) and LigPlot$^+$ (C and D), respectively.

panel C and D), it is evident that Lys71, Lys90, Leu201, Met272, Asp273, Ser274, and Lys277 of NSP15 can form hydrogen bonds with Juglanin. LigPlot+ analysis data show that there are 10 hydrogen bonds between Juglanin and NSP15, which establishes the strong binding of the ligand at the active site of NSP15. When we compared the binding of Apigenin, Herbacetin, and Juglanin with NSP15 and NSP16, the best free energy from the docking studies was found to be –8.44 kcal/mol when Juglanin interacts with NSP15 at the active site. Studies showed that Lys71, Lys90, Val166, Thr167, Ser198, Arg199, Asn200, Leu201, Gln202, Leu252, Phe259, Leu266, Ser274, Thr275, Lys277, Tyr279, Val295, Ile296 and Asp297 are important amino acid residues of NSP15 protein. Vitamin D can bind to this particular region to stall the activity of NSP15 [113]. Compound methyl 4-[(2R,8S)-6-cyclooctyl-4,7-dioxo-3,6,17-triazatetracyclo[8.7.0.03,8.011,16]heptadeca-1(10),11, 13,15-tetraen-2-yl]benzoate (PubChem ID 1777791) was reported to have similar binding affinity with NSP15, notably with Arg199, Asn200, Tyr279, and Asp297 [114]. In our study, Juglanin-NSP16 interaction predicts the appearance of 7 hydrogen bonds, and the best free energy of binding was found to be –6.77 kcal/mol. From this, it can be predicted that Juglanin binds more strongly to NSP15 and also binds to NSP16 at the active site.

van der Waals interactions or van der Waals forces are comparatively weaker intermolecular interactions that come into play due to the electronic fluctuations within molecules. These forces play a crucial role in several aspects of chemistry and biology [115]. Important amino acids of NSP15, viz., Thr167, Ser198, Asn200, Leu252, Leu266, Pro271, and Thr275, interact through van der Waals interactions with Juglanin, which re-confirms the strong binding of Juglanin at the active site of the NSP15 protein. Other important interactions include C-H bond, Pi-anion interactions, Pi-Alkyl interactions, hydrophobic interactions, etc. All these interactions contribute to the stability of protein-ligand complexes [108, 116, 117]. All these types of interactions can be observed when Juglanin interacts with NSP15.

The binding of Apigenin and Herbacetin to NSP15 and NSP16 shows that both these ligands bind to the active pockets of the proteins. Notably, LigPlot+ analysis describes that Herbacetin-NSP15 interaction can have 10 hydrogen bonds, and the free energy of interaction was found to be –7.65 kcal/mol. But, when it interacts with NSP16, only 3 hydrogen bonds are formed, and the free energy of binding was found to be –7.04 kcal/mol. Interactions of Apigenin with both NSP15 and NSP16 proteins are poor, compared with those of Herbacetin and Juglanin. Apigenin forms 3 and 4 hydrogen bonds while interacting with NSP15 and NSP16, respectively, and the values of free energies of interactions were –7.09 kcal/mol and –7.22 kcal/mol, respectively.

From the literature, no side effects are reported with the flavonoids Apigenin, Herbacetin, and Juglanin, which is also clear from Table 6.3 (although Herbacetin is predicted to have mutagenicity). Apigenin is a flavonoid that has been tested against several diseases, including cancer, diabetes, different metabolic syndromes, and different viral diseases. It is reported that Apigenin can interact with vital viruses, which include African swine fever virus [118], foot-and-mouth disease [55], Epstein-Barr virus [119], enterovirus-71 [120], hepatitis B and C viruses [121], HIV [122], dengue [123], H1N1 [124], and many more. A few of the aforementioned viruses (e.g., Epstein-Barr virus) can cause fever and throat infection, as observed in SARS-CoV-2. These catalyze scientists to validate their efficacy against SARS-CoV-2. During the last 10 years, 1647 experiments have been carried out on COVID-19. Most of the group has tested the drug against the main protease of SARS-CoV-2, i.e., NSP5. In this study, we went a step forward to prove its efficacy against NSP15 and NSP16. However, for Herbacetin, only 15 studies have been reported in the last decade, of which only 2 examined its inhibitory action against SARS-CoV-2 NSP5. Interestingly, Juglanin is known for its ROS-quenching activity. No direct evidence was found through a literature search, which proves this drug can be used against SARS-CoV-2. In our study, we predict that Juglanin may be an important molecule that interacts with the active sites of NSP15 and NSP16 of SARS-CoV-2.

6.9 Conclusions and Future Perspective

Molecular docking has played a central role in fighting against SARS-CoV-2 and similar viral diseases, offering insights into drug discovery and therapeutic intervention. By simulating interactions between viral proteins and potential inhibitors, docking studies have accelerated the identification of candidate molecules capable of blocking critical viral functions, such as spike protein binding, protease activity, and RNA polymerase function. In the case of SARS-CoV-2, molecular docking has identified several promising antiviral agents. Our study shows that Juglanin, a flavonoid, may be useful for treating SARS-CoV-2.

Despite its successes, molecular docking is not without limitations. Inaccurate scoring functions, oversimplified protein-ligand interactions, and the lack of dynamic flexibility in the models can sometimes lead to false positives. To address these challenges, integrations of docking with advanced methods like molecular dynamics simulations, artificial intelligence, and machine learning have emerged as powerful alternatives to traditional bench work.

Looking ahead, molecular docking will continue to be an essential tool in combating viral diseases, especially with the growing threat of viral mutations and emerging pathogens. Future perspectives include the development of more accurate algorithms, integration with multi-omics data, and high-throughput virtual screening platforms. As computational power increases, docking studies can be expected to provide faster, more reliable predictions, aiding in the design of new antiviral drugs and ensuring preparedness for future pandemics.

Acknowledgements

Authors received financial support from the Department of Biotechnology (DBT), Government of India. The infrastructural support from the DBT-funded Bioinformatics Infrastructure Facility Centre (Project Sanction no: BT/PR40162/BTIS/137/48/2022, dated 31.10.2022) and National Network Project (Project Sanction no: BT/PR40192/BTIS/137/69/2023, dated 19.12.2023) were sanctioned to Prof. Angshuman Bagchi and utilized in this work. The authors are thankful to the BIF Centre, University of Kalyani for providing the infrastructure. The authors are also thankful to the University of Kalyani for providing a plagiarism-checking facility. Dr. Dipanjan Guha was supported with a salary from DBT, Govt. of India.

References

1. Chaudhary, K. K., & Mishra, N. (2016). A review on molecular docking: A novel tool for drug discovery. *Databases, 3*(4), 1029.
2. Muhammed, M. T., & Aki-Yalcin, E. (2024). Molecular docking: Principles, advances, and its applications in drug discovery. *Letters in Drug Design & Discovery, 21*(3), 480–495.
3. Sahu, M. K., Nayak, A. K., Hailemeskel, B., & Eyupoglu, O. E. (2024). Exploring recent updates on molecular docking: Types, methods, applications, limitations, and future prospects. *International Journal of Pharmaceutical Research and Allied Sciences, 13*(2–2024), 24–40.
4. Okpo, E. A., Agboke, A. A., Udobi, C. E., John, G. E., & Andy, I. E. (2024). The synergy of molecular docking and bioinformatics: An in-depth review in drug discovery. *Biotechnology Journal International, 28*(4), 119–136.
5. Sahu, D., Rathor, L. S., Dwivedi, S. D., Shah, K., Chauhan, N. S., Singh, M. R., & Singh, D. (2024). A review on molecular docking as an interpretative tool for molecular targets in disease management. *Assay and Drug Development Technologies, 22*(1), 40–50.
6. Yu, H., Li, C., Wang, X., Duan, J., Yang, N., Xie, L., ... & Li, Y. (2020). Techniques and strategies for potential protein target discovery and active pharmaceutical molecule screening in a pandemic. *Journal of Proteome Research, 19*(11), 4242–4258.
7. Villas-Boas, G. R., Rescia, V. C., Paes, M. M., Lavorato, S. N., Magalhães-Filho, M. F. D., Cunha, M. S., ... & Oesterreich, S. A. (2020). The new coronavirus (SARS-CoV-2): A comprehensive review on immunity and the application of bioinformatics and molecular modeling to the discovery of potential anti-SARS-CoV-2 agents. *Molecules, 25*(18), 4086.
8. Patel, A. R., Patel, H. B., Mody, S. K., Singh, R. D., Sarvaiya, V. N., Vaghela, S. H., & Tukra, S. (2021). Virtual screening in drug discovery. *Journal of Veterinary Pharmacology and Toxicology, 20*(2), 1–9.

9. Gao, G., Liu, W., Liu, P., Lei, W., Jia, Z., He, X., ... & Wu, G. (2022). Surveillance of SARS-CoV-2 in the environment and animal samples of the Huanan seafood market.

10. Liu, W. J., Liu, P., Lei, W., Jia, Z., He, X., Shi, W., ... & Wu, G. (2024). Surveillance of SARS-CoV-2 at the Huanan seafood market. *Nature, 631*(8020), 402–408.

11. Kumar, S., Sharma, P. P., Shankar, U., Kumar, D., Joshi, S. K., Pena, L., ... & Rathi, B. (2020). Discovery of new hydroxyethylamine analogs against 3CLpro protein target of SARS-CoV-2: Molecular docking, molecular dynamics simulation, and structure: Activity relationship studies. *Journal of Chemical Information and Modeling, 60*(12), 5754–5770.

12. Kumar, D., Batra, L., & Malik, M. T. (2020). Insights of Novel Coronavirus (SARS-CoV-2) disease outbreak, management, and treatment. *AIMS Microbiology, 6*(3), 183.

13. Arya, R., Kumari, S., Pandey, B., Mistry, H., Bihani, S. C., Das, A., ... & Kumar, M. (2021). Structural insights into SARS-CoV-2 proteins. *Journal of Molecular Biology, 433*(2), 166725.

14. Roe, M. K., Junod, N. A., Young, A. R., Beachboard, D. C., & Stobart, C. C. (2021). Targeting novel structural and functional features of coronavirus protease NSP5 (3CLpro, Mpro) in the age of COVID-19. *Journal of General Virology, 102*(3), 001558.

15. Stobart, C. C., Sexton, N. R., Munjal, H., Lu, X., Molland, K. L., Tomar, S., ... & Denison, M. R. (2013). Chimeric exchange of coronavirus NSP5 proteases (3CLpro) identifies common and divergent regulatory determinants of protease activity. *Journal of Virology, 87*(23), 12611–12618.

16. de Freitas Amorim, V. M., de Souza, R. F., Guzzo, C. R., & de Souza, A. S. (2023). Molecular dynamics simulations suggest SARS-CoV-2 3CLpro mutations in beta and omicron variants do not alter binding affinities for cleavage sites of non-structural proteins. *COVID, 3*(4), 622–636.

17. Li, G., Hilgenfeld, R., Whitley, R., & De Clercq, E. (2023). Therapeutic strategies for COVID-19: Progress and lessons learned. *Nature Reviews Drug Discovery, 22*(6), 449–475.

18. Salukhe, I., Choi, R., Van Voorhis, W., Barrett, L., & Hyde, J. (2023). Regulation of coronavirus NSP15 cleavage specificity by RNA structure. *PLoS One, 18*(8), e0290675.

19. Thomas, J., Kumar, S., & Satija, J. (2024). Integrated molecular and quantum mechanical approach to identify novel potent natural

bioactive compound against 2′-O-methyltransferase (NSP16) of SARS-CoV-2. *Journal of Biomolecular Structure and Dynamics, 42*(4), 1999–2012.

20. Pavan, M., & Moro, S. (2023). Lessons learnt from COVID-19: Computational strategies for facing present and future pandemics. *International Journal of Molecular Sciences, 24*(5), 4401.
21. Gentile, D., Patamia, V., Scala, A., Sciortino, M. T., Piperno, A., & Rescifina, A. (2020). Putative inhibitors of SARS-CoV-2 main protease from a library of marine natural products: A virtual screening and molecular modeling study. *Marine Drugs, 18*(4), 225.
22. Junqueira Ribeiro, M. M. (2024). Computer-aided drug discovery approaches in the identification of natural products against SARS-CoV-2: A review. *Current Computer-Aided Drug Design, 20*(4), 313–324.
23. Silva Arouche, T. D., Reis, A. F., Martins, A. Y., S. Costa, J. F., Carvalho Junior, R. N., & JC Neto, A. M. (2020). Interactions between remdesivir, ribavirin, favipiravir, galidesivir, hydroxychloroquine, and chloroquine with fragment molecular of the COVID-19 main protease with inhibitor N3 complex (PDB ID: 6LU7) using molecular docking. *Journal of Nanoscience and Nanotechnology, 20*(12), 7311–7323.
24. Celik, I., Erol, M., & Duzgun, Z. (2021). In silico evaluation of potential inhibitory activity of remdesivir, favipiravir, ribavirin, and galidesivir active forms on SARS-CoV-2 RNA polymerase. *Molecular Diversity*, 1–14.
25. Eroglu, E., & Toprak, C. (2021). Overview of favipiravir and remdesivir treatment for COVID-19. *International Journal of Pharmaceutical Sciences and Research, 12*(4), 1950–1957.
26. Singh, T. U., Parida, S., Lingaraju, M. C., Kesavan, M., Kumar, D., & Singh, R. K. (2020). Drug repurposing approach to fight COVID-19. *Pharmacological Reports, 72*, 1479–1508.
27. Ng, Y. L., Salim, C. K., & Chu, J. J. H. (2021). Drug repurposing for COVID-19: Approaches, challenges, and promising candidates. *Pharmacology & Therapeutics, 228*, 107930.
28. Uzunova, K., Filipova, E., Pavlova, V., & Vekov, T. (2020). Insights into antiviral mechanisms of remdesivir, lopinavir/ritonavir, and chloroquine/hydroxychloroquine affecting the new SARS-CoV-2. *Biomedicine & Pharmacotherapy, 131*, 110668.
29. Gómez-Ríos, D., López-Agudelo, V. A., & Ramírez-Malule, H. (2020). Repurposing antivirals as potential treatments for SARS-CoV-2: From SARS to COVID-19. *Journal of Applied Pharmaceutical Science, 10*(5), 001–009.

30. Pirolli, D., Righino, B., & De Rosa, M. C. (2021). Targeting SARS-CoV-2 spike protein/ACE2 protein-protein interactions: A computational study. *Molecular Informatics, 40*(6), 2060080.

31. Muhseen, Z. T., Hameed, A. R., AlHasani, H. M., ul Qamar, M. T., & Li, G. (2020). Promising terpenes as SARS-CoV-2 spike receptor-binding domain (RBD) attachment inhibitors to the human ACE2 receptor: Integrated computational approach. *Journal of Molecular Liquids, 320,* 114493.

32. Gimeno, A., Mestres-Truyol, J., Ojeda-Montes, M. J., Macip, G., Saldivar-Espinoza, B., Cereto-Massagué, A., ... & Garcia-Vallvé, S. (2020). Prediction of novel inhibitors of the main protease (M-pro) of SARS-CoV-2 through consensus docking and drug reposition. *International Journal of Molecular Sciences, 21*(11), 3793.

33. Ferreira, L. G., Dos Santos, R. N., Oliva, G., & Andricopulo, A. D. (2015). Molecular docking and structure-based drug design strategies. *Molecules, 20*(7), 13384–13421.

34. Santos, L. H., Ferreira, R. S., & Caffarena, E. R. (2019). Integrating molecular docking and molecular dynamics simulations. *Docking Screens for Drug Discovery*, 13–34.

35. Hongmao, S. (2015). *A Practical Guide to Rational Drug Design*. Woodhead Publishing.

36. Mandal, S., & Mandal, S. K. (2009). Rational drug design. *European Journal of Pharmacology, 625*(1–3), 90–100.

37. Faisal, S., Badshah, S. L., Kubra, B., Sharaf, M., Emwas, A. H., Jaremko, M., & Abdalla, M. (2021). Computational study of SARS-CoV-2 RNA-dependent RNA polymerase allosteric site inhibition. *Molecules, 27*(1), 223.

38. Pantsar, T., & Poso, A. (2018). Binding affinity via docking: Fact and fiction. *Molecules, 23*(8), 1899.

39. Erickson, J. A., Jalaie, M., Robertson, D. H., Lewis, R. A., & Vieth, M. (2004). Lessons in molecular recognition: the effects of ligand and protein flexibility on molecular docking accuracy. *Journal of Medicinal Chemistry, 47*(1), 45–55.

40. Halperin, I., Ma, B., Wolfson, H., & Nussinov, R. (2002). Principles of docking: An overview of search algorithms and a guide to scoring functions. *Proteins: Structure, Function, and Bioinformatics, 47*(4), 409–443.

41. Korb, O., Olsson, T. S., Bowden, S. J., Hall, R. J., Verdonk, M. L., Liebeschuetz, J. W., & Cole, J. C. (2012). Potential and limitations of ensemble

docking. *Journal of Chemical Information and Modeling, 52*(5), 1262–1274.

42. Chen, L., & Chen, X. (2012). Results of molecular docking as descriptors to predict human serum albumin binding affinity. *Journal of Molecular Graphics and Modelling, 33*, 35–43.
43. Coleman, R. G., Carchia, M., Sterling, T., Irwin, J. J., & Shoichet, B. K. (2013). Ligand pose and orientational sampling in molecular docking. *PloS One, 8*(10), e75992.
44. Ghasemi, J. B., Abdolmaleki, A., & Shiri, F. (2017). Molecular docking challenges and limitations. In *Pharmaceutical Sciences: Breakthroughs in Research and Practice*, 770–794, IGI Global.
45. Wade, R., Pantsar, T., & Poso, A. (2018). Binding affinity via docking: Fact and fiction. *Molecules, 23*(8).
46. Llanos, M. A., Gantner, M. E., Rodriguez, S., Alberca, L. N., Bellera, C. L., Talevi, A., & Gavernet, L. (2021). Strengths and weaknesses of docking simulations in the SARS-CoV-2 era: The main protease (Mpro) case study. *Journal of Chemical Information and Modeling, 61*(8), 3758–3770.
47. Masimirembwa, C. M., Bredberg, U., & Andersson, T. B. (2003). Metabolic stability for drug discovery and development: Pharmacokinetic and biochemical challenges. *Clinical Pharmacokinetics, 42*, 515–528.
48. Lee, S., Seok, C., & Park, H. (2023). Benchmarking applicability of medium-resolution cryo-EM protein structures for structure-based drug design. *Journal of Computational Chemistry, 44*(14), 1360–1368.
49. Vakser, I. A. (1995). Protein docking for low-resolution structures. *Protein Engineering, Design and Selection, 8*(4), 371–378.
50. Russo, M., Spagnuolo, C., Tedesco, I., & Russo, G. L. (2010). Phytochemicals in cancer prevention and therapy: Truth or dare? *Toxins, 2*(4), 517–551.
51. Singh, J. (2023). Bioactive nutrients in vegetables for human nutrition and health. In *Vegetables for Nutrition and Entrepreneurship*, 57–72, Singapore: Springer Nature Singapore.
52. Russo, G. L., Tedesco, I., Spagnuolo, C., & Russo, M. (2017, October). Antioxidant polyphenols in cancer treatment: Friend, foe, or foil? In *Seminars in Cancer Biology, 46*, 1–13, Academic Press.
53. Spagnuolo, C., Moccia, S., & Russo, G. L. (2018). Anti-inflammatory effects of flavonoids in neurodegenerative disorders. *European Journal of Medicinal Chemistry, 153*, 105–115.

54. Lalani, S., & Poh, C. L. (2020). Flavonoids as antiviral agents for enterovirus A71 (EV-A71). *Viruses, 12*(2), 184.

55. Qian, S., Fan, W., Qian, P., Zhang, D., Wei, Y., Chen, H., & Li, X. (2015). Apigenin restricts FMDV infection and inhibits viral IRES-driven translational activity. *Viruses, 7*(4), 1613–1626.

56. Bafandeh, S., Khodadadi, E., Ganbarov, K., Asgharzadeh, M., Köse, Ş., & Samadi Kafil, H. (2023). Natural products as a potential source of promising therapeutics for COVID-19 and viral diseases. *Evidence-Based Complementary and Alternative Medicine, 2023*(1), 5525165.

57. Kumar, S., Sharma, P. P., Upadhyay, C., Kempaiah, P., & Rathi, B. (2021). Multi-targeting approach for NSP3, NSP9, NSP12, and NSP15 proteins of SARS-CoV-2 by Diosmin as illustrated by molecular docking and molecular dynamics simulation methodologies. *Methods, 195*, 44–56.

58. Chowdhury, N., & Bagchi, A. (2022). Identification of the binding interactions of some novel antiviral compounds against nsp1 protein from SARS-CoV-2 (COVID-19) through high-throughput screening. *Journal of Biomolecular Structure and Dynamics, 40*(14), 6634–6641.

59. Bank, P. D. (1971). Protein Data Bank. *Nature New Biol, 233*(223), 1010–1038.

60. UniProt Consortium. (2007). The universal protein resource (UniProt). *Nucleic Acids Research, 36*(suppl_1), D190–D195.

61. Bienert, S., Waterhouse, A., De Beer, T. A., Tauriello, G., Studer, G., Bordoli, L., & Schwede, T. (2017). The SWISS-MODEL Repository: New features and functionality. *Nucleic Acids Research, 45*(D1), D313–D319.

62. Varadi, M., Bertoni, D., Magana, P., Paramval, U., Pidruchna, I., Radhakrishnan, M., ... & Velankar, S. (2024). AlphaFold Protein Structure Database in 2024: Providing structure coverage for over 214 million protein sequences. *Nucleic Acids Research, 52*(D1), D368–D375.

63. Velankar, S., Best, C., Beuth, B., Boutselakis, C. H., Cobley, N., Sousa Da Silva, A. W., ... & Kleywegt, G. J. (2010). Protein Data Bank in Europe. *Nucleic Acids Research, 38*(suppl_1), D308–D317.

64. Kinjo, A. R., Suzuki, H., Yamashita, R., Ikegawa, Y., Kudou, T., Igarashi, R., ... & Nakamura, H. (2012). Protein Data Bank Japan (PDBj): Maintaining a structural data archive and resource description framework format. *Nucleic Acids Research, 40*(D1), D453–D460.

65. Pieper, U., Webb, B. M., Dong, G. Q., Schneidman-Duhovny, D., Fan, H., Kim, S. J., ... & Sali, A. (2014). ModBase, a database of annotated comparative protein structure models and associated resources. *Nucleic Acids Research*, *42*(D1), D336–D346.

66. Murzin, A. G. (1996). Structural classification of proteins: New superfamilies. *Current Opinion in Structural Biology*, *6*(3), 386–394.

67. Orengo, C. A., Michie, A. D., Jones, S., Jones, D. T., Swindells, M. B., & Thornton, J. M. (1997). CATH: A hierarchical classification of protein domain structures. *Structure*, *5*(8), 1093–1109.

68. Madej, T., Addess, K. J., Fong, J. H., Geer, L. Y., Geer, R. C., Lanczycki, C. J., ... & Bryant, S. H. (2012). MMDB: 3D structures and macromolecular interactions. *Nucleic Acids Research*, *40*(D1), D461–D464.

69. Kim, S., Thiessen, P. A., Bolton, E. E., Chen, J., Fu, G., Gindulyte, A., ... & Bryant, S. H. (2016). PubChem substance and compound databases. *Nucleic Acids Research*, *44*(D1), D1202–D1213.

70. Irwin, J. J., & Shoichet, B. K. (2005). ZINC: A free database of commercially available compounds for virtual screening. *Journal of Chemical Information and Modeling*, *45*(1), 177–182.

71. Wishart, D. S., Knox, C., Guo, A. C., Cheng, D., Shrivastava, S., Tzur, D., ... & Hassanali, M. (2008). DrugBank: A knowledgebase for drugs, drug actions, and drug targets. *Nucleic Acids Research*, *36*(suppl_1), D901–D906.

72. Gaulton, A., Hersey, A., Nowotka, M., Bento, A. P., Chambers, J., Mendez, D., ... & Leach, A. R. (2017). The ChEMBL database in 2017. *Nucleic Acids Research*, *45*(D1), D945–D954.

73. Gilson, M. K., Liu, T., Baitaluk, M., Nicola, G., Hwang, L., & Chong, J. (2016). BindingDB in 2015: A public database for medicinal chemistry, computational chemistry, and systems pharmacology. *Nucleic Acids Research*, *44*(D1), D1045–D1053.

74. Sitzmann, M., Weidlich, I. E., Filippov, I. V., Liao, C., Peach, M. L., Ihlenfeldt, W. D., ... & Nicklaus, M. C. (2012). PDB ligand conformational energies calculated quantum-mechanically. *Journal of Chemical Information and Modeling*, *52*(3), 739–756.

75. Pence, H. E., & Williams, A. (2010). ChemSpider: An online chemical information resource.

76. Marques, J. M., Da Roz D'Alessandre, N., Guardia, G. D., Craveiro, F. L., Grutzmacher, C., Silva, G. O., ... & Batista, R. L. (2024). 8738 Unveiling potential therapeutic candidates for ACTH-silent pituitary neuroendocrine tumors through high-throughput drug screening. *Journal of the Endocrine Society*, *8*(Supplement_1), bvae1363–1367.

77. Tripathi, N., Goel, B., Bhardwaj, N., Sahu, B., Kumar, H., & Jain, S. K. (2022). Virtual screening and molecular simulation study of natural products database for lead identification of novel coronavirus main protease inhibitors. *Journal of Biomolecular Structure and Dynamics, 40*(8), 3655–3667.

78. Gupta, P. P., Bastikar, V. A., & Chhajed, S. S. (2019). Chemical structure databases in drug discovery. In *Computer Applications in Drug Discovery and Development*, 47–61, IGI Global.

79. Wishart, D. S., Tzur, D., Knox, C., Eisner, R., Guo, A. C., Young, N., ... & Querengesser, L. (2007). HMDB: The human metabolome database. *Nucleic Acids Research, 35*(suppl_1), D521–D526.

80. Gupta, C. L., Khan, M. B., Ampasala, D. R., Akhtar, S., Dwivedi, U. N., & Bajpai, P. (2019). Pharmacophore-based virtual screening approach for identification of potent natural modulatory compounds of human toll-like receptor 7. *Journal of Biomolecular Structure and Dynamics.*

81. Morris, G. M., Goodsell, D. S., Huey, R., Hart, W. E., Halliday, S., Belew, R., & Olson, A. J. (2001). AutoDock automated docking of flexible ligands to receptor: User guide.

82. Trott, O., & Olson, A. J. (2010). AutoDock Vina: Improving the speed and accuracy of docking with a new scoring function, efficient optimization, and multithreading. *Journal of Computational Chemistry, 31*(2), 455–461.

83. Liu, N., & Xu, Z. (2019, February). Using LeDock as a docking tool for computational drug design. In *IOP Conference Series: Earth and Environmental Science, 218*, 012143, IOP Publishing.

84. Friesner, R. A., Banks, J. L., Murphy, R. B., Halgren, T. A., Klicic, J. J., Mainz, D. T., ... & Shenkin, P. S. (2004). Glide: A new approach for rapid, accurate docking and scoring. 1. Method and assessment of docking accuracy. *Journal of Medicinal Chemistry, 47*(7), 1739–1749.

85. Vilar, S., Cozza, G., & Moro, S. (2008). Medicinal chemistry and the molecular operating environment (MOE): Application of QSAR and molecular docking to drug discovery. *Current Topics in Medicinal Chemistry, 8*(18), 1555–1572.

86. Jones, G., Willett, P., Glen, R. C., Leach, A. R., & Taylor, R. (1997). Development and validation of a genetic algorithm for flexible docking. *Journal of Molecular Biology, 267*(3), 727–748.

87. Ruiz-Carmona, S., Alvarez-Garcia, D., Foloppe, N., Garmendia-Doval, A. B., Juhos, S., Schmidtke, P., ... & Morley, S. D. (2014). rDock: A fast, versatile and open-source program for docking ligands to proteins and nucleic acids. *PLoS Computational Biology, 10*(4), e1003571.

88. Rarey, M., Kramer, B., Lengauer, T., & Klebe, G. (1996). A fast, flexible docking method using an incremental construction algorithm. *Journal of Molecular Biology, 261*(3), 470–489.
89. Jain, A. N. (2007). Surflex-Dock 2.1: Robust performance from ligand energetic modeling, ring flexibility, and knowledge-based search. *Journal of Computer-aided Molecular Design, 21*, 281–306.
90. Grosdidier, A., Zoete, V., & Michielin, O. (2011). SwissDock, a protein-small molecule docking web service based on EADock DSS. *Nucleic Acids Research, 39*(suppl_2), W270–W277.
91. Guedes, I. A., Krempser, E., & Dardenne, L. E. (2017). DockThor 2.0: A free web server for protein-ligand virtual screening. *XIX SBQT–Simpósio Brasileiro de Química Teórica*, 2013–2014.
92. Kiss, R., Sandor, M., & Szalai, F. A. (2012). http://Mcule.com: A public web service for drug discovery. *Journal of Cheminformatics, 4*(Suppl_1), P17.
93. Swiss Institute of Bioinformatics. (2013). Click2Drug website. http://www.click2drug.org/
94. Williams, C. J., Headd, J. J., Moriarty, N. W., Prisant, M. G., Videau, L. L., Deis, L. N., ... & Richardson, D. C. (2018). MolProbity: More and better reference data for improved all-atom structure validation. *Protein Science, 27*(1), 293–315.
95. McNutt, A. T., Francoeur, P., Aggarwal, R., Masuda, T., Meli, R., Ragoza, M., ... & Koes, D. R. (2021). GNINA 1.0: Molecular docking with deep learning. *Journal of Cheminformatics, 13*(1), 43.
96. Shin, W. H., & Seok, C. (2012). GalaxyDock: Protein–ligand docking with flexible protein side-chains. *Journal of Chemical Information and Modeling, 52*(12), 3225–3232.
97. Gorgulla, C., Boeszoermenyi, A., Wang, Z. F., Fischer, P. D., Coote, P. W., Padmanabha Das, K. M., ... & Arthanari, H. (2020). An open-source drug discovery platform enables ultra-large virtual screens. *Nature, 580*(7805), 663–668.
98. Tovchigrechko, A., & Vakser, I. A. (2006). GRAMM-X public web server for protein–protein docking. *Nucleic Acids Research, 34*(suppl_2), W310–W314.
99. Kozakov, D., Hall, D. R., Xia, B., Porter, K. A., Padhorny, D., Yueh, C., ... & Vajda, S. (2017). The ClusPro web server for protein–protein docking. *Nature Protocols, 12*(2), 255–278.
100. Schneidman-Duhovny, D., Inbar, Y., Nussinov, R., & Wolfson, H. J. (2005). PatchDock and SymmDock: Servers for rigid and symmetric docking. *Nucleic Acids Research, 33*(suppl_2), W363–W367.

101. De Vries, S. J., Van Dijk, M., & Bonvin, A. M. (2010). The HADDOCK web server for data-driven biomolecular docking. *Nature Protocols, 5*(5), 883–897.

102. Liu, Y., Grimm, M., Dai, W. T., Hou, M. C., Xiao, Z. X., & Cao, Y. (2020). CB-Dock: A web server for cavity detection-guided protein–ligand blind docking. *Acta Pharmacologica Sinica, 41*(1), 138–144.

103. Meng, Q., Lai, X., Yan, Z., Wang, Y., & Wu, M. (2021). Position control with zero residual vibration for two degrees-of-freedom flexible systems based on motion trajectory optimization. *Information Sciences, 575*, 698–713.

104. Baur, M. E., Horsma, D. A., Knobler, C. M., & Perez, P. (1969). Dielectric constant and refractive index of weak complexes in solution. *The Journal of Physical Chemistry, 73*(3), 641–650.

105. Bosch, E., Rosés, M., Herodes, K., Koppel, I., Leito, I., Koppel, I., & Taal, V. (1996). Solute-solvent and solvent-solvent interactions in binary solvent mixtures. 2. Effect of temperature on the ET (30) polarity parameter of dipolar hydrogen bond acceptor-hydrogen bond donor mixtures. *Journal of Physical Organic Chemistry, 9*(6), 403–410.

106. Whitty, A., Zhong, M., Viarengo, L., Beglov, D., Hall, D. R., & Vajda, S. (2016). Quantifying the chameleonic properties of macrocycles and other high-molecular-weight drugs. *Drug Discovery Today, 21*(5), 712–717.

107. Carugo, O., & Djinović-Carugo, K. (2013). Half a century of Ramachandran plots. *Acta Crystallographica Section D: Biological Crystallography, 69*(8), 1333–1341.

108. Bhattacharya, A., Guha, P. S., Chowdhury, N., Bagchi, A., & Guha, D. (2023). Virtual screening and molecular docking of flavone derivatives as a potential anticancer drug in the presence of dexamethasone. *Biointerface Res. Appl. Chem, 13*(3), 19.

109. Yang, C. Y., Sun, H., Chen, J., Nikolovska-Coleska, Z., & Wang, S. (2009). Importance of ligand reorganization free energy in protein–ligand binding-affinity prediction. *Journal of the American Chemical Society, 131*(38), 13709–13721.

110. Gilson, M. K., & Zhou, H. X. (2007). Calculation of protein-ligand binding affinities. *Annu. Rev. Biophys. Biomol. Struct., 36*(1), 21–42.

111. Zeegers-Huyskens, T., & Huyskens, P. (1991). Intermolecular forces. In *Intermolecular Forces: An Introduction to Modern Methods and Results*, 1–30, Berlin, Heidelberg: Springer Berlin Heidelberg.

112. Dong, K., Zhang, S., & Wang, J. (2016). Understanding the hydrogen bonds in ionic liquids and their roles in properties and reactions. *Chemical Communications, 52*(41), 6744–6764.

113. Shalayel, M. H., Al-Mazaideh, G. M., Aladaileh, S. H., Al-Swailmi, F. K., & Al-Thiabat, M. G. (2020). Vitamin D is a potential inhibitor of COVID-19: In silico molecular docking to the binding site of SARS-CoV-2 endoribonuclease nsp15. *Pakistan Journal of Pharmaceutical Sciences, 33*(5).

114. Motwalli, O., & Alazmi, M. (2021). Analysis of natural compounds against the activity of SARS-CoV-2 NSP15 protein towards an effective treatment against COVID-19: A theoretical and computational biology approach. *Journal of Molecular Modeling, 27*(6), 160.

115. Roth, C. M., Neal, B. L., & Lenhoff, A. M. (1996). Van der Waals interactions involving proteins. *Biophysical Journal, 70*(2), 977–987.

116. Garg, S., & Roy, A. (2020). In silico analysis of selected alkaloids against the main protease (Mpro) of SARS-CoV-2. *Chemico-Biological Interactions, 332*, 109309.

117. Snoussi, M., Ahmad, I., Patel, H., Noumi, E., Zrieq, R., Saeed, M., ... & Kadri, A. (2022). Lapachol and (α/β)-lapachone as inhibitors of SARS-CoV-2 main protease (Mpro) and hACE-2: ADME properties, docking, and dynamic simulation approaches. *Pharmacognosy Magazine, 18*(79).

118. Hakobyan, A., Arabyan, E., Avetisyan, A., Abroyan, L., Hakobyan, L., & Zakaryan, H. (2016). Apigenin inhibits African swine fever virus infection in vitro. *Archives of Virology, 161*, 3445–3453.

119. Wu, C. C., Fang, C. Y., Cheng, Y. J., Hsu, H. Y., Chou, S. P., Huang, S. Y., ... & Chen, J. Y. (2017). Inhibition of Epstein-Barr virus reactivation by the flavonoid apigenin. *Journal of Biomedical Science, 24*, 1–13.

120. Zhang, W., Qiao, H., Lv, Y., Wang, J., Chen, X., Hou, Y., ... & Li, E. (2014). Apigenin inhibits enterovirus-71 infection by disrupting viral RNA association with transacting factors. *PloS One, 9*(10), e110429.

121. Yang, P. W., Chen, T. T., Zhao, W. X., Liu, G. W., Feng, X. J., Wang, S. M., ... & Zhang, S. H. (2021). Scutellaria barbata D. Don and Oldenlandia diffusa (Willd.) Roxb crude extracts inhibit hepatitis-B-virus-associated hepatocellular carcinoma growth through regulating circRNA expression. *Journal of Ethnopharmacology, 275*, 114110.

122. Schonhofer, C., Yi, J., Sciorillo, A., Andrae-Marobela, K., Cochrane, A., Harris, M., ... & Tietjen, I. (2021). Flavonoid-based inhibition

of cyclin-dependent kinase 9 without concomitant inhibition of histone deacetylases durably reinforces HIV latency. *Biochemical Pharmacology*, *186*, 114462.

123. Coulerie, P., Nour, M., Maciuk, A., Eydoux, C., Guillemot, J. C., Lebouvier, N., ... & Figadère, B. (2013). Structure-activity relationship study of biflavonoids on the dengue virus polymerase DENV-NS5 RdRp. *Planta Medica*, *79*(14), 1313–1318.

124. Zhang, X. X., Wu, Q. F., Yan, Y. L., & Zhang, F. L. (2018). Inhibitory effects and related molecular mechanisms of total flavonoids in Mosla chinensis Maxim against H1N1 influenza virus. *Inflammation Research*, *67*, 179–189.

Chapter 7

Role of Computational Methodologies to Find Better Potent Inhibitors Against Several Fatal Diseases: A Multi-Targeted Drug Repurposing Approach

Sourav Majumdar and Anup Pramanik
Department of Chemistry, Sidho-Kanho-Birsha University, Purulia, West Bengal, India
anup.pramanik@skbu.ac.in

Drug repurposing has become a popular approach in computational drug discovery to find out the possible interaction pattern with respect to a receptor of a compound as a drug. In recent years, for the treatment of several fatal diseases, e.g., COVID-19, Dengue, and Cancer (especially breast and lung), the multi-targeted drug repurposing approach has reached a new height due to its cost- and time-efficient nature. Researchers used several efforts for the treatment of these diseases but a few of them became successful, that is where the role of virtual screening, molecular docking (both

Computational Methods for Molecular Modelling and Docking
Edited by Arup Kumar Ghosh and Suban Kumar Sahoo

ISBN 978-981-5352-25-2 (Hardcover), 978-1-042-00204-7 (eBook)
www.jennystanford.com

DOI: 10.1201/9781042002047-7

standard and extra precision), conceptual density functional theory (DFT)-based electronic parameters like E_{HOMO}, E_{LUMO}, global softness (S), absolute hardness (η), etc., and molecular dynamics (MD) simulation come into play to reduce the extra wastage and extra cost-cutting program. Further to validate whether a lead compound could be used as a drug or not, one needs to study pharmacological features of the respective lead compounds, e.g., absorption, distribution, metabolism, and excretion (ADME), and toxicity via different computational servers. Thus, the present chapter provides significant insights into different polyphenolic compounds by using computational tools for the treatment of several fatal diseases using a multi-targeted drug repurposing approach.

7.1 Introduction

It is widely recognized that developing a compound into a potent drug involves extensive experimentation, rigorous testing, and thorough validation. Thus, the traditional process of drug discovery is a high-risk and time-consuming task. Therefore, to minimize the time requirement, a new era of drug discovery has come into play in recent years, popularly known as computational drug discovery or computer-aided drug discovery (CADD). CADD has a big role in saving time and costs in the production of different categories of drugs, such as anti-viral, anti-hypertensive, anti-dengue, anti-cancer, etc. [1, 2]. For the production and identification of the potential therapeutic agents as a drug for these diseases, the workflow of CADD can be distributed into three parts: (a) hit identification, (b) lead generation, and (c) lead optimization phases [3]. Nowadays, producing effective and cost-cutting drugs for the treatment of anti-viral, anti-hypertensive, and anti-cancer diseases is a demanding trend. The severity of COVID-19 disease from the end of 2019 to the middle of 2021 is well remembered by everyone till now [4]. To investigate probable drugs for the treatment of COVID-19 disease, in 2021, a group of scientists conducted a multimodal drug repurposing approach for COVID-19 treatment. To find out probable new inhibitors for COVID-19 treatment with respect to the proposed anti-COVID-19 inhibitors, they used two separate Restricted Boltzmann Machine (RBM) approaches [5]. Several investigations were done to search for potential drugs for probable treatment against the COVID-19 disease, but only a few of them became successful. Polyphenolic compounds have been found to be an effective source for treating COVID-19, hypertension, dengue, and several cancerous diseases (especially breast and lung cancer) in recent years. In 2024, Ananto et al. took the structural data for the main protease of SARS-CoV-2 (PDB ID 5R7Y) from the RCSB Protein Data Bank (PDB) [6] in PDB format and did an in silico molecular docking study with 225 newly designed polyphenolic compounds, the top-scored ligand showed the most favorable docking score of greater than –8 kcal/mol, higher than the reference compound, indicating better binding characteristics, and the stability of the complex was confirmed by an MD simulation study [7]. Apart from COVID-19, hypertension belongs to another class of fatal diseases. Several types of research were carried out, and a plethora of works are still in progress regarding the treatment of hypertension.

Normally, five classes of anti-hypertensive drugs are available in the market: (a) Angiotensin converting enzyme (ACE) inhibitors, (b) Calcium channel blockers (CCBs), (c) Angiotensin-2-receptor antagonists (AT2R), (d) Beta blockers, and (e) Loop diuretics [8]. Again, the role of polyphenolic compounds obtained from various natural sources has become significant for the treatment of hypertension. A group of scientists in 2024 isolated flavonoids, phenolic acids, and iridoids from the *Vinca soneri* species [9]. Apart from doing *in vitro* enzymatic kinetic assay, molecular docking, DFT, and molecular mechanics-based generalized born surface area (MM-GBSA) calculations were done to find out the most promising candidate.

Among various life-threatening diseases, cancer stands out as a particularly fatal condition. Among the numerous types of cancer worldwide, breast and lung cancer are currently the most significant. Regarding cancerous cell targets, epidermal growth factor receptor (EGFR) and cyclin-dependent kinase (CDK) inhibitors are two crucial therapeutic targets. Recently, the role of flavonoid compounds obtained from various medicinal plants has played a major role in the treatment of breast and lung cancer. In 2023, a group of scientists explored the potential of fluoro-flavonoid compounds as a potential anti-lung cancer agent via DFT, molecular docking, and MD simulation approaches. To assess the interaction profiles of the fluoro-flavonoid derivatives against six different proteins, e.g., HER2 (PDB ID 7JXH), EGFR (PDB ID 4UV7), FPPS (PDB ID 1YQ7), Prostaglandin D Synthase (PDB ID 1V40), Deoxycytidine Kinase (PDB ID 1P60), and ECH-associated protein 1 (PDB ID 1X2J), the authors used AutoDock Vina [10] to carry out the molecular docking study. To find out the stability of the protein–ligand complexes, an MD simulation study was carried out [11]. Additionally, A. Shah et al. conducted virtual screening of a library of 329 flavonoid compounds obtained from an online database of medicinal plants. All the aforementioned compounds were validated by using the absorption, distribution, metabolism, excretion, and toxicity (ADMET) study. Again, a DFT study was carried out to find out several global reactivity parameters, e.g., E_{HOMO}, E_{LUMO}, η, S, etc., for the mentioned compounds. Finally, to validate the stability of the protein–ligand complex, they carried out the MD simulation study and found a significant influence of the ligand on the respective protein [12].

Apart from COVID-19, hypertension, and cancer, dengue is another life-threatening disease. Until now, no potent standard drugs have been found to inhibit dengue disease. To overcome the situation, the role of polyphenolic compounds using a multi-targeted drug repurposing approach is crucial. Medicinal plants are the best sources from which we can easily obtain different polyphenolic compounds. To identify a more potent inhibitor against the dengue disease, a group of scientists carried out two phases of virtual screening using an ML-based QSAR model and a molecular docking study. Different phytochemicals obtained from four medicinal plants, e.g., *Aloe vera, Cannabis sativa* (Hemp), *Ocimum sanctum* (Tulsi), and *Zingiber officinale* (Ginger) were selected and screened against NS5 protein. To validate the results obtained from the docking study, the redocking experiment was carried out, and finally, they carried out an MD simulation study to find out the stability of protein–ligand complexes [13].

Thus, the present chapter focuses on the use of different computational approaches (DFT, virtual screening, molecular docking, QSAR, MD simulation, etc.) with respect to different polyphenolic compounds in finding better potent inhibitors against different fatal diseases (COVID-19, hypertension, different types of cancer, dengue, etc.) by using a drug repurposing approach. The overall workflow of the computer-aided drug design (CADD) method is given in Figure 7.1.

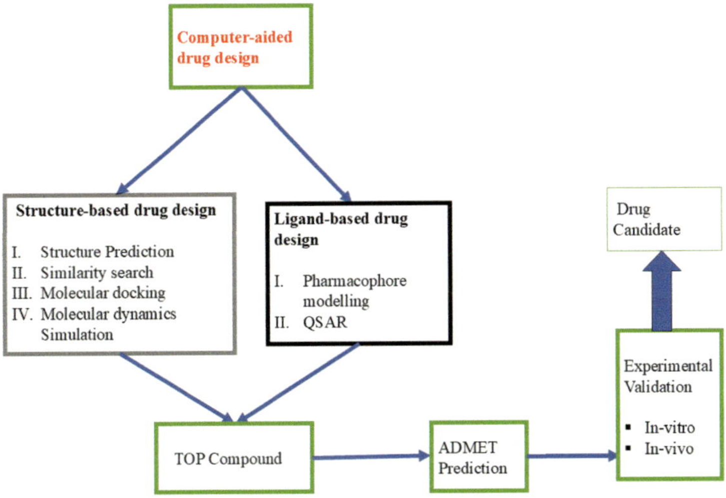

Figure 7.1 Workflow of computer-aided drug design [14].

7.2 CADD Approach

Structural modelling, binding site prediction through molecular docking, visualization and interaction analysis of docked structures, lead optimization, and quantitative structure-activity relationship (QSAR) model development are widely recognized as integral components of CADD [15]. Overall, CADD approaches can be categorized into two types: (a) structure-based drug design (SBDD) and (b) ligand-based drug design (LBDD).

7.2.1 SBDD Approach

In the context of the computational drug discovery process, one normally uses an SBDD approach when the 3D structure of the receptor molecule is well known. To understand the molecular-level mechanism of action of the ligand and several important biological interactions of the ligand with the receptor molecule, one normally uses molecular docking as one of the important SBDD approaches.

The SBDD approach can be divided into two cycles. 1st cycle contains virtual screening of a large scale of compounds through molecular docking concerning a certain receptor for binding site identification, and the 2nd cycle contains analysis and assessment of protein stability of the complex with the most positive leads of the 1st cycle, i.e., with the compounds with the highest binding affinities toward the targeted receptor [16]. Herein, several computational studies are presented on the potential inhibition of these compounds for the treatment of various fatal diseases using the SBDD approach.

7.2.1.1 Molecular docking study

Among the different components of the SBDD study, the molecular docking study is one of the important components. A molecular docking study can be carried out in a few steps as follows:

❖ **Target structure prediction and validation**

To carry out the molecular docking study of a compound potent enough to be a drug, one must first identify its target structure, e.g., protein, enzyme, receptors, etc., which play a critical role in the mechanistic action of the compound to be a drug. To gain deeper insights into the structure and function of target proteins, one has to take a close look at the 3D structure of the protein/receptor obtained from the RCSB PDB [6] in PDB format [17]. For various types of diseases and activity prediction, the 3D formats of the target structures are varied. Finally, to validate the structure of the protein, many web servers, e.g., Pro Check [18], and Saves v 6.0, etc., are used.

Table 7.1 Different kinds of polyphenolic compounds and their uses as a drug [25]

Serial no.	Compound name (Ligand)	PubChem ID	Uses as a drug
1.	Chalcone	637760	Anti-inflammatory, anti-tumor
2.	Hinokiflavone	5281627	Anti-neoplastic agent, neuroprotective agent
3.	Robustaflavone	5281694	Anti-microbial, anti-viral, anti-neoplastic and anti-oxidant
4.	Flavanone	10251	Anti-microbial, anti-inflammatory, anti-fungal

❖ Screening of a library of compounds

After the protein structure was selected as the target, one had to screen a library of compounds to identify which compounds had biological activity concerning that specific target. The traditional approach to drug discovery is widely recognized as time-consuming and costly. That's why the role of screening a library of compounds is very effective and essential [19]. With the help of virtual screening/ molecular docking study, we can easily screen those compounds for their use as a potential inhibitor against several fatal diseases, e.g., COVID-19, anti-hypertension, anti-cancer, etc. To download several small-molecule inhibitors for use as drugs, PubChem online databases (www.pubchem.ncbi.nlm.nih.gov), ChemBL database, etc., are normally used. Such well-known polyphenolic compounds, along with their specificity, are given in Table 7.1.

❖ Grid generation, docking calculation, and further studies

Now, to carry out the molecular docking study, after the preparation of the target and selection of the potential ligands, normally a grid box is generated with certain coordinates (different concerning site-specific or blind docking), and finally, with the help of Lamarckian genetic algorithm (LGA), molecular docking studies are carried out. Based on the different algorithms used in LGA, e.g., population size, number of genetic algorithm (GA) runs, number of generations, etc., molecular docking calculations can be divided into two main types: standard precision (SP) docking and extra precision (XP) docking. In recent years, several researches have been carried out concerning molecular docking studies of different polyphenolic compounds to inhibit several kinds of fatal diseases, e.g., COVID-19, hypertension, cancer (breast and lung types) dengue, etc. Very recently, to find a more potent inhibitor against the COVID-19 disease, Rabie et al. [4] took small libraries of different heterocyclic compounds and did *in vitro* anti-RdRp/ExoN and anti-SARS-CoV-2 bioassays. They found that ChloViD 2022, CoViTris2022, and Taroxaz-26 could effectively inhibit the replication of new virulent strains of SARS-CoV-2. Apart from *in vitro* assessment, in order to confirm the inhibition character of the said compounds theoretically, they took 3D structures of targeted SARS-CoV-2 RdRp protein (PDB ID 7BV2) and ExoN protein (PDB ID 7MC6) from the RCSB PDB in PDB format and performed molecular docking study with the said compounds along with the reference drug Molnupiravir with the help of MOE software. Upon doing a molecular docking study, they found that among the screened amino imidazoles ChloVid2022 showed greater docking scores (−7.1 kcal/mol with RdRp protein and −7.9 kcal/mol with ExoN protein) with respect to known reference drug Molnupiravir (−6.7 kcal/mol with RdRp protein and −7.1 kcal/mol with ExoN protein), which confirms its better inhibition characteristics [4]. As already discussed, hypertension also belongs to another class of fatal diseases. The role of different polyphenolic compounds has reached a new height in the treatment of hypertension through the molecular docking approach. To identify the anti-hypertensive activity of different kinds of 5,7-dihydroxyflavones, in silico simulations have been done, followed by *in vitro* validation. From an in silico molecular docking study, it has been found that chrysin shows better inhibition toward the Enos receptor (PDB ID 3NOS), and it has been reported that a large number of interacting amino acid residues are involved in stabilizing the protein–ligand complexes [20]. However, different kinds of software can be used in docking studies, e.g., AutoDock 4.2 [21], AutoDock Vina [10], generalized optimized ligand docking (GOLD), etc. After completion of the docking protocol, the protein–ligand interaction residual profiler can be shown via various kinds of software, e.g., Discovery Studio, Python Molecular Viewer (PyMol), etc., to analyze the docking results.

7.2.1.2 Conceptual DFT study

To assess different popular qualitative chemical concepts in order to analyze different molecular and/or structure-activity relationships within a molecular system, a conceptual DFT study always plays a crucial role in computational chemistry. Several types of electronic parameters related to the chemical reactivity of a molecule, e.g., highest occupied molecular orbital (HOMO) energy (E_{HOMO}), lowest unoccupied molecular orbital (LUMO) energy (E_{LUMO}), electronegativity (χ), chemical potential (μ), electrophilicity index (ω), etc., are normally generated from conceptual DFT study to find out the structure-activity relationship of a molecule, which can be correlated with the molecular docking and other parameters [22]. Among different chemical quantities, χ and η are the two most popular indices related to chemical reactivity. Regarding the concept of electronegativity, according to Pauling, it is the power of a molecule to attract electrons toward itself, and according to Pearson's concept, η quantifies the resistance against the polarization of the electron cloud of atoms, ions, and molecules. To describe this parameter, Pearson derived two important principles, e.g., hard soft acid base (HSAB) and maximum hardness principle (MHP). HSAB principle normally tells us that hard likes hard and soft likes soft molecules, whereas with reference to MHP, it is clear that molecules arrange themselves to become as hard as possible [23]. Apart from these two reactivity parameters, the energy gap between the energy of the highest occupied molecular orbital and the energy of the lowest unoccupied molecular orbital ($\Delta E = E_{\text{LUMO}} - E_{\text{HOMO}}$) is another quantity from which we can easily confirm that a molecule will be hard and stable or soft and reactive. The higher the energy gap between the HOMO and LUMO orbitals, the harder and more stable compounds will be; whereas, the lower the gap between the HOMO and LUMO orbitals, the softer and more reactive compounds will be [24]. Different kinds of global reactivity parameters and their equations are listed below.

$$\text{Electronegativity } (\chi) = (E_{\text{HOMO}} + E_{\text{LUMO}})/2$$

$$\text{Absolute hardness } (\eta) = (E_{\text{LUMO}} - E_{\text{HOMO}})/2$$

$$\text{Global softness } (S) = 1/\eta$$

7.2.1.3 MD simulation study

MD simulation is a powerful computational technique used in CADD to study the dynamic behavior of biomolecular systems, including drug-receptor interactions. It provides insights into the stability, flexibility, and conformational changes of biomolecules over time, enhancing the accuracy of drug discovery. In order to investigate the time evaluation of a set of molecules in a virtual environment, MD simulations are generally carried out [26, 27]. For finding out better potent inhibitors among different polyphenolic compounds with respect to several fatal diseases, e.g., COVID-19, hypertension, cancer (mainly breast and lung), and dengue, MD simulations have been shown to play a pivotal role concerning the stability of the protein–ligand complex. MD simulations help analyze the binding stability of drug molecules to their target receptors by evaluating parameters like binding free energy, hydrogen bonding, and molecular interactions over time. While Molecular docking predicts the static binding mode of a ligand to a protein, MD simulations refine these results by incorporating molecular flexibility and solvent effects. It allows observation of key interactions, such as hydrogen bonds, hydrophobic contacts, and salt bridges, which contribute to drug efficacy. A typical protein–ligand interaction profiler (PLIP) visualization, showing different types of interactions between flunarizine, a potent antihypertensive drug with two receptors, 1O8A and 3LV3, is shown in Figure 7.2. In general, proteins and ligands undergo structural fluctuations; however, MD simulations capture these dynamic conformational changes and help understand their impact on drug binding. By simulating physiological conditions (such as water environment, pH, and temperature), MD provides a realistic picture of how drug molecules behave *in vivo*. Advanced MD methods like Molecular Mechanics Poisson–Boltzmann Surface Area (MM-PBSA) and MM-GBSA can estimate binding free energy, guiding lead optimization. Furthermore, MD helps explore drug resistance mechanisms in mutated proteins, aiding in the design of more effective inhibitors. There are different varieties of software from which MD simulation can be carried out, e.g., Groningen Machine for Chemical Simulation (GROMACS), Schrodinger, etc. In order to find different potent inhibitors for the treatment of hypertension, the role of MD simulation study is essential, as it provides detailed conformational stability of the protein–ligand complex over time.

Very recently, to identify potential inhibitors against the two important SARS-CoV-2 targets, main protease (PDB ID 5RG1) and spike protein (PDB ID 1WYY), a group of 32 polyphenolic compounds were taken from *Kabasura Kudineer* and molecular docking studies were performed with the two receptors with the help of Glide module of Schrodinger software and finally to find out stability/ conformation changes of the top scored protein–ligand complex, MD simulations were performed with the help of system builder wizard of Schrodinger software which confirmed their stability [28]. Recently, Yue et al. took 680 diverse known molecules from the *Suaeda salsa* medicinal plant and conducted an ADMET study of these molecules. After the ADMET study molecular docking study of the molecules was carried out. Finally, the MD simulation study and post-MD simulation study of the top six compounds obtained from the molecular docking study were carried out to find out the top potent compound [29].

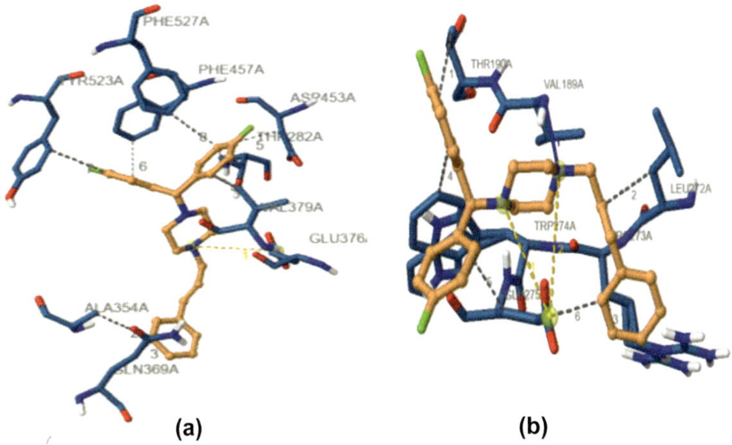

Figure 7.2 PLIP visualization of (a) 1O8A-flunarizine and (b) 3LV3-flunarizine complex [25].

To perform an MD simulation study, one needs to create the topology and parameter files for both the receptor and ligand. To create the topology files for both the protein and the ligand, one has to use several force fields, which can be found in different programs of MD simulation. Several tools and servers, e.g., PRODRG (http://prodrg1.dyndns.org/) and OPLS-AA (All-Atom), are used to create the topology of the drug molecule. After topology preparation, the overall system gets solvated within different types of boxes, e.g., cubic, dodecahedron, etc., as per system requirements. After the addition of suitable amounts of Na^+ and Cl^- in physiological conditions, energy minimization of the overall system is normally carried out. Finally, after NVT and NPT equilibration, the MD simulation study of the overall system is carried out. To analyze the results obtained from the MD simulation study, root mean square deviation (RMSD), root mean square fluctuation (RMSF), radius of gyration (Rg), and solvent accessible surface area (SASA) plots were generated and normally visualized using different Grace plotting tools. Further confirmation of the stability of the protein–ligand complex can be visualized with the help of trajectory movements using VMD software.

7.2.2 LBDD Approach

Apart from SBDD, LBDD belongs to another classical approach to CADD. In this method of drug discovery approach, compounds/ ligands potent enough to be drugs are chosen for screening based on the compounds with known biological significance. The LBDD approach is mainly based on the similarity principle, and it is commonly used to predict a compound concerning a known molecule with good biological similarity to the former one [30].

One normally uses the LBDD approach when severe complications are raised concerning the use of well-known drug molecules, such as low affinity toward the targeted receptor, adverse pharmacological effects, etc. In such cases, either pharmacophore-based modelling or structural modifications of that well-known compound could be done. In those cases, drug designing normally occurs based on the similarity of the known drugs, so that the binding interaction pattern of the former one does get altered [31].

7.2.2.1 Quantitative-structure activity relationship (QSAR) analysis

The main principle of QSAR is based on the study of the relationship between the bioactivity of different molecules with respect to their different structures. QSAR study is mainly used for hit-to-lead identification, lead optimization, and study analysis of different kinds of drugs. Finally, to study the biological activity of the molecules by constructing different models, multiple linear regression (MLR) analysis of QSAR is normally used [32]. In recent years, the QSAR approach has been used in several studies concerning the study of potent inhibitors for several fatal diseases, e.g., COVID-19, hypertension, cancer (especially breast and lung cancer), dengue, etc. In 2022, Santos Costa and his co-workers took a dataset consisting of 32 bicycloproline derivatives, and they did a QSAR-based molecular similarity search study to reproduce reliable new Mpro inhibitors of SARS-CoV-2 [33]. Similarly, in 2024, in order to find different inhibitors against SARS-CoV-2 disease, a group of scientists did a systematic review of advances in different cheminformatic modelling approaches. They reported different varieties of algorithms and different relevant studies corresponding to the QSAR approach to find probable inhibitors against SARS-CoV-2 [34]. Apart from COVID-19, in recent years, the QSAR approach has also been proven to be an effective approach to finding potent anti-hypertensive drugs. In 2023, for the treatment of hypertension and cardiovascular disease (CVD), S. Shah et al. did a genetic algorithm-multiple linear regression model (GA-MLR) of the QSAR approach for the 255 structurally diversified compounds and experimentally validated ACE inhibitors to find novel inhibitors that can also inhibit the ACE receptor [35]. In order to predict robust and effective QSAR models, both linear and non-linear approaches have been taken into consideration with strong emphasis on model validation [36]. The QSAR model validation procedure can be subdivided into several steps:

❖ **Dataset preparation and curation**

Among the different steps of the QSAR approach, dataset preparation and curation are essential and one of the most challenging parts of the QSAR process. Considering the dependency of the endpoint on experimental protocol and different experimental conditions, one normally uses the data endpoint for the prediction of QSAR models. Irrespective of that, in order to avoid false or non-predictive models, different chemical and biological data obtained from different sources require strict curation [37].

❖ **Molecular descriptors calculation**

Based on the chemical constitution of a compound, molecular descriptors are used to find different information about the specific compound with respect to several numerical values. Molecular descriptors normally allow the correlation of the chemical structure of a compound with the physical properties/biological activities of the compound. With respect to different dimensions of descriptors, e.g., 0D–3D, molecular descriptors can be divided into various types. Among them, 0D, 1D, and 2D molecular descriptors can be divided into large-scale 2D molecular descriptors, where large-scale 2D molecular descriptors consist of topological, structural, 2D matrix-based, and 2D atom pairs, etc. On the other hand, in the case of 3D molecular descriptors, electronic, spatial, and geometric structures are important parameters. In the QSAR model, these descriptors normally describe specific aspects of a molecule.

❖ Dataset division and feature selection

Normally, a dataset of compounds can be divided into two sets: (a) a training set and (b) a test set. Training sets are commonly used for the generation of the QSAR model. In order to encompass a wide chemical domain, one needs to select a training set in such a way that it can be used in developing the QSAR model. There are several methods that can involve dataset division, these are (a) Diversity-based Euclidean distance [38], (b) Kennard-Stone [39], (c) sorted response [40], etc. Irrespective of the dataset division, developing a correlation between the response and predictor variable feature selection process is a vital tool. Among a large pool of compounds, a dataset division can help select the most critical descriptors. Several feature selection techniques include genetic algorithms and genetic function approximation.

❖ Model development algorithms and domain of applicability determination

To build transparency in model building and make it reproducible to others, OECD guidelines explain using an "unambiguous algorithm" that explains the development of the QSAR model. To use the model development algorithm, there are several kinds of linear modelling techniques, such as multiple linear regression, principal component regression, and partial least squares. After the development of these algorithms, to find out a certain degree of reliability, the domain of applicability of the QSAR model is normally used.

❖ QSAR model validation and its interpretation

To interpret and predict biological responses, one needs to validate a QSAR model by reproducing the biological activities of untested compounds. Among several parameters of QSAR, goodness of fit and robustness are the two most important quantities. To check the model prediction quality of QSAR, one has to use both regression and classification modelling. Finally, to interpret the QSAR model, one has to identify different features of the QSAR variables that contribute to a more thorough understanding of the overall phenomenon. Different software tools are available to find out and calculate a wide variety of molecular descriptors, such as PaDEL, Chemdes, RDKit, and ChemAxon.

7.2.3 Role of Pharmacological Features in CADD

Irrespective of the different kinds of methods used in SBDD and LBDD, to confirm the drug-like character of the potent compound, the role of different pharmacological feature analyses is essential. To study pharmacological features, we normally have to study different molecules' ADMET properties. To identify the deviation of a compound from potent drug-like behavior, the role of the ADMET study is essential [41]. Normally, ADMET property evaluation of a compound can be done with the help of Lipinski's Ro5 (Rule of 5), which is MW < 500 g/mol, number of H bond donors ≤ 5, number of H bond acceptors ≤ 10, $\mathrm{LogP}_{o/w}$ (logarithm of partition coefficient between aqueous and lipophilic phase) < 5 [42]. Apart from Lipinski's Ro5, Ghose and Veber gave several important criteria to determine different compounds' pharmacological properties. The ADME properties of different ligands that are potent to be a drug can be calculated using the SwissADME online server (http://www.swissadme.ch) [43], and the toxicity study of different compounds can be calculated using Protox-II (https://tox-new.charite.de/protox_II/).

7.3 Conclusions

CADD is a revolutionary technique, with the help of which one can easily determine the probable interaction patterns between a compound/ligand (potent enough to be a drug) and a biological macromolecular system to find out their interaction patterns. During the last several decades, the complexity of different diseases has reached a new height, among them COVID-19, hypertension, cancer, and dengue are the important types. Several studies were carried out regarding different polyphenolic compounds' ability to inhibit these diseases. In this chapter, we tried to explain the overall scheme of different computational methodologies, e.g., conceptual DFT, molecular docking, MD simulation, QSAR, etc., which were used with respect to different polyphenolic compounds to find out some potent inhibitors that can prevent and control these diseases.

References

1. P. Katara, "Computational Approaches for Drug Target Identification BT - Computer-Aided Drug Design," D. B. Singh, Ed. Singapore: Springer Singapore, 2020, pp. 163–185. doi: 10.1007/978-981-15-6815-2_8.
2. S. O. Oselusi et al., "The role and potential of computer-aided drug discovery strategies in the discovery of novel antimicrobials," *Comput. Biol. Med.*, vol. 169, p. 107927, 2024, doi: https://doi.org/10.1016/j.compbiomed.2024.107927.
3. S. Morgan, P. Grootendorst, J. Lexchin, C. Cunningham, and D. Greyson, "The cost of drug development: A systematic review," *Health Policy (New. York).*, vol. 100, no. 1, pp. 4–17, 2011, doi: https://doi.org/10.1016/j.healthpol.2010.12.002.
4. A. M. Rabie and W. A. Eltayb, "Potent Dual Polymerase/Exonuclease Inhibitory Activities of Antioxidant Aminothiadiazoles Against the COVID-19 Omicron Virus: A Promising In Silico/In Vitro Repositioning Research Study," *Mol. Biotechnol.*, vol. 66, no. 4, pp. 592–611, 2024, doi: 10.1007/s12033-022-00551-8.
5. S. A. Hooshmand, M. Zarei Ghobadi, S. E. Hooshmand, S. Azimzadeh Jamalkandi, S. M. Alavi, and A. Masoudi-Nejad, "A multimodal deep learning-based drug repurposing approach for treatment of COVID-19," *Mol. Divers.*, vol. 25, no. 3, pp. 1717–1730, 2021, doi: 10.1007/s11030-020-10144-9.
6. P. W. Rose et al., "The RCSB Protein Data Bank: integrative view of protein, gene and 3D structural information," *Nucleic Acids Res.*, vol. 45, no. D1, pp. D271–D281, Jan. 2017, doi: 10.1093/nar/gkw1000.
7. A. D. Ananto and W. Haryadi, "Flavonoid compound of red fruit papua and its derivatives against SARS-CoV-2 mpro : An in silico approach," *J. Appl. Pharm. Sci.*, vol. 14, no. 12, pp. 90–97, 2024, doi: 10.7324/JAPS.2024.177392.
8. S. Majumdar and A. Pramanik, "Exploring the efficacy of some biologically active compounds as anti-hypertensive drugs: an insightful evaluation through DFT, molecular docking and molecular dynamics simulations," *In Silico Pharmacol.*, vol. 13, no. 1, p. 4, 2024, doi: 10.1007/s40203-024-00291-4.
9. E. Uyanır et al., "Alpha-Glucosidase Inhibitory Effects of Flavonoids, Phenolic Acids and Iridoids Isolated From Vinca Soneri: In Vitro and In Silico Perspectives," *Chem. Biodivers.*, vol. 21, no. 10, p. e202401386, Oct. 2024, doi: https://doi.org/10.1002/cbdv.202401386.

10. O. Trott and A. J. Olson, "AutoDock Vina: improving the speed and accuracy of docking with a new scoring function, efficient optimization, and multithreading," *J. Comput. Chem.*, vol. 31, no. 2, pp. 455–461, Jan. 2010, doi: 10.1002/jcc.21334.

11. N. J. I. Esha et al., "Exploring the potential of fluoro-flavonoid derivatives as anti-lung cancer agents: DFT, molecular docking, and molecular dynamics techniques," *Int. J. Quantum Chem.*, vol. 124, no. 1, p. e27274, Jan. 2024, doi: https://doi.org/10.1002/qua.27274.

12. A. Shah et al., "Discovery of novel anticancer flavonoids as potential HDAC2 inhibitors: virtual screening approach based on molecular docking, DFT and molecular dynamics simulations studies," *3 Biotech*, vol. 14, no. 3, p. 83, 2024, doi: 10.1007/s13205-023-03912-5.

13. F. Alotaibi et al., "Investigating the effects of four medicinal plants against dengue virus through QSAR modeling and molecular dynamics studies," *J. Biomol. Struct. Dyn.*, pp. 1–18, doi: 10.1080/07391102.2024.2301744.

14. A. Tiwari and S. Singh, "Chapter 13 - Computational approaches in drug designing," D. B. Singh and R. K. B. T.-B. Pathak, Eds. Academic Press, 2022, pp. 207–217. doi: https://doi.org/10.1016/B978-0-323-89775-4.00010-9.

15. D. B. Singh and R. K. Pathak, "Computational Approaches in Drug Designing and Their Applications BT - Experimental Protocols in Biotechnology," N. Gupta and V. Gupta, Eds. New York, NY: Springer US, 2020, pp. 95–117. doi: 10.1007/978-1-0716-0607-0_6.

16. P. Śledź and A. Caflisch, "Protein structure-based drug design: from docking to molecular dynamics," *Curr. Opin. Struct. Biol.*, vol. 48, pp. 93–102, Feb. 2018, doi: 10.1016/j.sbi.2017.10.010.

17. W. Yang, Y. Wang, D. Han, W. Tang, and L. Sun, "Recent advances in application of computer-aided drug design in anti-COVID-19 Virials Drug Discovery," *Biomed. Pharmacother.*, vol. 173, p. 116423, 2024, doi: https://doi.org/10.1016/j.biopha.2024.116423.

18. R. A. Laskowski, M. W. MacArthur, D. S. Moss, and J. M. Thornton, "PROCHECK: a program to check the stereochemical quality of protein structures," *J. Appl. Crystallogr.*, vol. 26, no. 2, pp. 283–291, Apr. 1993, doi: https://doi.org/10.1107/S0021889892009944.

19. C. M. Song, S. J. Lim, and J. C. Tong, "Recent advances in computer-aided drug design," *Brief. Bioinform.*, vol. 10, no. 5, pp. 579–591, Sep. 2009, doi: 10.1093/bib/bbp023.

20. Z.-H. Shen et al., "5,7-Dihydroxyflavone acts on eNOS to achieve hypotensive effects in spontaneously hypertensive rats," *Sci. Rep.*, vol. 15, no. 1, p. 594, 2025, doi: 10.1038/s41598-024-84259-6.
21. G. M. Morris et al., "AutoDock4 and AutoDockTools4: Automated docking with selective receptor flexibility," *J. Comput. Chem.*, vol. 30, no. 16, pp. 2785–2791, Dec. 2009, doi: 10.1002/jcc.21256.
22. P. Geerlings et al., "Conceptual density functional theory: status, prospects, issues," *Theor. Chem. Acc.*, vol. 139, no. 2, p. 36, 2020, doi: 10.1007/s00214-020-2546-7.
23. F. De Proft, P. W. Ayers, S. Fias, and P. Geerlings, "Woodward-Hoffmann rules in density functional theory: Initial hardness response," *J. Chem. Phys.*, vol. 125, no. 21, p. 214101, Dec. 2006, doi: 10.1063/1.2387953.
24. M. Roney et al., "Computational evaluation of quinones of Nigella sativa L. as potential inhibitor of dengue virus NS5 methyltransferase," *J. Biomol. Struct. Dyn.*, vol. 42, no. 16, pp. 8701–8711, Nov. 2024, doi: 10.1080/07391102.2023.2248262.
25. S. Majumdar, S. Karmakar, and A. Pramanik, "In silico analysis via molecular docking study to identify a better potent inhibitor for the main protease (Mpro) of SARS-CoV-2," *J. Sci. Enq.*, vol. 2, no. 1, pp. 1–10, 2022, doi: 10.54280/jse.222101.
26. T. Bereau, "Computational compound screening of biomolecules and soft materials by molecular simulations," *Model. Simul. Mater. Sci. Eng.*, vol. 29, no. 2, p. 23001, 2021, doi: 10.1088/1361-651X/abd042.
27. R. Garduño-Juárez et al., "Molecular Dynamic Simulations for Biopolymers with Biomedical Applications," *Polymers*, vol. 16, no. 13. 2024. doi: 10.3390/polym16131864.
28. S. K. Mandal et al., "In silico anti-viral assessment of phytoconstituents in a traditional (Siddha Medicine) polyherbal formulation – Targeting Mpro and pan-coronavirus post-fusion Spike protein," *J. Tradit. Complement. Med.*, vol. 14, no. 1, pp. 55–69, 2024, doi: https://doi.org/10.1016/j.jtcme.2023.07.004.
29. G. Yue, H. Gu, K. Zhang, Y. Song, and Y. Hao, "ACE inhibitors from Suaeda salsa: 3D-QSAR modeling, metabolomics, molecular docking and molecular dynamics simulations," *Silico Pharmacol.*, vol. 12, no. 2, p. 59, 2024, doi: 10.1007/s40203-024-00233-0.
30. E. Singh et al., "Ligand-Based Approaches for the Development of Drugs Against SARS-CoV-2 BT - In Silico Modeling of Drugs Against Coronaviruses: Computational Tools and Protocols," K. Roy, Ed. New York, NY: Springer US, 2021, pp. 117–134. doi: 10.1007/7653_2020_65.

31. T. S. 3rd Rush, J. A. Grant, L. Mosyak, and A. Nicholls, "A shape-based 3-D scaffold hopping method and its application to a bacterial protein–protein interaction," *J. Med. Chem.*, vol. 48, no. 5, pp. 1489–1495, Mar. 2005, doi: 10.1021/jm040163o.

32. P. Gupta et al., "Computer-aided Drug Design: Innovation and its Application in Reshaping Modern Medicine," *Curr. Artif. Intell.*, 2024, doi: https://doi.org/10.2174/0129503752321279241126091807.

33. A. S. Costa, J. P. A. Martins, and E. B. de Melo, "SMILES-based 2D-QSAR and similarity search for identification of potential new scaffolds for development of SARS-CoV-2 MPRO inhibitors," *Struct. Chem.*, vol. 33, no. 5, pp. 1691–1706, 2022, doi: 10.1007/s11224-022-02008-9.

34. Q. Wang et al., "Recent advances in chemometric modelling of inhibitors against SARS-CoV-2," *Heliyon*, vol. 10, no. 2, p. e24209, 2024, doi: https://doi.org/10.1016/j.heliyon.2024.e24209.

35. S. Shah et al., "In silico study to recognize novel angiotensin-converting-enzyme-I inhibitors by 2D-QSAR and constraint-based molecular simulations," *J. Biomol. Struct. Dyn.*, vol. 42, no. 5, pp. 2211–2230, Mar. 2024, doi: 10.1080/07391102.2023.2203261.

36. P. De, S. Kar, P. Ambure, and K. Roy, "Prediction reliability of QSAR models: an overview of various validation tools," *Arch. Toxicol.*, vol. 96, no. 5, pp. 1279–1295, 2022, doi: 10.1007/s00204-022-03252-y.

37. P. Ambure and M. N. D. S. Cordeiro, "Importance of Data Curation in QSAR Studies Especially While Modeling Large-Size Datasets BT - Ecotoxicological QSARs," K. Roy, Ed. New York, NY: Springer US, 2020, pp. 97–109. doi: 10.1007/978-1-0716-0150-1_5.

38. H. Golmohammadi, Z. Dashtbozorgi, and W. E. J. Acree, "Quantitative structure-activity relationship prediction of blood-to-brain partitioning behavior using support vector machine," *Eur. J. Pharm. Sci. Off. J. Eur. Fed. Pharm. Sci.*, vol. 47, no. 2, pp. 421–429, Sep. 2012, doi: 10.1016/j.ejps.2012.06.021.

39. R. W. Kennard and L. A. Stone, "Computer Aided Design of Experiments," *Technometrics*, vol. 11, no. 1, pp. 137–148, Feb. 1969, doi: 10.1080/00401706.1969.10490666.

40. K. Roy, P. Ambure, and S. Kar, "How Precise Are Our Quantitative Structure–Activity Relationship Derived Predictions for New Query Chemicals?," *ACS Omega*, vol. 3, no. 9, pp. 11392–11406, Sep. 2018, doi: 10.1021/acsomega.8b01647.

41. A. Singh, A. Mishra, A. Meena, N. Mishra, and S. Luqman, "Exploration of selected monoterpenes as potential TRPC channel family modulator in lung cancer, an in silico upshot," *J. Biomol. Struct. Dyn.*, vol. 42, no. 15, pp. 7917–7933, Oct. 2024, doi: 10.1080/07391102.2023.2241900.
42. C. A. Lipinski, F. Lombardo, B. W. Dominy, and P. J. Feeney, "Experimental and computational approaches to estimate solubility and permeability in drug discovery and development settings," *Adv. Drug Deliv. Rev.*, vol. 46, no. 1–3, pp. 3–26, Mar. 2001, doi: 10.1016/s0169-409x(00)00129-0.
43. A. Daina, O. Michielin, and V. Zoete, "SwissADME: a free web tool to evaluate pharmacokinetics, drug-likeness and medicinal chemistry friendliness of small molecules," *Sci. Rep.*, vol. 7, no. 1, p. 42717, 2017, doi: 10.1038/srep42717.

Glossary

List of Available Computational Software

1. **ABINIT:** Used in materials science for computing electronic properties (total energy, charge density, DFT calculations), model materials, examine molecular, crystal, surface, or interface systems, calculate and predict material properties (mechanical, vibrational, and optical), Geometry Optimization, and perform advanced calculations (compute band structures, density of states, molecular dynamic simulations, dielectric properties, and thermodynamic properties).
2. **ACES (Advanced Concepts in Electronic Structure Theory):** used for calculating atomic and molecular energies and properties accurately.
3. **AMPAC:** Helps in Quantum Mechanical Calculations on molecules. It calculates thermodynamic properties (heat of formation), vibrational spectra, dipole moments, polarizabilities and hyperpolarizabilities, electronic spectra, ESR charge distributions etc.
4. **Amsterdam Density Functional:** Helps calculate first principle electronic structure calculations using DFT.
5. **Avogadro:** It is a molecular editor and visualization platform usually used in computational chemistry, molecular modelling, bioinformatics, and materials science.
6. **CHARMM:** It is a molecular simulation program designed for studying complex macromolecular systems (such as biological macromolecules) and nanomaterials.
7. **CASINO:** Designed to perform Monte Carlo (QMC) electronic structure calculations.
8. **CP2K:** Open-source software used for atomic and molecular simulations in Quantum chemistry and solid-state physics.
9. **CPMD:** It is a useful for performing ab-initio molecular dynamic simulation by implementing Density functional theory (DFT) using a parallelized plane wave or pseudopotential approach.

10. **Dalton:** Used for advanced molecular electronic-structure calculations.
11. **DIRCCR12:** It performs coupled cluster calculations with explicit electron calculations.
12. **GAMESS:** It is a package that performs DFT and other semi-empirical calculations.
13. **Gaussian:** Helps in electronic structure calculations and predicting various molecular properties (molecular structures and energies, vibrational frequencies, IR and Raman spectra, bond energies and reaction pathways etc).
14. **GROMACS:** Open-source software used for molecular dynamic simulations particularly biological macromolecules and polymers.
15. **Hyperchem:** It is a molecular modelling software which is applicable for both macro and micro molecules. It combines 3D visualization and animation with quantum chemical calculations, molecular mechanics and molecular dynamics.
16. **Jaguar:** Software used for electronic structure calculations (molecules in both gas and solution phase), spectroscopic properties, geometry optimization, solvation calculations, molecular properties, pseudo spectral approximations etc.
17. **Molpro:** Useful for accurate ab-initio Quantum chemistry calculations such as electronic structure calculations, electron correlations, molecular properties etc.
18. **NAMD:** This software is used for high performance simulations of large biomolecular systems.
19. **NW Chem:** It is a software useful for performing quantum chemical and molecular dynamic calculations.
20. **OCTOPUS:** This software is used to perform Kohn-Sham density functional theory (DFT) and time dependent density functional theory (TDDFT) calculations. It calculates magnetic susceptibilities, static and dynamic polarizabilities, absorption spectra and can perform molecular dynamic simulations.
21. **Orca:** This software consists of electronic structure methods well suited for studying larger molecules, transition metal complexes and their spectroscopic properties.
22. **Psi3:** It is an open-source program used for ab-initio molecular electronic structure computations, implementation of standard quantum chemical methods, and development of new methods.

23. **Quantum ESPRESSO:** This software finds its use in electronic structure calculations and material modelling at nano-levels. It can perform ground state calculations, ab-initio molecular dynamics, optimize structures, calculate spectroscopic properties, generate pseudopotentials, find transition states and minimum energy paths etc.

24. **SAPT:** This is used in electronic structure theory to analyze and compute non-covalent interactions between atoms and molecules such as hydrogen bonding, van der Waals interactions, intermolecular forces, and other type of interactions.

25. **Schrödinger:** This software helps in molecular modelling, Quantum mechanics simulations, molecular dynamic simulations, and research in computational chemistry.

26. **Spartan:** This method is used in molecular modelling and quantum chemistry calculations such as molecular structure building, computational calculations, spectral analysis and many more.

27. **VASP:** It is used in modelling materials at atomic scale with attention on electronic structure calculations and quantum-mechanical molecular dynamics.

28. **WIEN2K:** This software helps to perform quantum mechanical calculations on periodic solids. It performs electronic structure calculations, geometry optimization. It calculates optical properties, electron and spin densities, density of states (DOS), X ray structure factors etc.

29. **Yambo:** It helps in calculating excited-state properties of solid-state systems.

Index